INTERNATIONAL SERIES IN
PURE AND APPLIED BIOLOGY
DIVISION: **ZOOLOGY**
GENERAL EDITOR: G. A. KERKUT

VOLUME 54

INSECT CLOCKS

Some titles of related interest

BERNHARD, C. G.
Functional Organization of the Compound Eye

BOYDEN, A.
Perspectives in Zoology

CAMPBELL, P. N.
The Structure and Function of Animal Cell Components

CLOUDSLEY-THOMPSON, J. L.
Desert Life

CLOUDSLEY-THOMPSON, J. L.
Spiders, Scorpions, Centipedes and Mites

COHEN, J.
Living Embryos

INGLIS, J. K.
A Textbook of Human Biology

MARSHALL, P. T.
The Development of Modern Biology

MUZZARELLI, R.
Chitin

PARSONS, T. R. & TAKAHASHI, H.
Biological Oceanographic Processes

ROGER, F.
Onchocerciasis in Zaire

WHITE, D. C. S. & THORSON, J.
The Kinetics of Muscle Contraction

INSECT CLOCKS

BY

D. S. SAUNDERS

Department of Zoology, University of Edinburgh

PERGAMON PRESS

OXFORD · NEW YORK · TORONTO
SYDNEY · PARIS · FRANKFURT

U. K.	Pergamon Press Ltd., Headington Hill Hall, Oxford, OX3 0BW, England
U. S. A.	Pergamon Press Inc., Maxwell House, Fairview Park, Elmsford, New York 10523, U.S.A.
CANADA	Pergamon of Canada Ltd., P.O. Box 9600, Don Mills M3C 2T9, Ontario, Canada
AUSTRALIA	Pergamon Press (Aust.) Pty. Ltd., 19a Boundary Street, Rushcutters Bay, N.S.W. 2011, Australia
FRANCE	Pergamon Press SARL, 24 rue des Ecoles, 75240 Paris, Cedex 05, France
WEST GERMANY	Pergamon Press GmbH, 6242 Kronberg/Taunus, Pferdstrasse 1, Frankfurt-am-Main, West Germany

First edition 1976

Library of Congress Cataloging in Publication Data

Saunders, David Stanley, 1935–
Insect clocks.

(International series in pure and applied biology: Division, Zoology)
Bibliography: p.
Includes index.
1. Insects—Physiology. 2. Insects—Behavior.
3. Biological rhythms. I. Title.
QL495.S28   1975   595.7′01   75–17512
ISBN 0-08-018211-9

Printed in Hungary

CONTENTS

PREFACE

TIME is one of the three fundamental "quantities" in terms of which a physicist can describe the Universe; however, unlike the other two (mass and length), it is difficult to define. In this book I have looked at time from a biologist's point of view, and in terms of the motion of the "heavenly bodies", particularly the rotation of the Earth on its axis and around the Sun, and the revolution of the Moon around the Earth, movements which give rise to the familiar successions of day and night, months, years and tides. Organisms on this planet have been exposed to such rhythmic changes since life began, and this aspect of time must be the most meaningful one as far as they are concerned!

All aspects of physiology have a "time course", and many phenomena — from heart beats and nerve impulses to the interactions between predator and prey — are rhythmic or oscillatory in nature. This book, however, is concerned only with those phenomena in which "environmental time" has a functional significance in the life of insects, enabling them to perform behavioural or physiological events at the "right time" of the day, month, year or tide. The majority of these rhythmic phenomena are endogenous, and when allowed to "free-run" in the absence of temporal cues, reveal a natural periodicity which is *close to* that of the solar day (or month, year or tidal cycle), accurate, and temperature compensated. They possess, in fact, the properties one normally attributes to man-made time measuring devices or clocks. It is the nature and functional significance of these insect clocks which is dealt with here: restriction of the examples to the Class Insecta merely reflects a life-long passion — or perhaps a prejudice — on my part.

The work has been organized so that the fundamental properties of circadian rhythms are presented first, followed by the longest sections on seasonal photoperiodism. This arrangement has been followed in order to present and discuss the problem of photoperiodism in terms of circadian rhythmicity. The alternative point of view that time measurement is accomplished by "hour glasses" has also been given full attention; my conclusion is that both forms of time measurement are to be seen in the insects, sometimes in the same species. Concluding chapters compare other insect clocks, rhythmic and non-rhythmic, with the circadian system, and describe what is known about the anatomical location of the circadian pacemakers and the photoreceptors which facilitate their entrainment to the environmental cycles of light and dark. Whilst much of the material and its interpretation has naturally been derived from the writings of others, I must accept all responsibility for reporting them faithfully, and for those opinions and suggestions which are mine.

The writing of this book has been greatly aided by stimulating discussions with my friends and colleagues, particularly C. S. Pittendrigh, J. Aschoff, A. D. Lees, J. N. Brady and W. Engelmann. These authors, together with many others, have given permission to reproduce copyrighted material, and in many cases provided me with original artwork or photographs of their published figures; these sources are individually acknowledged in the respective figure legends. Much of my own research in the field of Insect Clocks has been carried out in the University of Edinburgh with the continued interest and encouragement of Professor J. M. Mitchison, and with the generous financial support of the Science Research Council. I also gratefully acknowledge the technical assistance provided by Mrs. Helen MacDonald and Mrs. Margo Downie, the help from Mr. D. F. Cremer in making photographic copies of most of the original figures, and from Mr. J. J. Holmes for redrawing Figs. 3.15 and 3.17. Lastly I would like to thank Miss A. Keegan for typing the manuscript.

Edinburgh   D. S. SAUNDERS

CHAPTER 1

INTRODUCTION: RHYTHMS AND CLOCKS

EVER since life first appeared on this planet it has been subjected to daily cycles of light and dark, and to seasonal cycles of climatic change, caused by the rotation of the earth around its axis and around the sun. Marine and intertidal organisms have in addition been subjected to tidal and lunar periodicities. Only those animals which have invaded the depths of the ocean, or underground caves and rivers, have avoided this fluctuating environment. Other species—especially those on the land, where daily and seasonal changes may include violent fluctuations in temperature and humidity—have developed strategies to counteract or to exploit this periodicity. The majority of insects, for example, show daily and annual cycles of activity and development. They may be nocturnal, diurnal or crepuscular. They may hibernate or aestivate. Plants may produce leaves or flowers only at certain seasons, and flowers may open and close at particular times of the day.

Some of these phenomena are direct responses to environmental change, but many more are overt manifestations of an endogenous periodicity. These innate rhythms must have astounded early workers such as the French astronomer De Mairan who discovered (in 1729) that the daily leaf movements of *Mimosa* would persist in constant darkness. The oscillations underlying such phenomena are now known to provide a temporal organization for physiological and behavioural activities in practically every group of organisms apart from the prokaryotes. Of particular interest are those endogenous oscillations which have evolved with a periodicity close to 24 hours (circadian rhythms), and are used by animals and plants to "time" daily events and thus allow the organism to perform functions at the "right time of the day", or to attain synchrony with other individuals of the population. It is clear that these circadian oscillations in the cell and the organism have evolved to match almost exactly the oscillations in the physical environment. In *Drosophila melanogaster* the period of the pupal eclosion rhythm is inherited and the gene responsible has been located on the X chromosome (Konopka and Benzer, 1971). These rhythms, therefore, are not "imposed" on the organism by the environment, neither are they "learned". The natural cycles of light and temperature, however, do serve to entrain and phase-control these endogenous oscillators so that under natural conditions their periods become exactly 24 hours. In the absence of temporal cues from the environment (i.e. in darkness and constant temperature) the rhythms "free-run" and reveal their own natural period (τ) which is close to, but significantly different from, that of the solar day. The observation that this period is temperature-compensated, and that the rhythms are used by the organisms to measure the passage of time (Pittendrigh, 1954, 1960), justifies the use of the term "biological clock".

1

Apart from circadian rhythms which have evolved as a match to the 24-hour periodicity of the Earth's rotation around its axis, endogenous oscillations with tidal ($\sim$ 12.4 hour), semilunar ($\sim$ 14.7 day), lunar ($\sim$ 29.4 day) or annual ($\sim$ a year) periods are also to be found in organisms, including the insects. In many cases the endogenous nature of these rhythms has been demonstrated by allowing them to "free-run" in the absence of the environmental cues *(Zeitgebers)* which normally entrain them.

The brief account of these biological oscillations given above—and the more extensive description of their properties given later in this book—amply demonstrate their endogeneity. They are, in fact, every bit as much a part of the organism as its morphological organization. Some investigators, however—principally Brown (1960, 1965)—have held an alternative view, namely that all of the observed periodicities are in some way exogenously controlled by "subtle geophysical forces" associated with the solar day (such as air pressure, periodic fluctuations in gravity associated with the Earth's rotation in relation to the Sun and the Moon, or cosmic ray intensity) which remain unaccounted for in laboratory experiments in which the obvious periodicities (light, temperature, etc.) have been eliminated. This view will receive no further attention in this book even though, theoretically, it must remain an open question until unequivocal experiments (perhaps involving organisms travelling away from the influence of the Earth) have been performed. As a partial answer to the endogenous-exogenous controversy, Hamner *et al.* (1962) maintained a number of organisms at the South Pole on a turntable arranged to rotate once every 24 hours counter to the Earth's own rotation, thereby eliminating most of the diurnal variables. Under these conditions several rhythmic systems, including the pupal eclosion rhythm in *D. pseudoobscura*, continued to show a circadian periodicity apparently unaffected by either their location at the South Pole or by their rotation on the turntable. Therefore, as far as these experiments or the results allow, the data support the endogenous hypothesis.

Although the clock analogy should not be pursued too closely it is a useful one, and there is an interesting parallel between the development of man-made "time-pieces" and those "clocks" found in nature. Early man was aware of the passage of time by watching the movement of the Sun, Moon, and stars, or by observing the movement of the Sun's shadow on the ground or on a dial. Such methods, of course, have nothing to do with clocks. Neither have the *direct* responses of animals and plants to daily periodicities. These exogenous effects are widespread in nature and in some animals the observed rhythm of activity is related to the immediate effects of the daily changes in light intensity. Under field conditions most daily rhythms of activity—although innate—are nearly always strongly modulated by the immediate character of the environment, particularly the rapid changes in light intensity at dawn and dusk. These effects will be discussed only where they modify an endogenous periodicity.

The first man-made time-measuring devices were probably sand-glasses, clepsydras (water clocks) and candles. These "clocks" did not oscillate and had to be reset or "turned over" once all the water or sand had run out, or the candle burnt to the bottom. This type of device finds its equivalent in some of the "hour-glass" timers performing night-length measurement in aphids which, after measuring the duration of the dark period, require to be "turned over" by light before they can function again (Lees, 1968).

Mechanical clocks introduced in the fourteenth and fifteenth centuries were either weight-driven or spring-driven and incorporated oscillatory devices which ran continuously so long as the weight was raised or the spring wound up. These find their counterpart in the biological oscillations mentioned above. The escapement in these early clocks consisted of a crown wheel and a verge and foliot. The system was not isochronous and the clocks so constructed tended to lose or gain up to 15 minutes every day. In the seventeenth century the incorporation of a pendulum with an escapement to maintain a constant amplitude introduced isochrony to the clock, and brought the error down to about 10 seconds per day. Although the concrete nature of biological oscillations remains unknown, this pendulum analogy and sine-wave representation of the oscillation's time course are often instructive and useful in model building.

For really accurate time measurement temperature-compensation is required. In man-made clocks an uncompensated pendulum lengthens as the temperature rises and therefore swings more slowly. By the eighteenth century George Graham had compensated for such temperature changes by using a mercury-vial pendulum. When the quantity of mercury was correctly adjusted its thermal expansion raised the centre of oscillation to compensate for the lengthening of the pendulum rod. Graham's clock varied by as little as 1 second per day; Harrison's grid-iron pendulum, which operated on a similar principle, later cut this error down to less than 1 second. In biological systems most physiological processes more than double their rate with every 10° rise in temperature, and such temperature effects would render time measurement impossible. However, during evolution this challenge has been met: most biological oscillators with a "clock" function have a Q_{10} between 0.85 and 1.1. This property is an absolute functional prerequisite for a clock mechanism. It is also essential for effective entrainment by a natural (24-hour) Zeitgeber because if the oscillator had a Q_{10} of 2.0 or more it would, at some temperatures, fall outside the limits within which the light-cycle could hold it. The manner in which temperature-compensation is achieved in biological clocks, however, remains as obscure as the nature of the oscillations themselves.

The "clock" analogy begins to founder at this point. In man-made clocks hour-glasses clearly antedate oscillators, but the reverse would seem to be the case in biological systems. Pittendrigh (1966) suspects that circadian oscillations—which occur in all eukaryotes and possess the common but somewhat "improbable" properties of accuracy and temperature-compensation—are monophyletic in origin and therefore very ancient. Although it is far from clear what their original functional significance was, they are now widely used for the purposes of chronometry. In many species of animals and plants the circadian system is causally involved in the measurement of day- or night-length in "classical" photoperiodism. In some insects, on the other hand, this function is performed by means of an "hour-glass" rather than by circadian oscillations—which they surely must possess. Evidence of an evolutionary convergence such as this suggests that the adoption of an hour-glass for photoperiodic time measurement is a comparatively recent event.

Many aspects of insect physiology and behaviour are "clock-controlled". There are, for example, daily rhythms of general locomotion, feeding, mating, oviposition, pupation, and pupal eclosion, in which these activities are restricted to a particular part of the day or night. Photoperiodism also involves a clock which measures day- or night-length, the most frequent response being the seasonal appearance of a dormant stage in the life cycle. The adaptive significance of diapause is clear, but it is not always easy to see the adaptive significance of daily rhythms, and in the absence of concrete experimental evidence most conclusions must remain conjectural. However, adults of Drosophila pseudoobscura emerge from their puparia close to dawn when the relative humidity of the air is at its highest, and it is known that success in the act of eclosion is greatest under these conditions (Pittendrigh, 1958). Cycles of feeding may be correlated with the supply of food: the classical example of this is probably the "time-memory" *(Zeitge-dächtnis)* of bees. Bees can be "trained" to visit a food source at a particular time of the day (Beling, 1929), this mechanism ensuring that they visit nectar sources every day at the same time. The significance of this behaviour lies in the observation that not only do flowers open and close at particular hours, but that nectar production is also a circadian event (Kleber, 1935). In many cases the selective advantage of an event being clock-controlled lies in the synchrony attained between individuals of the population. Mating rhythms of certain Diptera, for example, ensure that all sexually active individuals in the population are looking for mates at the same time and thereby increase the likelihood of successful encounters between the sexes. Differences in mating times between different species are also known to provide effective mechanisms for genetic isolation (Tychsen and Fletcher, 1971).

Biological clocks have been classified in a number of ways. Pittendrigh (1958) differentiated (1) "Pure" rhythms, such as colour change in the crab *Uca pugnax* (Brown *et al.*, 1953), from (2) Interval timers, in which a particular event such as pupal eclosion occurs at a particular time of the day, and (3) Continuously consulted clocks such as the bees' *Zeitgedächtnis* and time-compensated sun orientation in which time may be "recognized" at any time of the day. Lees (1960a) has also used the term interval timer to describe some of the non-oscillatory timing devices in aphids.

Truman (1971d) has recently proposed that animal clocks fall into two well-defined groups. In Type I, such as the rhythm of pupal eclosion in *Drosophila* spp. (Pittendrigh, 1966) and *Antheraea pernyi* (Truman, 1971a), the compound eyes (or other "organized" photoreceptors) are not involved, the photoreceptors lying in the brain itself. These rhythms are also damped out by moderate intensities of continuous light, and the magnitude of the phase-shifts generated by quite short light perturbations may be in the order of 10 hours (for *D. pseudoobscura*). These clocks are generally associated with developmental rhythms such as hatching, moulting, eclosion or release of brain hormone. Truman also places photoperiodism in this category. In Type II clocks, such as those controlling locomotor activity rhythms, the compound eyes are the principal and sometimes the only photoreceptors involved and the brain itself is insensitive to light. These rhythms "free-run" in both DD and *LL* of quite high intensity and the phase-shifts generated by light perturbations are usually much smaller than in Type I. Truman includes *Zeitgedächtnis* and time-compensated sun orientation in this category solely because of their association with locomotor activity.

Although this scheme is attractive and points out several possibly fundamental differences between biological clocks, too few species have been examined in sufficient detail and it may be premature to adopt the classification. Nevertheless, the distinction between developmental rhythms which can only be appreciated in mixed-age *populations*, and those such as general locomotor activity which are performed repeatedly by *individual* insects often over quite long periods of time, is certainly a useful one. Consequently this distinction is used in the present book, and forms the basis for the first two chapters.

Annotated Summary

1. Insects, like other organisms, have evolved in an environment dominated by daily, monthly, annual and, in some cases, tidal periodicities.

2. This environmental periodicity is frequently matched by an appropriate endogenous rhythmicity which is a constituent and characteristic physiological feature of living tissue. Organisms may possess circadian ($\sim$24 hours), circa-tidal ($\sim$12.4 hours), semi-lunar or circasyzygic ($\sim$14.7 days), circa-lunar ($\sim$29.4 days) or circannual ($\sim$a year) periodicities.

3. These oscillations have natural periods which are approximately equal to that in the environment, are accurate, and are temperature-compensated. They provide the organisms with a "temporal organization" allowing them to perform functions with a selective advantage at the "right time of the day", or to "measure time" as "biological clocks".

4. Biological clocks control a wide variety of behavioural and physiological activities in insects. These include daily rhythms of locomotion, feeding, mating, oviposition, pupation and eclosion. These rhythms may be operational either in individual insects or in populations which behave, in this respect, like "super-organisms". Clocks also control cuticle deposition, metabolism and the seasonal control of alternate developmental pathways (photoperiodism).

CHAPTER 2

CIRCADIAN RHYTHMS OF ACTIVITY IN INDIVIDUAL INSECTS

INSECTS, like other organisms, usually restrict their activity to certain times of the diel cycle. In natural conditions, or in the artificial light and temperature cycles provided in the laboratory, they may be—with respect to a particular activity—either night-active (nocturnal), day-active (diurnal) or twilight-active (crepuscular). The mechanisms controlling these activity rhythms may be exogenous (i.e. a direct response to environmental changes) or endogenous (i.e. controlled by an underlying circadian oscillation, or oscillations, which are a part of the physiological make-up of the organism). Most activity rhythms have proved to be a "mixture" of endogenous and exogenous components, the overt rhythm of activity, although controlled by an endogenous oscillation, being continuously modulated by the *direct* effects of the environmental cycles of light and temperature, particularly the abrupt changes in light intensity at dawn and dusk. Here we are mainly interested in the endogenous aspects of rhythmic phenomena because the intrinsic and self-sustained physiological oscillations controlling them function as "biological clocks", and provide temporal organization for a wide array of behavioural activities.

This chapter is concerned with those aspects of activity such as general locomotion, flight, feeding, and oviposition which are performed repeatedly by *individual* insects and may persist as a daily rhythm for quite long periods of time. "Once-in-a-lifetime" events such as egg hatching, moulting, pupation and adult eclosion—which may also be governed by an on-going circadian oscillation—will be discussed in Chapter 3: questions of photoreception and the location of the "clock" will be dealt with in Chapter 10.

Activity rhythms have been studied in a wide range of insect types. Here we will illustrate the general properties of the circadian organization of these rhythms using some of the most intensively investigated systems. These include the general locomotor activity of cockroaches (Harker, 1956; Roberts, 1960), crickets (Lutz, 1932; Nowosielski and Patton, 1963), beetles (Lohmann, 1964; Birukow, 1964) and stick insects (Eidmann, 1956; Godden, 1973), and flight activity in mosquitoes (Jones *et al.*, 1967; Taylor and Jones, 1969) and other species of Diptera (Roberts, 1956; Brady, 1972). Emphasis is placed on the more recent or on the more fully investigated species, but a more comprehensive list of those insects exhibiting activity rhythms of this kind is presented in an appendix.

A. Activity in Light/Dark-cycles

Cockroaches are almost entirely nocturnal in their habits. Under natural and laboratory conditions activity generally commences at or soon after dusk and continues more or less throughout the dark period; the insects become inactive during the day (Gunn, 1940; Mellanby, 1940; Harker, 1954; Roberts, 1960). In the laboratory their large size makes the recording of their activity a relatively easy task. Harker (1956, 1960a), for example, recorded the locomotor activity of *Periplaneta americana* in rocking actographs or in phototransistors using very dim red light, or by attaching a fine wire to the pronotum which wrote on a smoked drum when the insect moved. Roberts (1960) and a number of other authors have used running wheels.

In an artificial cycle of 12 hours light and 12 hours dark (*LD 12* : 12) most cockroaches commence activity shortly after the onset of darkness. Considerable variation between individual insects, and between sex, age and physiological state is apparent, however. Roberts (1960) used males of *Leucophaea maderae*, *Byrsotria fumigata* and *P. americana* in preference to females because their activity was "less erratic". Leuthold (1966) found that the activity rhythm of female *L. maderae* varied with the insect's reproductive state, locomotion being suppressed when mature eggs were present in the lower reproductive tract. Working with mature adult females of *P. americana*, Lipton and Sutherland (1970) found no activity rhythm that was obviously related to the lighting regime, and similarly concluded that the reproductive cycle interfered with the normal expression of the rhythm. Virgin females, on the other hand, exhibited an entrained rhythm very similar to that shown by adult males. Amongst the males they also found considerable variability. The majority showed a clearly entrained rhythm of activity, or at least a weak nocturnal rhythm or "pattern", but about 4 per cent were apparently random in their activity. Most of those with a well-marked rhythm showed the typical onset of activity within the first few hours of dark, but over 30 per cent showed a secondary active phase in the first few hours of light. A similar variation in rhythmicity has been recorded by Nishi-itsutsuji-Uwo *et al.* (1967); Ball (1972) has also described individuals of *Blaberus craniifer* with secondary activity after dawn.

Harker (1956) showed that most of the feeding took place during the active period. Nevertheless, when food was offered in the light period only, the insects became active at this time as well as during the night. When feeding was discontinued, however, the daytime feeding peak did not persist and she concluded that the activity rhythm in *P. americana* was not an expression of a hunger cycle.

The locomotor activity rhythm in the house cricket *Acheta domestica* is similar to that in cockroaches. Lutz (1932) showed that activity commenced soon after dark and continued for about 4 to 6 hours. As with cockroaches, however, the pattern of activity varied between individuals and with age. Nowosielski and Patton (1963), for example, showed that last instar larvae rarely showed a pronounced rhythm, and that some adults were biphasic with a second peak *prior* to the onset of dark. Cymborowski (1973) demonstrated three types of individual in *LD 12* : 12: some commenced activity *at* the light/dark transition, some commenced activity up to 3 hours *after* dark, and some began their period of intensified activity as much as 1 hour *before* light-off.

The activity rhythms of several mosquito species have been recorded by automatic devices in which the flight noise is amplified (Jones, 1964; Nayar and Sauerman, 1971). In *LD 12* : 12 *Anopheles gambiae* is nocturnal but with an intense activity lasting 20 to

30 minutes following both light-off and light-on (Jones *et al.*, 1966; Jones *et al.*, 1967). *Aëdes taeniorhynchus* is also a night-active insect with a similar bimodal pattern (Nayar and Sauerman, 1971). In this species the activity pattern originates in the adult instar and is not carried over from the developmental stages. The yellow-fever mosquito *A. aegypti*, however, is a diurnal insect with a main peak of activity about 1 to 2 hours before light-off and little or no activity in the dark portion of the cycle (Taylor and Jones, 1969). A bimodal pattern with a smaller peak following dawn is also apparent in this species.

The tsetse-fly *Glossina morsitans* is strictly diurnal. In a rocking actograph at LD *12* : 12 its activity occurs during the light in short bursts of about 1 minute duration separated by long intervals (Brady, 1970, 1972). Nevertheless, the mean hourly activity of groups of insects (teneral unfed males) reveals a clear V-shaped diurnal pattern with peaks in the morning and evening, similar to that observed in the field; activity during the dark is almost negligible. The question of bimodality in activity rhythms, especially in mosquitoes and tsetse-flies, will be re-examined in later sections, particularly with respect to its endogeneity.

B. The Endogenous Nature of Activity Rhythms

1. *"Free-running" behaviour in the absence of temporal cues*

Rhythms of activity in a light/dark-cycle provide few clues as to the physiological nature of the controlling mechanism, which might have both endogenous and exogenous components. The endogenous nature of a rhythm, however, is usually revealed when the organism is transferred from a light/dark-cycle (*LD*) into continuous dark (DD) or continuous light (*LL*), provided that temperature and other possible *Zeitgebers* are also held constant (Aschoff, 1960). Under these conditions an endogenous oscillation controlling a rhythmic activity will "free-run" and reveal its natural periodicity (τ). In this state τ often deviates slightly from 24 hours so that the onset or peak of activity appears either earlier or later by a few minutes every day. The fact that τ is close to but rarely equal to 24 hours is powerful evidence for an endogenous oscillator which is uncoupled from the environment and not being "driven" by any uncontrolled *Zeitgeber* associated with the solar day.

Roberts (1960) studied the free-running rhythms of locomotor activity in the cockroaches *Leucophaea maderae*, *Byrsotria fumigata* and *Periplaneta americana*. When transferred to DD he found that the rhythms persisted for at least 3 months at a constant temperature of 25°C. In the three species studied τ for individual cockroaches varied between about 23 and 25 hours. The value of τ for an individual was not absolutely fixed, however, and in some instances was observed to change abruptly and spontaneously (Fig. 2.1). Nevertheless, the *range* of realizable τ values in an individual insect is probably genotypic.

Earlier authors (Gunn, 1940; Harker, 1956) had reported a gradual loss of rhythmicity in cockroaches after a few days in constant light (*LL*). Roberts (1960), however, found no such loss for at least 20 days in *L. maderae* and for up to 7 weeks in *B. fumigata*. The difference between these results was attributed to the type of recorder used; in Harker's work, for example, tying the cockroach to a kymograph might have promoted a breakdown in activity not observed in a running wheel. Roberts (1960) also found

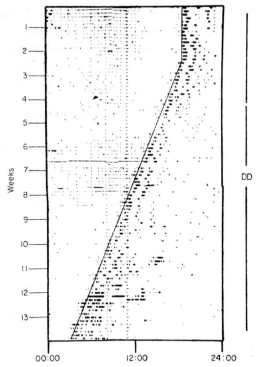

FIG. 2.1. The activity record of an individual of the cockroach *Leucophaea maderae* in constant darkness (DD) for 13 weeks showing an abrupt change in its free-running period (τ) in the third week from 24 hours to 23 hours 48 minutes. (From Roberts, 1960.)

that a transfer of the insects to *LL* caused (1) τ to *lengthen* by between 20 and 60 minutes, and (2) a *second* peak of running to occur about 10 hours after the onset of primary activity. After transfer to DD this secondary peak disappeared (Fig. 2.2).

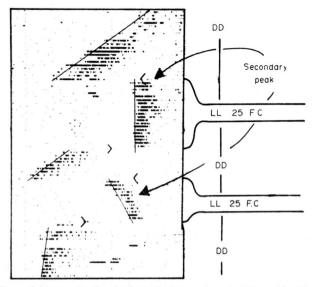

FIG. 2.2. The activity record of an individual of *L. maderae* in DD and in *LL*, showing the lengthening of the period (τ) in *LL*, and the appearance of a secondary activity peak in *LL* about 10 hours after the first. (From Roberts, 1960.)

The appearance of bimodality in *LL* and the spontaneous changes of τ in DD, both observed by Roberts (1960) in the cockroaches *L. maderae* and *B. fumigata*, are of particular interest because they indicate the probable multioscillator nature of the circadian system controlling activity rhythms (see Chapter 4). Both probably represent changes in the internal phase-relationship, or in the coupling, of the constituent subsystems.

The free-running behaviour of a circadian rhythm in constant conditions has been observed in a large number of insect species. A few of these will be mentioned here. Working with the house cricket *Acheta domestica*, Lutz (1932) observed the persistence of the rhythm of locomotor activity in DD and concluded that it was endogenously controlled. Nowosielski and Patton (1963) later observed that the rhythm persisted for at least 2 weeks in both DD and *LL*. Further examples have been recorded for mayfly nymphs (Harker, 1953), beetles (Park and Keller, 1932; Bentley *et al.*, 1941; Lohmann, 1964), the pond skater *Velia currens* (Rensing, 1961), and for the ant *Camponotus clarithorax* (McCluskey, 1965).

2. *"Aschoff's rule" and the "circadian rule"*

The range of τ values observed for a group of *Leucophaea maderae*, and the increase in τ on transfer from DD to *LL* are well illustrated in Fig. 2.3 (Lohmann, 1967). This change in circadian period is in accordance with "Aschoff's rule" (Pittendrigh, 1960),

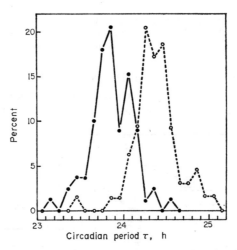

FIG. 2.3. Relative frequency distributions of free-running periods (τ) of *L. maderae* in DD (closed-circle lines) (N = 77) and in *LL* (open-circle lines) (N = 64). (Redrawn from Lohmann, 1967.)

which states that τ lengthens with an increase in light intensity, or on transfer from DD to *LL*, for dark-active animals (i.e. $\tau_{DD} < \tau_{LL}$, nocturnal), but shortens for light-active animals (i.e. $\tau_{DD} > \tau_{LL}$, diurnal). For the insects, however, the general applicability of this "rule" is in doubt (Table 2.1). In some nocturnal species such as cockroaches (Roberts, 1960), the house cricket *Acheta domestica* (Nowosielski and Patton, 1963), the pond skater *Velia currens* (Rensing, 1961), and the flour beetle *Tenebrio molitor* (Lohmann, 1964) the results for a transfer from DD to *LL*, and vice versa, are generally in agreement with "Aschoff's rule". In cockroaches, however, Roberts (1960) found no

TABLE 2.1. CIRCADIAN PERIOD (τ) OF OVERT LOCOMOTOR ACTIVITY IN INSECTS SHOWING CHANGES IN τ ON TRANSFER FROM DD TO LL, OR WITH AN INCREASE IN LIGHT INTENSITY. Asterisks mark those examples which violate "Aschoff's rule". N = nocturnal; D = diurnal.

Species	N/D	τ in DD, hr	τ in LL, hr	References
Leucophaea maderae	N	23.4–24.0	24.0–24.75	Roberts (1960)
Byrsotria fumigata	N	23.9–24.4	24.4–25.5	Roberts (1960)
Periplaneta americana	N	23.8	24.5	Roberts (1960)
Velia currens	N	24.2	26.0–27.7	Rensing (1961)
Acheta domestica	N	<24	>24	Nowosielski and Patton (1963)
Teleogryllus commodus	N	23.5	25.6	Loher (1972)
Tenebrio molitor	N		0.01 lux 24.3 2 lux 25.08 100 lux 26.07	Lohmann (1964)
Geotrupes sylvaticus*	D	24	>24	Geisler (1961)
Aëdes aegypti*	D	22.5	26	Taylor and Jones (1969)
Myrmeleon obscurus*	N	24–24.2	23.75–24	Youthed and Moran (1969a)

obvious correlation between τ and the intensity of illumination (in LL) and, in V. currens, Rensing (1961) failed to observe a systematic change of τ when the intensity was raised from 0.1 to 700 lux. Furthermore, in the mainly light-active dung beetle, Geotrupes sylvaticus, Geisler (1961) found that τ lengthened in LL—although the data are not compelling since the insects were only held in the recording apparatus for a 24-hour period. More recently, Youthed and Moran (1969a) have described a circadian rhythm of nocturnal pit-building activity by larvae of the ant-lion Myrmeleon obscurus with a peak in activity soon after dusk. This rhythm persisted in both LL and DD for at least 1 month. In continuous light τ varied from 23 hours 44 minutes to 23 hours 59 minutes; in DD it was between 24 hours 2 minutes and 24 hours 17 minutes. This lengthening of τ in DD for a nocturnal species clearly violates "Aschoff's rule". Taylor and Jones (1969) also demonstrated that τ increased in LL in the day-active mosquito Aëdes aegypti.

Aschoff (1960) later extended the so-called "Aschoff's rule" to cover two more parameters: (1) the ratio of activity-time (α) to rest-time (ϱ), and (2) the total amount of activity per circadian cycle. For a number of vertebrate species kept in constant conditions, Aschoff showed that both of these parameters increased with increasing light intensity in day-active animals, but decreased with light intensity in night-active animals. The generalization based on these observations—called the "circadian rule"—has been used in a model for circadian oscillations which suggests that the rhythm of locomotor activity is based on a continuous function which crosses a threshold twice during each daily cycle, and that activity only occurs when the function is above the threshold (Wever, 1965). Nevertheless, although this "circadian rule", like "Aschoff's rule", holds good for a number of birds and mammals (Hoffmann, 1965), it is violated by several examples from the insects. In particular, Lohmann (1964) showed—for the nocturnal beetle Tenebrio molitor—a positive correlation between the $\alpha : \varrho$ ratio and light intensity—despite the fact that the increase of τ on transfer from DD to LL agreed with "Aschoff's rule". Other violations of the circadian rule were observed for the cockroaches Byrsotria fumigata and Leucophaea maderae (Roberts, cited in Hoffmann, 1965).

3. *Temperature-compensation of the circadian period*

Roberts (1960) showed that the period of the locomotor rhythm in *Leucophaea maderae* was temperature-compensated. In one individual, for example, τ shortened from 25 hours 6 minutes at 20°C to 24 hours 24 minutes at 25°C and to 24 hours 17 minutes at 30°C. Although the calculation of a temperature coefficient was complicated by the known lability of τ in individual cockroaches, the Q_{10} was estimated to be about 1.04 for the rise in temperature from 20° to 30°C. This low value reveals the virtual "independence" of temperature which gives the circadian rhythm its functional significance as a time-measuring system (Pittendrigh, 1960).

4. *Bimodal activity patterns*

Circadian flight activity of the mosquito *Anopheles gambiae* has been intensively studied by automatically recording flight sounds (Jones *et al.*, 1967). This insect is a nocturnal species but with pronounced peaks of activity at both light-off and light-on. When transferred to DD the rhythm free-ran with a period (τ) close to 23 hours; constant light inhibited flight activity (Fig. 2.4). In DD it was the light-off peak which persisted and recurred with circadian frequency; the light-on peak appeared to be an

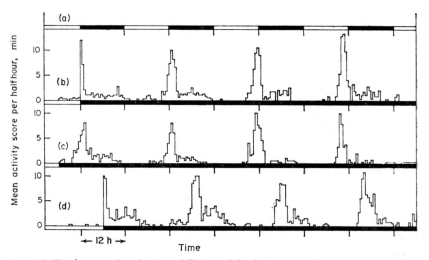

Fig. 2.4. The free-running rhythm of flight activity in the mosquito *Anopheles gambiae* when transferred to conditions of constant darkness (DD). (a) The rearing regime (*LD 12 : 12*); (b) transfer from *LD 12 : 12* to DD at the normal light-off time; (c) transfer from *LD 12 : 12* to DD 6 hours before normal light-off time; (d) transfer from *LD 12 : 12* to DD after 6 hours extra light. (From Jones *et al.*, 1967.)

exogenous or "startle" reaction associated with the sudden onset of light. For example, this peak disappeared in DD or when the abrupt onset of light was replaced by a more gradual artificial "dawn" (Jones *et al.*, 1972b). Flight activity in the diurnal species *Aëdes aegypti* was also bimodal with a small peak at light-on in addition to the main peak of activity about 1 to 2 hours before light-off. The rhythm persisted in DD with a low level of activity and a period (τ) of about 22.5 hours (Taylor and Jones, 1969). When adults were transferred from *LD 4 : 20* to DD both the light-on and the light-off peaks persisted at their expected times indicating that, unlike the bimodal activity pattern of *A. gambiae*, both peaks were endogenously controlled. Since they appeared to be independently phase-set by the "on" and "off" signals they may represent two independent ("dawn" and "dusk") oscillators (Taylor and Jones, 1969). The bimodal pattern of flight activity in *Aëdes taeniorhynchus* also persisted in DD with a period of about 23.5 hours (Nayar and Sauerman, 1971).

In the tsetse-fly *Glossina morsitans* the mean hourly activity of a group of teneral males showed a clear V-shaped pattern in *LD 12* : 12 with peaks in the morning and the evening (Brady, 1972). In conditions of constant darkness the free-running pattern

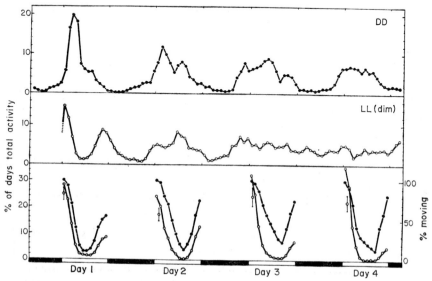

FIG. 2.5. The spontaneous locomotor activity of teneral males of the tsetse-fly *Glossina morsitans*. Upper panel: mean activity of nineteen flies in DD with individual records synchronized and showing free-running circadian rhythm. Middle panel: mean activity of eighteen flies in *LL* showing damping of rhythm within two or three cycle. Lower panel: mean activity of ten flies in *LD 12* : 12 showing bimodal activity pattern. Open circles are the mean hourly amounts of activity expressed as a percentage of each 24-hour total activity (left-hand ordinate); closed circles (right-hand ordinate) are the proportion of flies moving in any one hour. (From Brady, 1972.)

compiled by synchronizing the individual records to the time of appearance of their first major activity peak, showed that the rhythm persisted for at least 4 days (Fig. 2.5). The endogenous nature of the rhythm in *LL* was less clear.

C. Exogenous Effects

At the beginning of this chapter it was stated that most daily rhythms of activity and behaviour involve a mixture of endogenous and exogenous effects, and that the "pattern" of activity is always strongly modulated by the immediate character of the environment. Such an effect is to be seen, for example, in the "light-on" activity peak in *Anopheles gambiae* which is a direct response to the abrupt transition from dark to light. This "startle" reaction disappears in DD or if a gradual increase in light intensity is substituted for the abrupt change (Jones *et al.*, 1972).

In some species activity patterns in a light-cycle appear to be largely, if not entirely, exogenous. Odhiambo (1966), for example, showed that the locust *Schistocerca gregaria* was active during the light but inactive during the dark of an *LD 12* : 12 cycle. When the locusts were transferred to conditions of continuous darkness (DD) they became almost totally inactive, however, whereas in continuous light (*LL*) there was almost constant but erratic non-rhythmic activity. When they were transferred to a reversed cycle of light and dark (i.e. from *LD 12* : 12 to *DL* 12 : *12*) they adopted the new activity pattern immediately with none of the transient cycles (p. 18) expected from the re-entrainment of an oscillatory system.

Godden (1973) has recently demonstrated that the locomotor activity and oviposition rhythms in the stick insect *Carausius morosus* also have a pronounced exogenous component. In populations of this insect maintained at *LD 12* : 12 the egg-laying rhythm showed a peak of activity during the first 2 to 3 hours of darkness and very little activity during the light. When the insects were transferred to DD the population became apparently arrhythmic with activity occurring throughout the whole 24-hour period. In some —but not all—individuals, however, a persistent rhythm of egg-laying and locomotor activity was observed in DD, but the phase of the rhythm was reversed by about 12 hours with respect to previous entrainment. In one such example a free-running rhythm of locomotion was observed in DD with a period (τ) of about 24.3 hours. This persisted for 18 days until activity became random. The 12-hour time lag between the final *LD* cycle and the first activity peak in DD suggested that the initial response to light-off was a "rebound effect of release from photo-inhibition". The peak of activity following light-off must, therefore, be regarded as an exogenous response.

It is likely that most forms of crepuscular activity contain a strong exogenous component because the period of the day at which the activity can occur is restricted by environmental factors, particularly the light intensity. Tychsen and Fletcher (1971), for example, showed that mating in the Queensland fruit-fly *Dacus tryoni* occurred at an

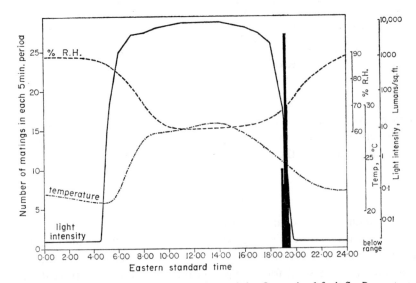

FIG. 2.6. Histogram of the number of matings of the Queensland fruit-fly *Dacus tryoni* occurring in each 5-minute period during a day under natural conditions. The changes in relative humidity, temperature, and light intensity are also shown. (From Tychsen and Fletcher, 1971.)

optimal light intensity of about 0.8 lux/ft². Under constant low light of this intensity mating followed a circadian rhythm which free-ran with an endogenous periodicity (τ) of about 28 hours; this endogenous component persisted for about 4 days. Conditions of constant darkness (DD) strongly depressed mating activity, however, and constant light (LL) of high intensity (900 lux/ft²) completely inhibited it. Nevertheless, "test dusks" given to samples of flies which had been kept in high-intensity continuous light revealed an endogenous rhythm of "readiness to mate" which persisted for two or three cycles before damping out. In the laboratory an instantaneous step-down from high to optimal light intensity applied at the normal time of dusk was an effective stimulus to mate. Under natural photoperiods the fact that mating was suppressed by both high light intensity and by darkness, restricted mating activity to a remarkably short period of 30 minutes each day at dusk when the light intensity reached its optimum (Fig. 2.6). The adaptive value of limiting mating to dusk is presumably because it synchronizes the sexual behaviour of all individuals in the population and thereby increases mating efficiency. Differences in mating times between different species of *Dacus* are also known to constitute effective barriers to hybridization.

D. Entrainment by Light and Temperature

1. *Entrainment by light-cycles*

When an endogenous oscillator (with period τ) is subjected to an environmental light-cycle (with period T) the period of the oscillator becomes the same as that of the *Zeitgeber*, provided that the latter is within the oscillator's range of entrainment. In natural light-cycles the oscillator therefore adopts a period which is exactly that of the solar day, namely 24 hours, and is considered to be entrained. In its entrained steady state the overt phase of the rhythm also adopts a fixed phase-relationship to the environmental cycle; entrainment thus constitutes both period-control and phase-control. In examining entrainment of a circadian rhythm to an environmental light cycle both the period of the *Zeigeber* (T) and the number of hours of light per cycle (the photoperiod) are important.

Roberts (1962) studied the entrainment of the locomotor activity rhythms in *Leucophaea maderae* and *Periplaneta americana* to a variety of light cycles in which both T and photoperiod were altered. When exposed to 24-hour cycles with a 12-hour photoperiod (LD *12* : 12, $T = 24$) the rhythms always attained a 24-hour period to match that of the *Zeitgeber*, with activity commencing at dusk. Figure 2.7 shows how a free-running rhythm of activity in *L. maderae* was "captured" and entrained by a cycle of LD *12* : 12. For the first 28 days of the experiment the rhythm free-ran in DD with a period (τ) of about 23.5 hours; subsequently the rhythm adopted the exact 24-hour period of the environmental cycle with sharp onsets of activity at light-off. The locomotor activity rhythm in *L. maderae* also became entrained to 24-hour cycles in which the photoperiod was other than 12 hours. When the photoperiod was long (e.g. LD *16* : 8 or LD *23* : 1) the overt phase of the rhythm and its pattern altered. With these very long light periods, as with LL (p. 9), there was a pronounced secondary activity which followed the primary onset at dusk by about 7 or 8 hours (Fig. 2.8). On return to DD this secondary peak of activity disappeared.

Roberts (1962) also demonstrated that the locomotor rhythm in *L. maderae* was entrained to a 24-hour period by some LD cycles whose periods were submultiples of

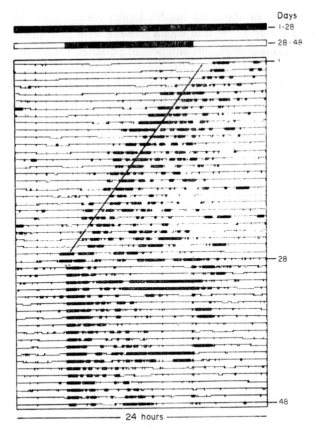

FIG. 2.7. Entrainment of the activity rhythm in an individual of *Leucophaea maderae* by a light/dark cycle (*LD 12* : 12) following free-run in DD. The natural period of the rhythm (τ) during the first 28 days is about 23 hours 30 minutes; subsequently the rhythm is entrained to a 24-hour period to match the *Zeitgeber*. (From Roberts, 1962.)

24 hours. Some individuals, for instance, became entrained to a cycle of *LD 2* : 2 (*T* = 4) or *LD 4* : 4 (*T* = 8), but entrainment failed with *LD 1* : 1 (*T* = 2). This phenomenon which is recognized as "frequency demultiplication" has also been recorded for the rhythm of flight activity of *Aëdes taeniorhynchus* in a cycle of *LD 6* : 6 (Nayer and Sauerman, 1971), and for the beetle *Carabus cancellatus* in *LD 1* : 1 and *LD 6* : 6 (Lamprecht and Weber, 1973). Data produced by Lohmann (1967) suggested that the range of entrainment for *L. maderae* was about the same as the range of free-running periods as shown in Fig. 2.3 (i.e. roughly between 23 and 25 hours). In *Carabus cancellatus* the primary range of entrainment was from 17 to 36 hours (Lamprecht and Weber, 1973).

2. *Phase-shifting by light-cycles and pulses, and the phase-response curve*

Following the experimental protocols established by Pittendrigh (1960) for the eclosion rhythm in *Drosophila pseudoobscura* (Chapter 3), Roberts (1962) showed that single or repeated light signals applied to the rhythm of locomotor activity in *L. maderae* "reset" the rhythm to a new phase, causing either advance phase-shifts ($+\Delta\phi$) or delay phase-shifts ($-\Delta\phi$) according to the circadian phase of the oscillator subjected to the light perturbation. Attainment of a new phase by the overt rhythm, however, was not instanta-

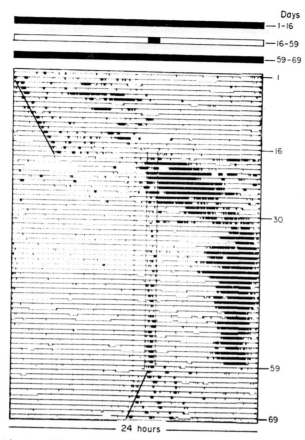

FIG. 2.8. Entrainment of the activity rhythm in *L. maderae* by LD 23 : 1. During the first 16 days the rhythm free-runs in DD with $\tau > 24$ hours. After the onset of the light cycle the phase of the rhythm is shifted, it becomes entrained to 24 hours, and a marked secondary peak of activity occurs about 7 hours after the first. On return to DD (day 59) the rhythm again free-runs, this time with $\tau < 24$ hours. (From Roberts, 1962.)

neous: the ultimate steady state was reached via a series of non-steady-state or "transient" cycles. Characteristically the number of transients was greater in the case of phase-advances than with phase-delays.

A single 12-hour light signal commencing during the middle of cockroach's "subjective day" caused a phase-delay of about 2 hours. A similar pulse applied during the "subjective night" caused a 1-hour phase-advance (Roberts, 1962). The magnitude of these phase shifts was therefore considerably less than those obtained for *D. pseudoobscura* (Pittendrigh, 1960, 1965) in which a light pulse of 15 minutes can induce up to a 10- or 12-hour advance or delay according to the circadian time at which the pulse is seen (Chapter 3). The effects of *repeated* signals applied out-of-phase were also investigated in *L. maderae*. When the insects were transferred from DD into *LD 12* : 12 so that the first 12-hour light pulse occurred during the insect's subjective night, the rhythm of activity phase-advanced by about 8 hours, but required about eleven transient cycles to effect the shift (Fig. 2.9A). Conversely, when the light regime started during the insect's subjective day, the rhythm phase-delayed by about $2\frac{1}{2}$ hours, but required many fewer transients to reach steady-state entrainment (Fig. 2.9B). Subsequent transfer to DD demonstrated that the phase of the endogenous rhythm had indeed been shifted by the light treatment.

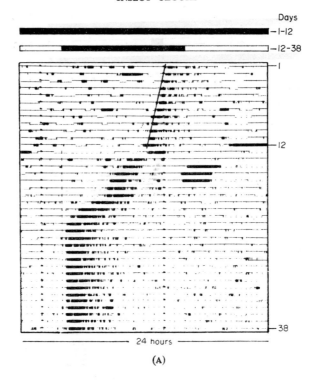

(A)

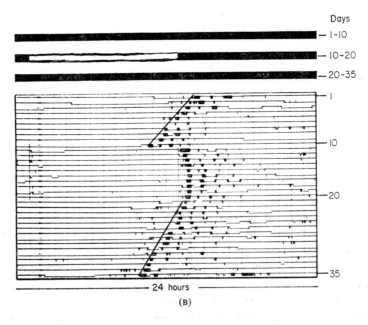

(B)

Fig. 2.9. Resetting the activity rhythm in *L. maderae* by light-cycles. A. Resetting with a light-cycle (*LD 12* : 12) initially out-of-phase with the free-running rhythm and with "dusk" falling during the insect's subjective day. The onset of running is gradually phase-advanced (+$\Delta\phi$) until entrainment is achieved after about eleven transient cycles. B. Resetting with a light-cycle (*LD 12* : 12) initially out-of-phase and with "dusk" falling during the insect's subjective night. The onset of running is rapidly phase-delayed (−$\Delta\phi$) and the rhythm becomes entrained. Subsequent transfer to DD (on day 20) shows that the phase of the endogenous rhythm has been shifted by the light treatment. (From Roberts, 1962.)

Re-entrainment of the locomotor rhythm in *P. americana* after a reversal in the *LD* cycle also required a series of transient cycles before its steady-state was achieved (Harker, 1956). A systematic analysis of the resetting pattern of the locomotor rhythm in *L. maderae* to single light pulses was not pursued; a phase response curve* for this species is therefore not available (Roberts, 1962).

The phase-setting and entrainment of flight rhythms by light pulses and cycles have also been investigated in the mosquitoes *Anopheles gambiae* and *Aëdes aegypti*. After an advance in the light-cycle achieved by shortening either one light or dark period, Jones *et al.* (1967) showed that the rhythm of flight activity in *A. gambiae* required several transient cycles before reaching its final steady state. In *A. aegypti*, a reversal of the *LD 12 : 12* light cycle by prolonging either one dark or one light period to 24 hours caused a resetting of the activity rhythm by a change in the time of light-off (Taylor and Jones, 1969).

Although a phase-response curve is not available for *L. maderae*, Jones *et al.* (1972a) have provided one for *A. gambiae*. Mosquitoes were raised at 25°C and *LD 12 : 12* and individual females were then subjected to 1-hour pulses of white light (70 lux) at different times in the first circadian cycle following a transfer to DD; the steady-state phases of the subsequent activity peaks were then measured at the end of the second cycle when the transients had subsided. The phase response curve so produced (Fig. 2.10) showed that, like *L. maderae*, the mean $\Delta\phi$ obtained was about 2 hours. Pulses applied late

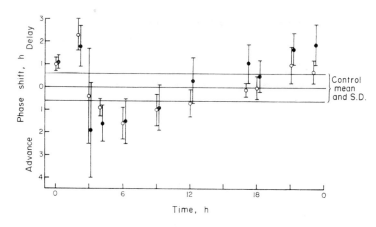

FIG. 2.10. The rhythm of flight activity in the mosquito *Anopheles gambiae*: phase response curve for 1-hour light pulses applied at different circadian times in the first cycle in DD after transfer from *LD 12 : 12*, showing maximum phase-delay ($-\Delta\phi$) of about 2 hours at 2 hours after normal light-off, and a maximum phase-advance ($+\Delta\phi$) of about 2 hours at 3 to 4 hours after normal light-off. Open circles: phase changes at the end of cycle 1; closed circles: phase changes at the end of cycle 2. (From Jones *et al.*, 1972a.)

in the subjective night caused phase advances ($+\Delta\phi$); those applied in the subjective day or early in the subjective night caused phase delays ($-\Delta\phi$). The phase response curve also showed the abrupt change between $-\Delta\phi$ and $+\Delta\phi$, which is also seen in *D. pseudoobscura* (p. 36) and all other species so examined (Aschoff, 1965). In *A. gambiae*, the

---

* A full account of the phase-response curve, the manner in which it is obtained, and its significance, will be given in Chapter 3 with reference to the fruit-fly *Drosophila pseudoobscura*.

magnitude of $\Delta\phi$ was found to be a function of signal energy. Five minutes at 70 lux was insufficient; with a 1-hour pulse $\Delta\phi$ increased up to 500 lux, but was insignificant below 10 lux.

3. *Entrainment by temperature cycles and pulses*

In the natural environment the diel cycle of light intensity is associated with a concomitant cycle of temperature, the normal phase relationship being when dawn falls somewhere near the low point of the temperature cycle. It is not surprising, therefore, that a daily temperature cycle, as well as a light-cycle, can act as a *Zeitgeber* and entrain circadian rhythms. Entrainment of such rhythmicities by temperature will be illustrated by reference to cockroach locomotor activity.

Roberts (1962) demonstrated that a 24-hour temperature cycle could entrain the locomotor rhythm in cockroaches (*Leucophaea maderae* and *Periplaneta americana*) held in otherwise constant conditions, despite the fact that earlier reports by Cloudsley-Thompson (1953) and Harker (1956, 1958) were conflicting or contradictory on this point. Figure 2.11, for example, shows an individual of *L. maderae* maintained throughout in DD. For the first 17 days the temperature was held at a constant 25°C and the locomotor activity rhythm free-ran with an endogenous periodicity less than 24 hours. From day 17 to day 44 the insect was subjected to a 24-hour sinusoidal temperature cycle fluctuating

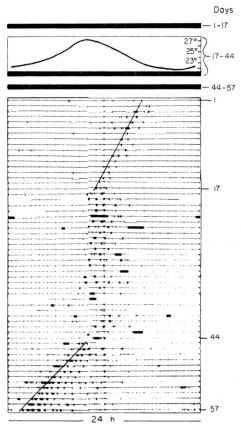

FIG. 2.11. Entrainment of the activity rhythm in *L. maderae* by a 24-hour sinusoidal temperature cycle (22–27°C) in constant darkness (DD). (From Roberts, 1962.)

between 22° and 27°C; the rhythm assumed the 24-hour period of the *Zeitgeber* with the onset of activity close to the high point of the cycle. After day 44 the temperature cycle was discontinued and the rhythm again free-ran.

Just as a temperature cycle can entrain a circadian rhythm as well as a light-cycle, a temperature pulse will cause a phase-shift in a manner comparable to a light pulse. For cockroaches *(P. americana)*, this was first demonstrated by Bünning (1959): phase-shifts occurred when the temperature was lowered for an interval of 12 hours or less, the magnitude of the phase-shift depending on the timing of the temperature treatment. Roberts (1962) showed that a 12-hour low-temperature pulse (7° or 12°C) applied to *L. maderae* otherwise maintained in DD at 25°C would also effect an alternation in phase. These phase shifts were similar to those effected by light although it was more difficult to differentiate unequivocally between phase-advances and phase-delays.

Since both light and temperature cycles can act as *Zeitgebers* in entraining the rhythm of locomotor activity, an interesting situation occurs when the phase angle between the two environmental driving cycles is altered (Roberts, unpublished, cited in Pittendrigh, 1960). In the natural environment, as already mentioned, the coolest part of the temperature cycle occurs close to dawn whereas the warmest part occurs late in the light period. Under this situation locomotor activity begins soon after dark. However, as the phase angle between sunset and the high point of the temperature cycle was experimentally widened, the onset of activity moved steadily to the right relative to the light regime until a discrete phase-jump occurred and the onset of activity "leaped" by about 12 hours to the next dusk (Fig. 2.12). It seems that at least 180° of the 360° of conceivable phase relative to the light-cycle constitutes a zone of "forbidden" phase relations, and indicates

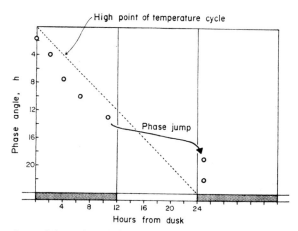

FIG. 2.12. The phase of the cockroach locomotor activity rhythm (onsets) as a function of the phase angle between simultaneously entraining light and temperature cycles. Ordinate shows the phase angle (in hours) between sunset and the high point of the temperature cycle. Note phase-jump that develops at a critical phase angle between the two *Zeitgebers*. (From Pittendrigh, 1960.)

that the light-cycle is a "stronger" *Zeitgeber* than temperature. A similar experiment with the rhythm of adult eclosion in *Drosophila pseudoobscura* will be described in Chapter 3.

Annotated Summary

1. Most activity rhythms in individual insects have a strong endogenous component, although they may be, to a variable extent, modulated by the direct effects of the environment.

2. The endogenous nature of circadian rhythms is revealed when the insects are transferred from a light/dark cycle into constant dark (DD) or constant light (LL) when the rhythm will "free-run" and present its natural periodicity (τ) provided that temperature cycles and other potential *Zeitgebers* are carefully controlled.

3. The endogenous circadian periodicity (τ) is rarely exactly 24 hours, and normally varies (between species or between individuals of the same species) from about 22 to 27 hours.

4. In many vertebrate species τ in LL is longer than in DD and lengthens as the light intensity increases if the animal is nocturnal, but is shorter or shortens with intensity if the animal is diurnal. This relationship—called "Aschoff's rule"—is not generally applicable to insects, however.

5. The period of the circadian oscillator in "free-run" is temperature-compensated, frequently showing a Q_{10} only a little greater than 1. (In the cockroach *Leucophaea maderae*, for example, it is 1.04.) This remarkable biological feature is an absolute functional prerequisite for a "clock".

6. When the free-running oscillator (with a period τ) is subjected to an environmental light or temperature cycle (with a period $T = 24$), the former adopts the period of the latter ($\tau = T$), or is entrained by it, undergoing phase-shifts, usually with one or more transient cycles.

7. The endogenous oscillator will also entrain to a light cycle whose period (T) is different from 24 hours, provided that it is within the oscillator's primary range of entrainment. The oscillator will also entrain to a periodicity where T is a submultiple of 24 hours (frequency demultiplication).

CHAPTER 3

CIRCADIAN RHYTHMS OF ACTIVITY
IN POPULATIONS OF INSECTS

ALTHOUGH many aspects of development occur but once in the life-cycle of an individual insect, their timing may be controlled by an on-going circadian oscillation in such a way that they occur at a particular time of the day or the night. In such cases a clear rhythm of activity is apparent only in a *population* of mixed developmental ages.

Probably the most spectacular of these rhythms is the emergence of adult insects from their pupae or puparia. In *Drosophila* spp., for example, pupal eclosion occurs during the hours close to dawn (Bünning, 1935; Kalmus, 1935, Pittendrigh, 1954, Brett, 1955). It also occurs close to dawn in the yellow dung fly *Scopeuma stercoraria* (Lewis and Bletchley, 1943), the Queensland fruit-fly *Dacus tryoni* (Myers, 1952; Bateman, 1955), and in the moths *Pectinophora gossypiella* (Pittendrigh and Minis, 1964) and *Heliothis zea* (Callahan, 1958). In the flour moth *Anagasta kühniella*, however, eclosion occurs in the late afternoon and early evening (Bremer, 1926; Scott, 1936; Moriarty, 1959), and in some Chironomids it occurs at night (Palmen, 1955). In still other insects such as the mosquito *Aëdes aegypti* pupal eclosion is not a rhythmic event (Haddow *et al.*, 1959).

Other events which are clock-controlled include egg hatching in *P. gossypiella* (Minis and Pittendrigh, 1968), pupation in *Aëdes* spp. (Nielsen and Haeger, 1954; McClelland and Green, 1970), and puparium-formation in the seaweed fly *Coelopa frigida* (Remmert, 1961) and the fruit-fly *Drosophila victoria* (Rensing and Hardeland, 1967; Pittendrigh and Skopik, 1970). Remmert (1962) gives an account of many of these rhythms.

Rhythms of oviposition have been extensively studied in the yellow-fever mosquito *Aëdes aegypti* (Haddow and Gillett, 1957) and in the pink bollworm *Pectinophora gossypiella* (Pittendrigh and Minis, 1964; Minis, 1965). Strictly speaking, since each female insect deposits many eggs on a number of occasions, these rhythms are not comparable to those of eclosion or pupation. Nevertheless, in *A. aegypti* a single bloodmeal leads to a single batch of eggs (= gonotrophic concordance), and each females lays her eggs in one or, in the most, two consecutive daily peaks (Gillett, 1962). No further eggs are then developed until the next bloodmeal is taken. For these reasons a rhythm is clearly expressed only in a population of mixed developmental ages; hence the inclusion of oviposition rhythms in this chapter.

The best-known "population" rhythm is undoubtedly that of pupal eclosion in *Drosophila* spp. The timing of this event attracted several early workers such as Kalmus (1935) and Bünning (1935), and has since been extensively studied by C. S. Pittendrigh

(1954, 1960, 1965, 1966) and his associates at Princeton and Stanford Universities. Much of what we know about *any* circadian system arises from Pittendrigh's work, and this example will form the basis of the present chapter.

A. General Properties of the Pupal Eclosion and Other Rhythms

Mixed-age populations of *Drosophila pseudoobscura* raised in alternating cycles of light and dark emerge as adults in a well-defined rhythm with peaks close to dawn. In its entrained steady state the rhythm assumes a definite phase-relationship to the driving light-cycle (Pittendrigh, 1965), the positions of the peaks depending on the length of the photoperiod. With very short photoperiods (less than 6 to 7 hours), for example, the eclosion peaks lie before dawn, whereas with longer photoperiods they occur after dawn. In LD *12* : 12 the median of the eclosion peak occurs about 2 to 3 hours after the onset of light (Fig. 3.1). This relationship is not linear, however: with photoperiods in excess of 18 hours there is a distinct change of phase, and in very long photoperiods, or in continuous light of sufficiently high intensity, the rhythm becomes inapparent.

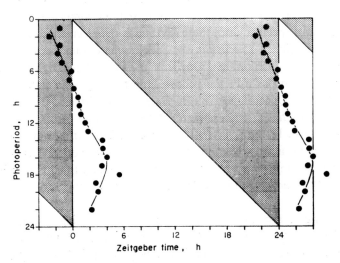

FIG. 3.1. The phase of the *Drosophila pseudoobscura* eclosion rhythm as a function of photoperiod. The plotted points are the medians [ϕ_r (R)] of the steady-state distributions of eclosion. (From Pittendrigh, 1965.)

Populations raised from the egg stage in DD and constant temperature show an arrhythmic pattern, but transfer to LD at any stage of larval or intra-puparial development will result in a rhythm being generated. Bünning (1935) showed that this could also take place after cultivation in constant dim light for fifteen generations. A rhythmic emergence of adult flies can also be generated by transferring populations from LD into DD, from LL into DD, or by means of a single unrepeated light signal—even as short as 1/2000 second—applied to an otherwise DD culture. The rhythm so produced "free-runs" in the further absence of temporal cues and reveals its natural periodicity (τ) which, in *D. pseudoobscura*, is very close to 24 hours. If a culture is transferred from DD to LL with an intensity of 0.3 to 3000 lux the rhythm so initiated damps out after three to four cycles (Chandrashekaran and Loher, 1969a): the period (τ) of the oscillation in *LL* is *less* than 24 hours (Bruce and Pittendrigh, 1957; Chandrashekaran and Loher, 1969a).

One of the important fundamental properties of endogenous circadian rhythms is that the period of the oscillation is temperature-compensated in the absence of temporal cues (Pittendrigh, 1954; Bruce, 1960). In *D. pseudoobscura*, for example, Pittendrigh (1954) demonstrated that temperature effects on period were very slight indeed. Three separate cultures raised in *LD 12* : 12 and at 16°, 21° and 26°C were transferred to DD whereupon

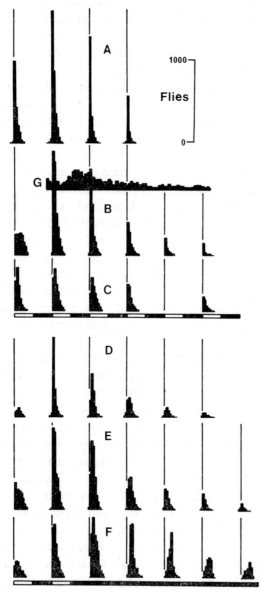

FIG. 3.2. The rhythm of pupal eclosion in *D. pseudoobscura*. A (26°C), B (21°C) and C (16°C) are from cultures maintained throughout their development under *LD* conditions, as indicated by the light regime represented under C. D (26°C), E (21°C) and F (16°C) are from cultures transferred, without temperature change, from *LD* to DD, as indicated by the light regime bar under F. G shows the aperiodic distribution of eclosion that is seen in cultures raised in DD conditions. (From Pittendrigh, 1954.)

the rhythms persisted for a further four or five cycles until all the flies had emerged. However, the period of the rhythm was not absolutely "temperature-independent": at 26° it was about 24 hours, whereas at 16° it has slowed to about 24.5 hours (Fig. 3.2). Zimmerman *et al.* (1968) have since shown that the period is 24.7 hours at 10°, 24.0 hours at 20°, and 23.7 hours at 28°. This slight positive dependence of τ on temperature ($Q_{10} = 1.02$) is important because it demonstrates that the rhythm is not being entrained by an *uncontrolled* aspect of the 24-hour environmental cycle (in which case the Q_{10} would have been 1.0), and amply demonstrates the endogenous nature of the rhythm.

Although the period of the rhythm is almost unaffected by temperature, certain other aspects are markedly temperature-sensitive. For example, since the rate of development varies with temperature, the population raised at 26°C utilized a smaller number of peaks than that at 16°C, but produced a larger proportion of the total population in each. In other words, although the period of the rhythm is temperature-compensated, its "*amplitude*" is not. The overt phase of the eclosion peaks also occurs later, relative to dawn, as the temperature is lowered. A third effect of temperature may be seen when a marked temperature *change* occurs at the same time as the transfer from LD to DD. Kalmus (1940) claimed that a rise in temperature at this point caused a shortening of the period, whereas a drop in temperature caused the period to be lengthened. On the basis of these observations he suggested that the endogenous rhythm controlling adult eclosion in *Drosophila* was temperature-dependent. Pittendrigh (1954), however, was able to demonstrate that it was only the first peak after the temperature change that was so affected, a fact apparently overlooked by Kalmus. Immediately after the initial delay or acceleration the system reverts to an essentially 24-hour periodicity only a little out-of-phase with that obtaining before the temperature shock. In 1954 Pittendrigh attributed the advance or delay of the first peak to a temperature-sensitive "terminal clock", separate from the temperature-compensated clock generally controlling eclosion. In later papers, however, these effects were recognized as "transient" cycles of the B-oscillator (see p. 36).

The pupal eclosion rhythm of *D. pseudoobscura* and *D. melanogaster* can be initiated by a light pulse or by the transfer of the population from LL to DD at any stage of larval or intra-puparial development (Bünning, 1935; Pittendrigh, 1954; Brett, 1955; Zimmerman and Ives, 1971). The oscillation governing eclosion is thus observed to operate continuously throughout post-embryonic development and to be unaffected by the extensive morphological reorganization that occurs during metamorphosis. Larval and pupal sensitivity is also the rule for the Mediterranean flour moth *Anagasta kühniella* (Moriarty, 1959), but in *Dacus tryoni* (Bateman, 1955) and *Sarcophaga argyrostoma* (Saunders, unpublished) entrainment and phase control can only be achieved with the larval stages, sensitivity to light coming to an end at puparium-formation. Presumably in these insects the clock becomes "uncoupled" from the environmental light cycle at this point. It is interesting, however, that temperature cycles applied to the intra-puparial stages of *Dacus tryoni* can continue to entrain the eclosion rhythm, and that a transfer of circadian phase in this species can take place through the egg (Bateman, 1955). In *S. argyrostoma* the fact that sensitivity to light ceases at puparium-formation makes biological "sense" in that puparia of this insect are subterranean and presumably not exposed to light cycles.

A number of other "population" rhythms will be reviewed here to illustrate their general properties and their similarities to the *D. pseudoobscura* case. The "giant" silkmoths, *Hyalophora cecropia* and *Antheraea pernyi*, for example, emerge from their pupae at species-specific times of the day. In *H. cecropia* emergence occurs in a broad peak 1 to 9 hours after dawn, whereas in *A. pernyi* it occurs later in the afternoon (Fig. 3.3) (Truman and Riddiford, 1970). As with *D. pseudoobscura* the rhythms attain a steady-state phase relationship to the light cycle which is a function of the photoperiod. In

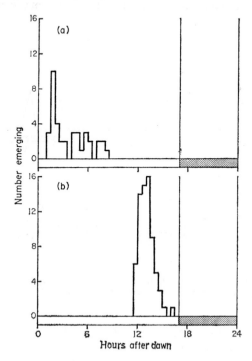

FIG. 3.3. The distributions of eclosions of the silkmoths (a) *Hyalophora cecropia*, (b) *Antheraea pernyi* in LD *17*:7. (Redrawn, from Truman and Riddiford, 1970.) (Copyright 1970 by the American Association for the Advancement of Science.)

A. pernyi, for instance, the eclosion peak is late in the afternoon in LD *12*: 12, LD *17* : 7 and LD *20* : 4, but "moves" to occupy the early hours of the night in LD *8* : 16 and LD *4* : 20. In very long photoperiods the eclosion rhythm begins to break down, emergence being scattered over a wide range of clock hours; in *LL* it becomes aperiodic (Truman, 1971b). Finally, when transferred from LD *17*: 7 into DD, the rhythm free-runs, revealing its endogenous periodicity (τ) of 22 hours.

Oviposition rhythms have been examined in at least two species: the yellow-fever mosquito *Aëdes aegypti* (Haddow and Gillett, 1957) and the pink bollworm *Pectinophora gossypiella* (Pittendrigh and Minis, 1964; Minis, 1965). The analysis of the oviposition rhythm in *A. aegypti* by A. J. Haddow and his co-workers deserves particular attention because it was carried out—at least during its early stages—in isolation from similar studies elsewhere (Gillett, 1972).

Haddow and Gillett (1957) showed that oviposition in caged populations of *A. aegypti* occurred in well-defined peaks towards the end of the light period in a normal tropical day (*LD 12* : 12)—both in the field and under laboratory conditions. Once again the steady-state phase of the rhythm was a function of the photoperiod: in LD *8* : 16, LD *12* : 12 and LD *16* : 8 the peaks occurred at the end of the light period, whereas with a shorter photoperiod (*LD 4* : 20) the peaks moved into the dark. Populations raised in the dark showed a weak periodicity, but with as little as 5 minutes of light per "day" a distinct rhythm became manifest (Gillett *et al.*, 1959). Those exposed to *LL*, however, were completely arrhythmic (Gillett *et al.*, 1959).

Larvae of mixed developmental ages raised in *LD 12* : 12 and transferred to DD as adults showed a periodic oviposition pattern for at least eleven cycles (Fig. 3.4). A similar

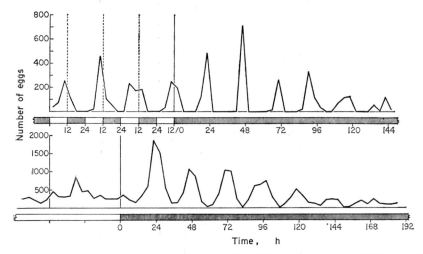

FIG. 3.4. The rhythm of oviposition in the mosquito *Aëdes aegypti* showing its "free-run" in constant darkness (DD). (Redrawn from Haddow *et al.*, 1961.)

rhythm was generated by the transfer of a population from *LL* into DD, or by the exposure of an otherwise DD population to as little as five seconds of light (Gillett *et al.*, 1961). Constant light with a single dark exposure did not elicit such a rhythm, however (Haddow *et al.*, 1961).

In the pink bollworm, *Pectinophora gossypiella*, the peak of oviposition occurred during the early part of the night and continued for about 7 hours (Pittendrigh and Minis, 1964; Minis, 1965). Unlike *D. pseudoobscura*, *A. pernyi* and *A. aegypti*, the duration of the photoperiod had little effect on the phase-relationship of the peaks to the light cycle; Minis (1965) attributed this to the complete suppression of oviposition by the light. Transfer of the moths to DD, however, revealed an endogenous periodicity (τ) of 22 hours 40 minutes (Fig. 3.5).

The rhythm of egg hatching in *P. gossypiella* has been investigated by Minis and Pittendrigh (1968). At 20° egg development takes about 9 to 10 days, and the initial 4-hour span of oviposition (one night's egg laying) is amplified in this time to about 52 hours. Eggs raised in either *LL* or DD show an aperiodic hatching pattern, but populations raised in *LD 12* : 12 show a distinct rhythm in which hatching is partitioned into discrete "packets" with medians just after dawn. This rhythm can also be initiated by a transfer from *LD 14* : 10 to DD, *LL* to DD, or by exposure to a single non-recurrent light pulse of 15 minutes. The last two treatments consist of signals giving no information on period; this rhythm therefore, like the others reviewed, is fully innate. A single temperature pulse of 28° will also generate rhythmicity.

One of the most interesting aspects of the egg-hatching rhythm in *P. gossypiella* was revealed by a systematic transfer of cultures from *LL* into DD every $5\frac{1}{2}$ hours during

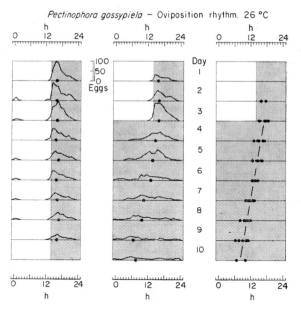

Pectinophora gossypiela — Oviposition rhythm. 26 °C

FIG. 3.5. The oviposition rhythm in the pink bollworm *Pectinophora gossypiella*. Left panel: the entrained steady state in *LD 14* : 10. Middle panel: a "free-run" in DD following *LD 14* : 10. Right panel: daily medians for seven free-runs, for which the average value of is 22 hours 40 minutes. (From Minis, 1965.)

development. This procedure showed that the rhythm could not be initiated until after the midpoint of embryogenesis (132 hours from deposition) (Fig. 3.6). Similar data for systematic transfers from *LD 12* : 12 to DD, or systematic exposures to 15 minutes of light, were also obtained. These results suggested that either the oscillator controlling egg hatching was not "differentiated" until 132 hours, or that it was present from the outset but was not coupled to the light cycle. The second of these alternatives was considered possible because a pinkish pigment appeared in the egg at about the time when initiation could first be achieved. On the other hand, the first alternative was thought to be the more plausible because a 12-hour temperature rhythm also failed to initiate rhythmicity when applied to the eggs during the first half of their development. If the oscillator does not develop until this point, *P. gossypiella* differs significantly from the Queensland fruit-fly, *Dacus tryoni*, in which a transfer of circadian phase can take place through the egg from adult to larva (Bateman, 1955).

The rhythm of pupation in the mosquito *Aëdes taeniorhynchus* has been intensively studied by Nayer (1967 a, b, 1968), Provost and Lum (1967) and Lum *et al.* (1968). In this species the daily peaks of adult emergence are *not* clock-controlled but merely reflect the antecedent rhythm of pupation; the interval between the pupal ecdysis and eclosion, for example, is dependent on temperature (Nielsen and Haeger, 1954; Provost and Lum, 1967). Nayar (1967 a, b) showed that larval cultures raised from the egg stage in DD showed a faint sinusoidal rhythm of pupation with a period (τ) of about 21.5 hours at both 27°C and 32°C. When an unrepeated light pulse of 4 hours was applied at the beginning of the last larval instar the peaks became more pronounced—as though the pulse had served to synchronize previously random oscillators—and τ became 22.5 hours. Since a single light pulse contains no information on period, and the frequency was unaffected by temperature, the rhythm of pupation was considered to be endogenous

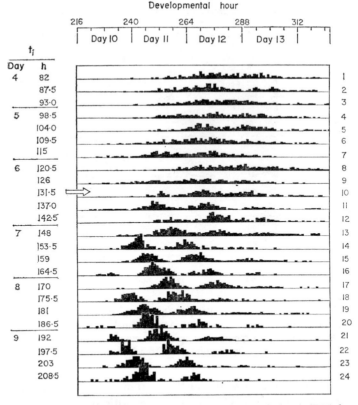

FIG. 3.6. Normalized distributions of egg hatch in continuous darkness (DD) for twenty-four populations of *Pectinophora gossypiella* eggs reared in 20°C and constant light (220 lux/m²) and systematically transferred to DD at 5½-hour intervals, from hour 82 to 208.5. The time of the transition is indicated by t_i. Note that a rhythm of egg hatch is initiated only when the *LL*/DD transition occurs after about the mid-point of development. (From Minis and Pittendrigh, 1968.) (Copyright 1968 by the American Association for the Advancement of Science.)

and temperature-compensated. A similar effect was observed following a single tempera-ture pulse of 4 hours at 32°C in an otherwise 27°C culture. The most curious feature of this rhythm was that it could not be entrained to a 24-hour cycle of light (*LD 12* : 12) or temperature (12 hours at 27°C, 12 hours at 32°C) and continued to run with a perio-dicity of about 22.5 hours (Provost and Lum, 1967; Nayar, 1967a). Only when the larvae were reared in the "stress" conditions of crowding, high salinity and "basic" ration was the period of the rhythm close to 24 hours (Nayar, 1967b). One explanation for this inability to lock onto the exact 24-hour periodicity of the solar day could be that the clock becomes "uncoupled" from the light cycle before the pupal ecdysis so that the pupation rhythm is actually free-running in *LD 12* : 12 as though it were in DD. Nevertheless, the larvae were clearly sensitive to light signals since reversed light-cycles or single light per-turbations caused phase-shifts with observable transients, continuous light (*LL*) caused arrhythmicity within 48 hours, and the rhythm could apparently be "entrained" to environmental light cycles shorter than 24 hours. In *LD 11* : 11, *LD 8* : 8 and *LD 6* : 6, for example, the rhythm adopted a period of 22, 20, 16 and 12 hours, respectively (Nayar, 1968). How far this can be considered normal entrainment, as opposed to an exogenous effect, is not clear.

B. Population Rhythms as "Gating" Phenomena

The general properties of the pupal eclosion rhythm in *D. pseudoobscura* suggest the existence of a self-sustained oscillator which partitions a mixed-age population into daily activity peaks. This further suggests that certain hours of the day constitute "forbidden zones" for eclosion, and that the "allowed" zones are dictated by the circadian clock. Pittendrigh (1966) has called these allowed zones "gates".

The validity of using a mixed-age population of insects to demonstrate an endogenous rhythm has been questioned, however. Harker (1964, 1965 a, b) related the timing of certain developmental events (head eversion, eye and wing pigmentation) in the pupa and pharate adult of *D. melanogaster* to the time of adult emergence on the one hand and to the environmental light cycle on the other. According to Harker the timing of these developmental events varied widely according to the circadian time at which the individual commenced an earlier stage in its development. For example, the interval between head eversion and the appearance of the yellow eye pigment was shorter when the earlier event occurred during the hours shortly after dawn than at later times in the cycle. She concluded that a fly simply emerged when its development was completed, and that its developmental time was only the sum of all the intermediate developmental stages. She further concluded that the observed rhythm of eclosion was not a function of an oscillation in an individual. However, her results would suggest—even if eclosion was not controlled by a circadian clock—that the intermediate "steps" were.

Recent work by Skopik and Pittendrigh (1967) and Pittendrigh and Skopik (1970) with both *D. pseudoobscura* and *melanogaster*, however, has failed to substantiate Harker's

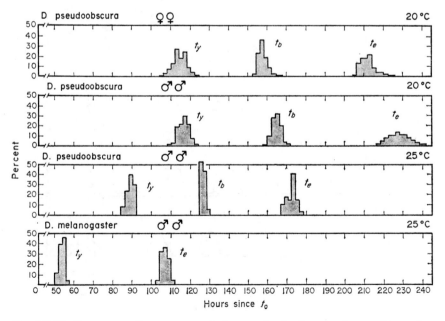

FIG. 3.7. The time course of emergence and of some prior developmental events in populations of *Drosophila pseudoobscura* which formed puparia within a known 1-hour interval ($t_0 \pm 0.5$ hour) and maintained throughout at 20°C or 25°C and continuous bright white light (*LL*). t_y = time of appearance of yellow eye pigmentation; t_b = black pigment appearing in ocellar bristles; t_e = time of eclosion. Note that the distribution of these events become wider, but remain unimodal. (From Pittendrigh and Skopik, 1970.)

findings, and these authors arrived at quite a different conclusion. Using developmentally
synchronous populations of flies collected as newly formed puparia in a 1-hour "collec-
tion window" (at t_0), Pittendrigh and Skopik (1970) examined the times of the following
events within the puparium: head eversion (t_h), yellow eye pigmentation (t_y), black
pigmentation in the ocellar bristles (t_b), and the final act of eclosion (t_e). In the *absence*
of a circadian oscillation (i.e. in continuous high intensity light) the distribution of each
development stage occurred as a single peak (Fig. 3.7), although the initial range of
1 hour during which puparium-formation had occurred had widened to 20 hours or
more by the time of emergence. The rate of development was, of course, temperature-
dependent, and the females completed their development before the males.

In the presence of a circadian oscillation, generated either by a transfer from *LL* to
DD, or by an alternating cycle of *LD 18* : 6, the picture was quite different. Figure 3.8
shows the fate of twenty-seven initially synchronous populations which experienced the
LL/DD transition at different developmental ages. The appearance of yellow eye pigmentation and black ocellar bristles clearly occurred at fixed time intervals after puparium
formation. The time of eclosion, however, depended on the circadian time at which the
transfer to DD occurred, and in some populations the distribution was clearly "split"
into two distinct peaks, almost 24 hours apart. In an *LD 18* : 6 regime these peaks occurred, as might be expected, in the hours immediately following "dawn". These results
showed that if the developing adults were not at the "correct" morphogenetic stage to
utilize one particular allowed zone or gate, they were required to remain within the

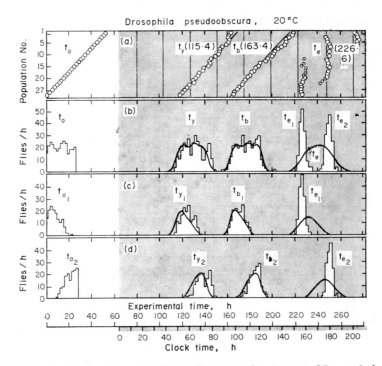

FIG. 3.8. Gated and ungated events in the development and emergence of *D. pseudoobscura*
(adult males) after a single step from *LL* to DD at different developmental times. The top
panel (a) plots the medians for twenty-seven individual populations of developmentally
synchronous pupae. Note that in certain populations the distribution of eclosion is
partitioned into two discrete peaks about 24 hours apart, whereas those for the intermediate
developmental steps are not. (From Pittendrigh and Skopik, 1970.)

puparium until the next, the intervening hours constituting a "forbidden zone" for eclosion. In *D. pseudoobscura* the gates recur with circadian frequency (modulo $\tau + 15$ hours) after the *LL*/DD transition.

These results demonstrate the reality of an endogenous oscillation in each developing fly which dictates the circadian time of eclosion but not that of the intervening developmental stages. In an earlier paper (Skopik and Pittendrigh, 1967; see also Pittendrigh,

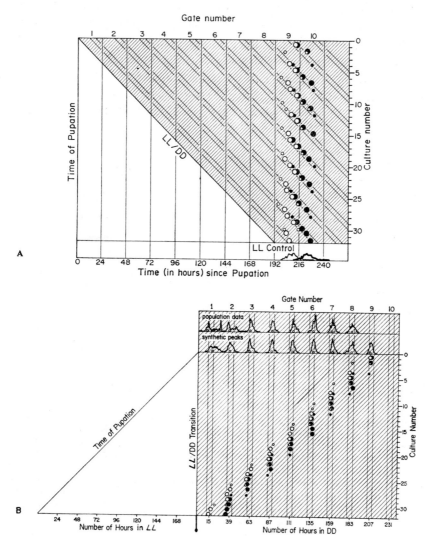

FIG. 3.9. (A) The effect of an *LL*/DD transition on the time of emergence of adults of *D. pseudoobscura* from populations of the *same* developmental age. The medians of the emergence peaks are given as points; open (female) and solid (males). For some cultures (e.g. No. 1, female) there is only one emergence peak. For others there are two (e.g. No. 2, female): the median points for these cultures are drawn at half size. (B) Simulation of a population of mixed developmental ages, by rearrangement of the data from (A). The uppermost panel (labelled "population data") gives the observed distribution of emergence peaks and their medians for a population of mixed developmental age that had been cultured in *LL* and transferred to DD at the point indicated. The next panel ("synthetic peaks") was obtained by summing, and normalizing to equal areas, all the distributions whose medians are given below as open and solid points. Thus the "synthetic peaks", like those marked "population data", include both sexes. (From Pittendrigh, 1966.)

1966) it was clearly demonstrated how a series of developmentally synchronous populations, in which the *LL/DD* transition was systematically varied, could be rearranged to simulate a population of *mixed* developmental age (Fig. 3.9).

The gating of certain developmental stages by such a mechanism is probably ubiquitous in insect "population" rhythms.* It is evident, for example, in pupal eclosion of *Antheraea pernyi* (Truman, 1971a), egg hatch (Minis and Pittendrigh, 1968) and oviposition of *Pectinophora gossypiella* (Pittendrigh and Minis, 1964) and in the oviposition rhythm of *Oncopeltus fasciatus* (Rankin *et al.*, 1972). McClelland and Green (1970) have described a similar situation for the mosquito *Aëdes vittatus*, in which the rhythm of pupation was so controlled. In these experiments virtually synchronous larval cultures were obtained by immersing eggs in filter paper into water for a 3-hour period. Eight such cultures were established, the start of each being staggered over a 24-hour period in 3-hour intervals. Figure 3.10 shows that larvae reared in *LL* showed no periodicity in pupation,

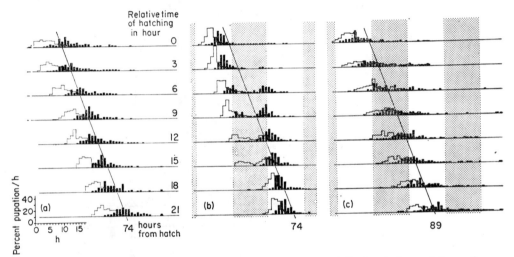

FIG. 3.10. The distribution of pupation times in sequentially hatched populations of mosquito larvae. Each successively lower line represents a population hatched 3 hours later relative to the one above and reared at 31°C under the light regime indicated by shading. The hollow histograms represent the percentage of total males in each population pupating in a single hour; the solid bars similarly represent females. The oblique line cuts each base line at the same age for each population and represents the mean time of pupating of females in the eight populations pooled (in (b) the value for (a) is used). (a) *Aëdes vittatus* in *LL*; (b) *A. vittatus* in LD *12* : *12*; (c) *A. aegypti* in LD *12* : *12*. Note how the distribution of pupation of *A. vittatus* in LD *12* : *12* is partitioned in some populations into discrete peaks or "gates". Pupation in *A. aegypti* is not a gated event. (From McClelland and Green, 1970.)

as expected, males developing in about 67 to 69 hours and the females in 72 to 76 hours. Under a regime of *LD 12* : *12*, however, the larvae were clearly forced to utilize pupation "gates" only available at certain phases of an on-going oscillation. Under certain circumstances, therefore, the populations were partitioned into two discrete pupation peaks. In *A. vittatus* the width of each gate is about 17 hours for males and about 15 hours for females; this contrasts with a gate width of only 6 hours for both sexes in *D. pseudo-obscura* (Skopik and Pittendrigh, 1967). It is of interest, however, that the females with

---

* See also the gating of the circannual rhythm in *Anthrenus verbasci* (Blake, 1958, 1959), Chapter 9.

their longer developmental time were sometimes forced to utilize a later gate than the males. In *D. pseudoobscura* the converse was true because the males showed a longer development. A similar periodicity of pupation has been described for *Aëdes taeniorhynchus* (Neilsen and Haeger, 1954), but is not apparent in *A. aegypti* (McClelland and Green, 1970).

Although pupation in *A. aegypti* is not a periodic phenomenon the deposition of eggs is a rhythmic event (Haddow and Gillett, 1957) and is therefore "gated". Gillett (1962) studied the "contribution" made by individual females to the population rhythm by enclosing them in small lamp-glass cages and observing the times of feeding and oviposition for each insect. In a natural tropical light cycle of *LD 12* : 12 Gillett showed that each mosquito that had completed egg development "waited" for the first available egg-laying period. If all her eggs were mature when the first gate arrived all were deposited in a single batch. On the other hand, some females were forced to lay some of their eggs in one gate and the remainder in the next, a full 24 hours later. A similar situation was observed for mosquitoes transferred from *LL* to DD, but in *LL* eggs were deposited when mature and with no sign of periodicity. When transferred to DD the first gate occurred 22 hours after the *LL*/DD transition and at circadian intervals thereafter. Following the terminology adopted for *D. pseudoobscura* (Pittendrigh, 1966), therefore, the gates were observed to occur at intervals of modulo $\tau + 22$ hours, where τ for *A. aegypti* was close to 24 hours. Since the rate of egg development was temperature-dependent whereas the period of the oscillation controlling the rhythm was not, mosquitoes kept at a lower temperature were forced to utilize later gates (Gillett, 1972).

C. Entrainment of Population Rhythms by Light and Temperature

1. *The phase response curve*

When the eclosion rhythm of *D. pseudoobscura* is allowed to free-run in DD after exposure to *LD 12* : 12 or to *LL* it is regarded as passing through alternate half-cycles of "subjective night" and "subjective day", with the beginning of subjective night (called hereafter circadian time, Ct 12) occurring at the *LL*/DD transition and at intervals (modulo τ) thereafter (Pittendrigh, 1965). Since τ for *D. pseudoobscura* is essentially 24 hours, subjective night and day each occupy $\tau/2$ or 12 hours. In such a free-running state the peaks of eclosion occur at intervals of modulo $\tau + 15$ hours after the *LL*/DD transfer. However, single pulses of white light applied at different phase points of the oscillation, i.e. at different circadian times will produce substantial phase-shifts in the subsequent steady state of the rhythm. For example, 15-minute light pulses applied early in the subjective night cause significant phase-delays ($-\Delta\varphi$), whereas those applied late in the subjective night or early in the subjective day produce significant phase-advances ($+\Delta\phi$). Pulses applied between Ct 4 and Ct 12, however, have little or no effect on the phase of the resulting rhythm of eclosion, and at Ct 18 there is an abrupt 360° change from phase-delay to phase-advance. These responses to single light perturbations provide a phase-response curve for the oscillation (Fig. 3.11). Similar responses to light pulses of 12 hours, 4 hours and 1/2000 sec have also been published (Pittendrigh, 1960). They differ mainly in the magnitude of the $\Delta\phi$ produced and in the circadian time of the phase-inversion. Although the concrete nature of the oscillation in *D. pseudoobscura* remains unknown, as it does in all other organisms, the phase-response curve provides one experi-

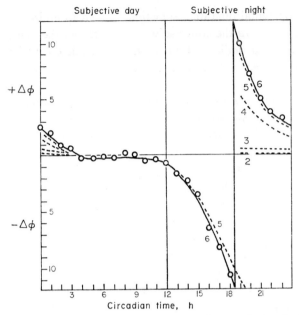

FIG. 3.11. The phase-response curve for the *D. pseudoobscura* eclosion rhythm based on 15-minute signals of white light. The solid curve, and the points plotted as open circles, describe the steady-state phase-shifts measured on day 6 after the signal. The curves plotted as dotted lines are based on the observed shifts on day 2, 3, 4 and 5. The delays ($-\Delta\phi$) are essentially completed immediately, but the advances ($+\Delta\phi$) develop slowly and are not complete until day 5 or even day 6. The phase-response curve has been plotted with advances above and delays below the control (unperturbed) line to comply with the new convention adopted by Pittendrigh (1967). (Redrawn from Pittendrigh, 1965.)

mental assay of the oscillation's phase, and is the best available characterization of its time-course.

The phase-response curve as represented in Fig. 3.11 is based on steady-state phase shifts ($\Delta\phi$) of the rhythm which, especially in the case of phase-advances, are only reached after several so-called transient (i.e. non-steady-state) cycles. Although some theoretical models for the circadian clock show that a *single* oscillator can undergo such transients after perturbation (see, for example, Klotter, 1960; Johnson and Karlsson, 1972a), transients can be accounted for by the *two*-oscillator model proposed by Pittendrigh and Bruce (1957, 1959) and Pittendrigh *et al.* (1958). The essential features of this model are that the physiological mechanism immediately underlying eclosion is governed by one oscillator (the B-oscillator) which is distinct from a second and light-sensitive A-oscillator. The A-oscillator is envisaged as a central pacemaker directly coupled to the environmental light cycle. The B-oscillator, on the other hand, is regarded as a *driven* element, coupled to the driver (A) and whose phase more immediately controls eclosion. It is further envisaged that the A-oscillator is immediately reset by the light pulse but the B-oscillator requires several cycles before it attains a steady-state relationship to the driver; hence the series of transients. The B-oscillator is not light-sensitive, but may be sensitive to temperature pulses and cycles. B therefore derives phase-control from two sources—directly from the temperature and indirectly via A from the light cycle. The two-oscillator model receives support from several types of experiment which will be examined later. Initially, however, the entrainment by light in the absence of temperature changes will be considered.

For several reasons it is convenient to represent the phase-response curve in a mathematically equivalent form in which the advances are displaced 360° (24 hours of circadian time) on the ordinate and treated as phase delays, thus yielding a monotonic curve (Pittendrigh, 1967) (Fig. 3.12). Figure 3.13 shows three cycles of the oscillation following an LL/DD transition in this monotonic form, and illustrates the terminology adopted by Pittendrigh (1965, 1966, 1967) to describe the parameters of the system.

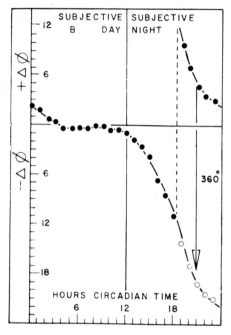

FIG. 3.12. The phase-response curve for *Drosophila pseudoobscura* showing its monotonic form. (Pittendrigh, 1967.)

As a test of Pittendrigh's hypothesis that light causes virtually instantaneous resetting of the A-oscillator, Chandrashekaran (1967a) transferred cultures of D. *pseudoobscura* from LD 12 : 12 to DD and then exposed them to *two* pulses of light (15 minutes at 3000 lux) in a single circadian cycle. In one experiment the first pulse (P_1) was timed to occur 27.5 hours after entry into DD and the second (P_2) at 34.0 hours after DD. According to the phase response curve for 15-minute perturbations (Fig. 3.11), the first pulse would fall at Ct 15.5 and cause a phase-delay ($-\Delta\phi$) of 5.0 hours in steady state; the second pulse, in isolation, would fall at Ct 22.0 and cause a phase advance ($+\Delta\phi$) of 4.8 hours. The result, however, indicated that a virtually instantaneous resetting had indeed occurred after P_1. The 5-hour $-\Delta\phi$ caused by P_1 resulted in P_2 falling, not at Ct 22.0, but at Ct 17.0 where it caused a further $-\Delta\phi$ of 8 hours; the net observed phase-delay of 13.5 hours was very close to the theoretically expected delay of 13.0 hours. In a second experiment, the opposing influences of pulses occurring 27.5 hours and 38.5 hours after DD cancelled themselves out because after the instantaneous $\Delta\phi$ caused by P_1 ($-\Delta\phi = 5.0$ hours) P_2 fell at Ct 21.5 and caused an advance ($+\Delta\phi$) of roughly the same magnitude. Pittendrigh (1974) has also verified the assumption that light pulses cause an instantaneous resetting by examining the phase-response curve following a light perturbation.

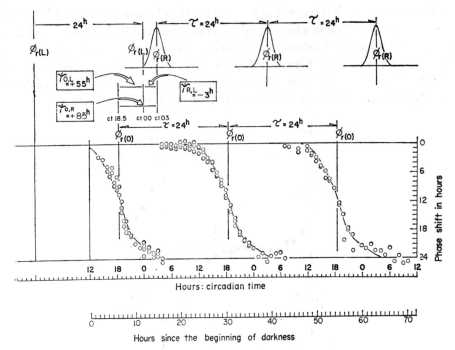

FIG. 3.13. The phase-response curve for the *D. pseudoobscura* eclosion rhythm, in its monotonic form, for three full cycles (72 hours) of the oscillation's free-run in DD following *LD 12 : 12*. The oscillation begins at Ct 12 when the lights go out. Each point plots the phase-shift as $-\Delta\phi$ caued by a single 15-minute light pulse at the times indicated on the abscissa. ϕ_r = phase reference point; $\phi_r(L)$ = phase reference point of light cycle; $\phi_r(R)$ = phase reference point of rhythm (median of eclosion peak); $\phi_r(O)$ = phase reference point of oscillation; $\psi_{R,L}$ = phase-angle (time) difference between the reference points of rhythm and light-cycle; $\psi_{R,O}$ = phase-angle difference between the reference points for rhythm and oscillation; $\psi_{O,L}$ = phase-angle difference between reference points for oscillation and light. (From Pittendrigh, 1967.)

2. *Entrainment by single recurrent light pulses of short duration*

The derivation of the phase response curve for the pupal eclosion rhythm in *D. pseudo-obscura* has been described above; its use in the prediction and interpretation of entrainment by light-cycles will now be examined.

When a self-sustained oscillation such as that controlling the eclosion rhythm (whose natural period is τ) is entrained by an external periodicity (whose period is T), the oscillation assumes the period of the driving cycle and maintains a fixed phase-relationship to it. This entrainment is effected by discrete and apparently instantaneous phase-shifts of the A-oscillator ($\tau - T = \Delta\phi_{ss}$) although, of course, there is usually a number of transient cycles before the driven system controlling eclosion "catches up". The magnitude and the sign of these phase-shifts depend on the circadian time at which the light pulse is seen. When T is shorter than τ the oscillation phase-lags the driver. Consequently the light pulse will fall in each cycle in the late subjective night and cause a phase-advance ($+\Delta\phi$). Conversely when $T > \tau$ the oscillation will phase-lead the driver, the pulse will fall in the early subjective night and generate a phase-delay ($-\Delta\phi$) (Fig. 3.14). For single 15-minute light pulses we can refer to the phase response curve (Fig. 3.11) to determine the phase-shifts so generated.

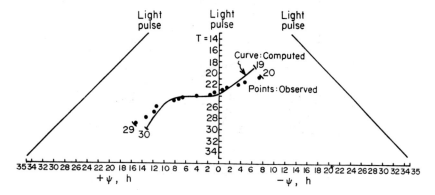

Drosophila pseudoobscura – Pupal Eclosion rhythm. 20° C.

FIG. 3.14. *Drosophila pseudoobscura*. The steady-state phases of the pupal eclosion rhythm ($\psi_{R,L}$) and its oscillation ($\psi_{0,L}$) to the period (T) of the light cycle defined by a single 15-minute light pulse per cycle. The solid curve marks $\psi_{0,L}$; it is a predicted curve (see Pittendrigh, 1965) which has been confirmed experimentally (see Pittendrigh, 1967). The points mark the phase (distribution means) of the rhythm. When $T > \tau$ the rhythm phase leads the oscillation; when $T < \tau$ the rhythm phase lags the oscillation. (From Pittendrigh and Minis, 1971.)

With recurrent 15-minute light pulses the interval between them defines the "environmental cycle" (T). For example, pulses 23 hours apart define a 23-hour environmental period ($T = 23$), whereas pulses 27 hours apart define $T = 27$. By effecting discrete phase-shifts the oscillation will become entrained to such cycles (Fig. 3.15). However, the oscillation is only able to assume the period (T) of a limited range of cycles. Steady-state entrainment is only possible when the light pulse falls on the response curve at points where the slope is less than -2.0 (Pittendrigh, 1966). The maximum utilizable phase-advance for Drosophila is $+5.9$ hours, and the maximum utilizable phase-delay is $-5.6f$ hors; the range of entrainment is, therefore, from about 18 to 30 hours. In practice it is lrom about 19 to 29 hours. Outside this range entrainment breaks down, but with still onger or shorter T values the oscillation may lock on to multiples or submultiples of τ.

3. *Entrainment by "skeleton" photoperiods*

Pittendrigh and Minis (1964) reported the important fact that two brief pulses of light *n* hours apart in each 24-hour cycle entrained the *D. pseudoobscura* rhythm to a steady-state phase-relation to such cycles that was identical to the phase-relation it assumed when entrained by a single long-duration pulse of *n* hours in each cycle. They described the two pulses as a "skeleton" photoperiod (PP$_s$) which simulates in its action on the oscillator the effect of a "complete" photoperiod (PP$_c$). In the later chapters on photoperiodism (Chapter 7) we will see further examples of how skeleton photoperiods may simulate the action of complete photoperiods in the induction of diapause. This phenomenon demonstrates the importance of the "on" and "of" signals of the photoperiod in the entrainment and phase-control of oscillations; the time interval between them is also important, but the fact that there is or is not light between the pulses is in some cases immaterial.

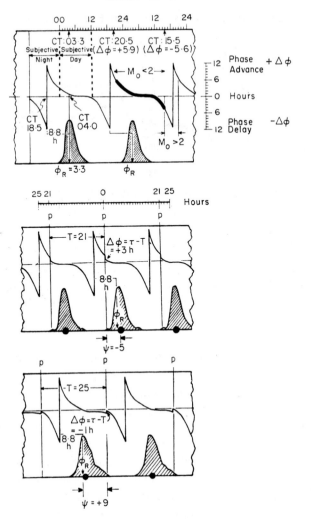

FIG. 3.15. The entrainment of the *D. pseudoobscura* oscillation by recurrent 15-minute light pulses. Upper panel: the phase-response curve (new convention). The subjective night is the half-cycle lying between Ct 12 and Ct 24; the subjective day lies between Ct 24 and Ct 12. Entrainment is only possible when the light pulse in the steady state falls at points on the response curve where its slope (M_0) is less than -2 (heavy line). Middle panel: entrainment by light cycles in which a 15-minute light pulse recurs every 21 hours ($T = 21$). In $T = 21$ the light pulse causes a phase advance of 3 hours ($\Delta\phi_{ss} = \tau - T$) in each cycle, and it falls at Ct 23.3 in the late subjective night. In $T = 25$ (lower panel) the light pulse causes a phase delay of 1 hour and falls at Ct 12.3 in the early subjective night. (From Pittendrigh, 1966.)

Pittendrigh and Minis (1964) described two types of skeleton photoperiod. "Symmetrical" skeletons are defined by two pulses of equal duration (say 15 minutes each), whereas "asymmetrical" skeletons are defined by a longer, or "main", photoperiod (say 4 to 14 hours in length) and a secondary pulse scanning the "night". The entrainment of the oscillation by symmetrical skeletons will be considered first.

In the *D. pseudoobscura* case symmetrical skeleton photoperiods consisting of two 15-minute pulses of white light simulate almost perfectly the action of complete photoperiods (at $T = 24$) up to about 11 hours (Fig. 3.16) (Pittendrigh, 1965). The phenomenon of entrainment by these two-point skeletons may be "explained" by the use of the phase-

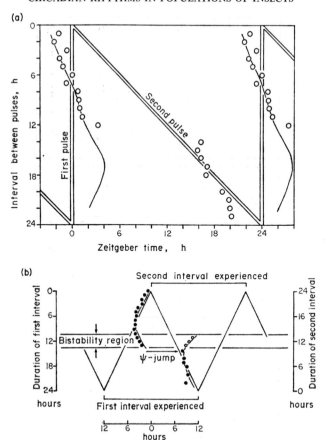

FIG. 3.16. (a) The phase of the *D. pseudoobscura* eclosion rhythm as a function of symmetrical skeleton photoperiods. The plotted points are medians [ϕ_r (R)] of the steady-state distributions of eclosion. The solid curve is that fitted to the medians for complete photoperiods (see Fig. 3.1). Note the phase-jump that occurs at about PP_s *14* : 10. (From Pittendrigh, 1965.) (b) Ditto, showing the "region of bistability" (between PP_s 10.3 and 13.7) for which there are two possible steady states with radically different ψ-values. (From Pittendrigh, 1966.)

response curve, if the final steady-state phase is computed from a sum of the two phase-shifts caused by the two separate pulses. Thus in steady state when the oscillator assumes the period of the driving light cycle (τ is changed to T), the following relation holds:

$$\tau - T = (\varDelta\phi_1) + (\varDelta\phi_2)$$

where $\varDelta\phi_1$ is the phase shift caused by one light pulse (p_1) in the skeleton photoperiod and $\varDelta\phi_2$ is the phase shift caused by the second pulse. Figure 3.17 shows how the phase shifts derived from the PRC may be used to calculate the steady-state phase relationship.

Symmetrical skeleton photoperiods, however, differ from complete photoperiods in one important respect: there is no continuous light between the "on" and "off" signals. For this reason a skeleton can be interpreted in two ways. A skeleton of PP_s 10 : 14, for example, is also a skeleton of PP_s 14 : 10, the only difference between the two being one of phase. Which of the two interpretations the oscillator accepts depends on a number of factors, the most important of which is the duration of the interval between the two pulses.

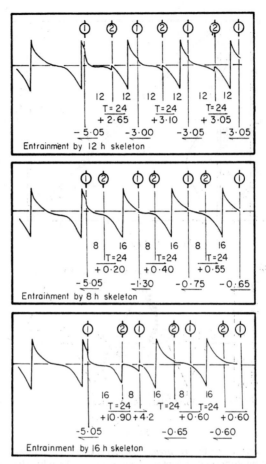

FIG. 3.17. The entrainment of the *D. pseudoobscura* oscillation by 2-pulse (symmetric) skeleton photoperiods, showing how the phase-response curve (new convention) may be used to predict the resulting phase of the eclosion rhythm. The panels show the $\Delta\phi$'s generated by the two 15-minute pulses (1 and 2) in PP_s *12* : 12 (upper), PP_s *8* : 16 (middle) and PP_s *16* : 8 (lower). In the case of PP_s *16* : 8 (lower panel) the steady-state phase of the oscillation eventually becomes identical to that of PP_s *8* : 16. (Redrawn from Pittendrigh, 1965.)

When the interval between the pulses is less than about 11 hours the simulation of PP_c by PP_s is almost perfect. With PP_s 12 : 12 simulation is less good in the sense that the phase relationship (ψ) of the rhythm to the pulse acting as dawn is no longer the same as that under a complete photoperiod of 12 hours. With skeletons longer than 14 hours (PP_s 14 : 10 and over), however, the light fails completely to simulate a 14-hour photoperiod and the oscillator "leaps" to the second "interpretation", namely that of PP_s 10 : 14. In *D. pseudoobscura* the general rule is that when the two pulses have an interval between them of longer than 13.7 hours, the rhythm *always* assumes a phase-relation which is characteristic of the shorter duration (Pittendrigh, 1966). The change in phase is called a "phase-jump" (ψ-jump) (Fig. 3.16).

Skeleton photoperiods close to $\tau/2$ show an additional property only revealed by computer simulation (Pittendrigh, 1966). All such cycles between PP_s 10.3 and PP_s 13.7 are open to two distinct interpretations, and the steady state ultimately adopted depends on two variables: (1) the phase-point in the oscillation which experiences the *first* pulse

·of the entraining cycle, and (2) the value of the first *interval*. Figure 3.18 compares the positions of the eclosion peaks adopted in four skeleton photoperiods: PP_s 9 : 15, PP_s 15 : 9, PP_s 11 : 13 and PP_s 13 : 11. For the first two regimens the resulting steady states are unique and the phase (ψ) adopted is always that of the shorter interval (ψ_9 in both). For the other two, however (both of which lie in the zone from PP_s to 10.3 to

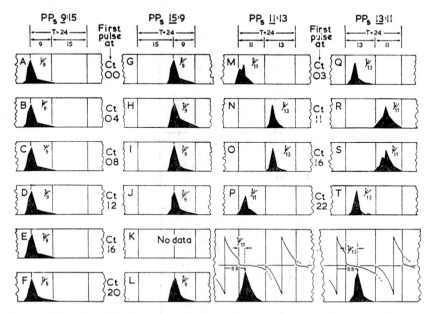

FIG. 3.18. The "bistability" phenomenon in the *D. pseudoobscura* eclosion rhythm. A unique phase-relationship (ψ_9) between the rhythm and light-cycle develops for PP_s 9 : 15 and PP_s 15 : 9, no matter where the first pulse that initiates the entraining cycle falls on the Ct scale. On the other hand, for both PP_s 11 : 13 and PP_s 13 : 11 there are two alternative steady states, ψ_{11} and ψ_{13}. Which develops depends on where the first pulse that initiates the entraining cycle falls on the Ct scale. The steady states of ψ_{11} and ψ_{13} are shown in terms of the phase-response curve (new convention) in the two lower right-hand panels. (Redrawn from Pittendrigh, 1966.)

PP_s 13.7), the phase adopted depends on the circadian time at which the first pulse is seen. For example, a PP_s 11.13 assumes the phase characeristics of PP_s 11 : 13 (ψ_{11}) when the first interval seen is 11 hours and the first pulse falls at Ct 03 or Ct 22, but assumes a phase characteristic of PP_s 13 : 11 (ψ_{13}) if the first pulse falls at Ct 11 or Ct 16. Conversely, a PP_s 13 : 11 assumes ψ_{13} when the first interval seen is 13 hours and the first pulse falls at Ct 03 or Ct 22, but assumes ψ_{11} if the first pulse falls at Ct 11 or Ct 16. This region close to $\tau/2$ is called the region of "bistability" (Pittendrigh, 1966). In later sections we will see apparently similar behaviour of oscillator(s) involved in photoperiodic phenomena.

Asymmetrical skeleton photoperiods also show the phenomenon of a phase-jump. Figure 3.19 shows the steady-state phase of the eclosion rhythm in *D. pseudoobscura* to asymmetric ($T = 24$) skeletons comprising a 4-hour main photoperiod and a 15-minute pulse scanning the accompanying dark period. As the "night interruption" falls later and later, from Zt 05 to Zt 12 the phase of the rhythm moves to the right. When the pulse falls at Zt 14 and later, however, a ψ-jump occurs and the peak of eclosion moves to later clock hours thus accepting the shorter interpretation. It is clear that

Drosophila pseudoobscura Pupal Eclosion Rhythm 20°C

hrs, ZT

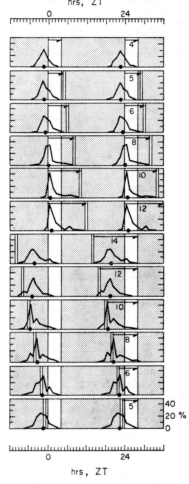

FIG. 3.19. The entrainment of the *D. pseudoobscura* eclosion rhythm by asymmetric skeleton photoperiods comprising a main photoperiod of 4 hours and 15-minute night interruptions. The effective skeleton is indicated by arrows and its duration by the number (of hours) below it. Note the phase-jump when the skeleton exceeds about 14 hours. (From Pittendrigh, 1965.)

when the pulse falls early in the night it is "read" as a terminator of a skeleton photoperiod (a new "dusk"), but when the pulse falls late in the night it is read as an initiator (new "dawn"). With longer and longer main photoperiods the position of the ψ-jump becomes later and later. In all cases except those close to the ψ-jump simulation of the corresponding complete photoperiod is good. Similar effects have been described for the action of asymmetric skeleton photoperiods on the oviposition rhythm in *Pectinophora gossypiella* (Pittendrigh and Minis, 1964, 1965) and on the pupal eclosion rhythm in the same species (Pittendrigh and Minis, 1971).

Asymmetric skeleton photoperiods are of particular interest in the analysis of the photoperiodic clock and will be examined further in Chapter 7. At this juncture, however, one difference between symmetrical and asymmetrical skeletons deserves attention: with asymmetrical skeletons it is possible to entrain to longer simulated photoperiods.

Therefore, with a main photoperiod in the region of 10 to 14 hours, the pulse scanning the night simulates "long-day" photoperiods—in the classical photoperiodic sense—at two points, one on either side of the ψ-jump.

The importance of both "on" and "off" signals of the photoperiod in entrainment and phase control was recognized by Pittendrigh (1960, 1965) and evident, for example, in the simulation of complete by skeleton photoperiods. Several authors have since attempted to separate the effects of these signals by examining light steps as opposed to light pulses. Engelmann (1966) attempted to explain the eclosion rhythm in terms of "on" and "off" oscillators. Chandrashekaran (1967b) transferred populations of *D. pseudoobscura* from *LD 12* : 12 to DD and then exposed the cultures to either a single 15-minute light pulse or to a light step (light "on" without light "off") at Ct 15.5 or at Ct 21.5. It was found that the pulses caused, as expected, a $-\Delta\phi$ of 5.0 hours at Ct 15.5 and a $+\Delta\phi$ of 4.4 hours at Ct 21.5. The light steps, however, caused an appreciable *advance* in phase at both circadian times. Chandrashekaran postulated that the light "off" signal in the early part of the subjective night acted as a "new dusk" and was effective in causing a delay phase-shift and the light "on" signal in the second half of the night was interpreted as a "new dawn" and caused a phase advance. The light step,

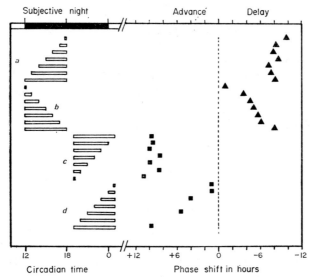

FIG. 3.20. Shifting the phase of the *D. pseudoobscura* eclosion rhythm with light pulses of 1000 lux and varying duration given in the first and second halves of the subjective night. The light pulses are represented by the unfilled bars and are arranged in four batches on the Ct scale of Pittendrigh and Minis (1964). In batch *a* the different populations experienced the "on" transitions of pulses at different hours but experienced the "off" transition at the same phase (Ct 18). In batch *b* the populations experienced the "on" transitions at the same circadian hour (Ct 12) but the "off" transition occurred for each population at a different hour. The pulses of batches *a* and *b* scan the first half of the subjective night. Batches *c* and *d* scan the hours of the second half of the subjective night. In batch *c* the "on" transitions of all the pulses were in alignment (at Ct 19) with the "off" transition occurring at a different hour for each population. In batch *d*, on the other hand, the "on" transitions were systematically staggered and the "off" transitions of pulses aligned. The filled triangles represent averaged median values of eclosion peaks of experimental populations 4–5 days after light treatment, which responded with $-\Delta\phi$. The filled squares represent averaged mean values of peaks 4–5 days after light treatment showing $+\Delta\phi$. Note that the "off" transitions determine direction and degree of $\Delta\phi$ during the first half of the night and the "on" transitions determine $\Delta\phi$ during the second half of the night. (From Chandrashekaran *et al.*, 1973.)

however, had no "off" component and only produced its effect (phase advance) when light extended into the second half of the night. A simpler alternative, however, might be that continuous light merely caused τ to shorten (see, for example, Bruce and Pittendrigh, 1957; Chandrashekaran and Loher, 1969b).

In a later paper, Chandrashekaran et al. (1973) subjected the subjective night to light pulses of between 15 minutes and 6 hours. These pulses were arranged in four groups. In the first half of the night (Ct 12 to 18) there were two groups with either "on" signals or "off" signals synchronized respectively; in the second half of the night (Ct 19 to 01) the other two groups were similarly arranged (Fig. 3.20). The results showed that the "off" transitions determined both the direction and the magnitude of $\Delta\phi$ during the first part of the night, whereas the "on" transitions determined $\Delta\phi$ in the latter half. Similar effects were noted when phase response curves for pulses between 15 minutes and 12 hours were plotted for either the "on" or the "off" transitions. The "off" signal in the first half of the night clearly simulated a "new dusk" and caused a delay phase-shift whereas the "on" signal in the second half of the night simulated a "new dawn" and caused an advance phase shift, as in the asymmetrical skeleton photoperiods first described by Pittendrigh and Minis (1964).

4. Entrainment by temperature pulses and cycles

Although the period of the oscillation controlling pupal eclosion in D. pseudoobscura is almost the same over a wide range of constant temperatures ($Q_{10} = 1.02$), temperature changes (e.g. steps, pulses and cycles) will cause phase-shifts and entrain (Pittendrigh, 1954, 1960). Temperature cycles will also entrain the eclosion rhythms of Dacus tryoni (Bateman, 1955) and Anagasta kühniella (Scott, 1936; Moriarty, 1959).

Working with Drosophila pseudoobscura, Zimmerman et al. (1968) showed that a square-wave temperature cycle, or thermoperiod, consisting of 12 hours at 28°C and 12 hours at 20°C ($T = 24$) will entrain the eclosion rhythm in DD. They also showed that single, non-recurrent, temperature pulses and temperature steps (up or down) will cause $\Delta\phi$'s similar in principle to those generated by light. The use of temperature steps is particularly interesting since the effects of a sharp rise and a sharp fall in temperature defining a pulse can be separated in a way which is difficult for light "on" and light "off" effects. Zimmerman et al. showed that temperature steps-up from 20° to 28°C caused phase advances ($+\Delta\phi$), the magnitude of which depending on the phase of the oscillator exposed to the temperature signal. Figure 3.21 shows that the maximum $+\Delta\phi$ was obtained when the step-up occurred at about Ct 22. Conversely, temperature steps-down caused phase-delays ($-\Delta\phi$); once again the magnitude of the response depending on phase. The amplitude of the $+\Delta\phi$'s were always greater than that for the $-\Delta\phi$'s, and the data strongly suggested that the minimum for both was close to Ct 10.

Just as the $\Delta\phi$'s for the "on" and the "off" signals in a light-cycle, as simulated by separate 15-minute light pulses in a skeleton photoperiod n hours apart, may be summed to calculate the net $\Delta\phi$ for a light pulse n hours long, so the $\Delta\phi$'s for temperature steps-up and steps-down may be summed to determine the net $\Delta\phi$ for temperature pulses. Thus the net $\Delta\phi$ for high-temperature pulses (HTP), consisting of a 12-hour period a 28° in an otherwise 20° culture, was a sum of the advance and delay phase-shifts caused by the two signals. The net $\Delta\phi$ was, of course, a function of the circadian time at which the pulse was seen (Fig. 3.22), with phase advances occurring when the HTP was initiated

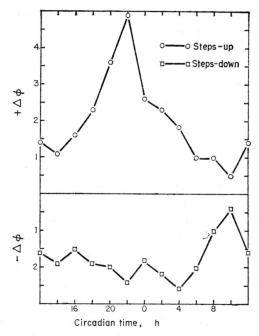

FIG. 3.21. Phase shift response curves for temperature steps-up (20/28°C) and temperature steps-down (28/20°C) in the *D. pseudoobscura* eclosion rhythm. Pupae were systematically exposed to temperature steps during the first 24 hours of DD free run (Ct 12 of day 0 to Ct 12 of day 1) after prior entrainment to *LD 12* : 12 at either constant 20° or 28°C. *Δφ* values for temperature steps-down are the average of *Δφ*'s (median of experimental minus that of free-run control) for days 6, 7 and 8; for temperature steps-up, the *Δφ* values are the average of *Δφ*'s for days 4 and 5. (From Zimmerman *et al.*, 1968.)

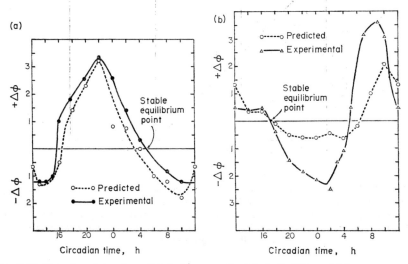

FIG. 3.22. Experimental and predicted *Δφ* curves for (a) single 12-hour high temperature pulses (HTP's) (20:28:20°C) and (b) single 12-hour low temperature pulses (LTP's) (28/20/28°C) in the *D. pseudoobscura* eclosion rhythm. Pupae were systematically exposed to temperature pulses during the first 24 hours of DD free-run (Ct 12 of day 0 to Ct 12 of day 1) after prior entrainment to *LD 12* : 12 at a constant 20° or 28°C. *Δφ* values for HTP's are the average of *Δφ*'s (median of experimental minus that of free-run control) for days 7 and 8; that for LTP's the *Δφ* values are the average of *Δφ*'s for days 4 and 5. Predicted *Δφ* values were derived from the response curves for temperature steps as described in Fig. 3.21. (From Zimmerman *et al.*, 1968.)

at points between Ct 17 and Ct 05, and phase delays when the HTP was initiated between Ct 05 and Ct 17. Conversely, the phase response curve for low-temperature pulse (LTP) showed phase delays between Ct 17 and Ct 05 and phase advances between Ct 05 and Ct 17. The fact that separate $\Delta\phi$'s for the two steps may be used to determine the net $\Delta\phi$ for the pulse suggests that the phase-shifts are accomplished quite rapidly (within a few hours). These results, therefore, show that temperature *changes* act as *Zeitgebers* in a manner comparable to changes from dark to light, or changes in light intensity.

Simultaneous cycles of light and temperature are particularly interesting because the two-oscillator model conceived for *D. pseudoobscura* (Pittendrigh and Bruce, 1957, 1959) includes one temperature-compensated oscillator (A) directly coupled to the light cycle and a temperature-sensitive driven element (B) whose phase controls eclosion. Therefore, concurrent cycles of light and temperature might be expected to produce a "conflict" in the steady state finally achieved. Using a sinusoidal temperature cycle (19–29°C) in conjunction with a light cycle of *LD 12* : 12, such an effect was observed (Pittendrigh, 1960). When the low point of the temperature cycle occurred close to dawn, as in the "natural" situation, no disturbance of phase was observed. But as the phase angle between the two *Zeitgebers* was changed, the peak of pupal eclosion moved to the right relative to the fixed light cycle until, at about 15 to 16 hours after dawn, a discrete 180° ψ-jump occurred (Fig. 3.23). The results showed that out of the total conceivable 360° of phase, only 180° is realizable: there is a 180° zone of "forbidden" phase relations during which eclosion cannot occur.

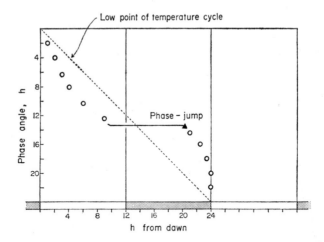

FIG. 3.23. The phase of the eclosion rhythm in *D. pseudoobscura* as a function of the phase angle between simultaneously entraining light and temperature cycles. Ordinate shows the phase angle (in hours) between dawn and the low point of the temperature cycle. Note the phase-jump develops at a critical phase angle between the two *Zeitgebers*. (From Pittendrigh, 1960.)

The explanation for this abrupt change of phase lies in the two-oscillator model for the clock underlying eclosion. For instance, when the system attempts to entrain to the two conflicting *Zeitgebers*, one part of the system (A) strictly follows the phase of the light-cycle, whereas another (B) was more influenced by the temperature cycle. There is clearly a limit to the phase-angle difference between these two subsystems: when that limit is reached the driven system is forced to leap back to its original phase close to dawn. A very similar result has been obtained for the cockroach, *Leucophaea maderae* (Roberts, cited in Pittendrigh, 1960) and for the eclosion rhythm in *Pectinophora gossypiella* (Pittendrigh and Minis, 1971).

D. Genetic Experiments

The non-steady-state (transient) cycles which follow perturbation with light pulses, and the "conflict" generated when the phase-angle between opposing light and temperature cycles is varied, suggest the participation of at least two oscillators in the eclosion clock of *D. pseudoobscura*. The existence and full significance of these subsystems have been made clear by certain types of genetic experiment. The first of these involved the selection for "early" and "late" strains (Pittendrigh, 1967), and the second the production of period mutants by chemical mutagens (Konopka and Benzer, 1971).

Pittendrigh (1967) systematically bred from parents that emerged either very early or very late in the daily distribution of emergence activity in populations kept at *LD 12* : *12*.

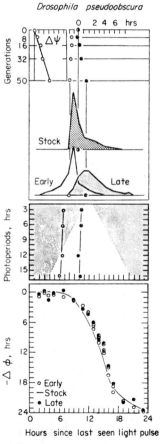

FIG. 3.24. The effect of selection for "early" and "late" eclosion strains of *D. pseudoobscura*. Top panel: the change in $\psi_{R,L}$ in the two strains through fifty generations of selection (both selection and monitoring were performed in *LD 12* : *12*). Middle panel: the $\psi_{R,L}$ for the two strains in five photoperiods. Lower panel: the phase-response curve for "early" (open circles) and "late" (solid circles) compared with that from the stock culture (solid line), showing that selection had affected $\psi_{R,L}$ but not $\psi_{0,L}$. (From Pittendrigh and Minis, 1971.)

After fifty generations of such selection the difference in eclosion time between the two artificially selected strains was about 4 hours (Fig. 3.24). This difference in phase angle was maintained at all photoperiods. However, when the phase response curves for the two strains were measured they were found to be identical. The phase response curve is a measure of the phase relation of the A-oscillator to the light cycle ($\psi_{O, L}$) whereas the overt rhythm of eclosion is a measure of the phase relationship of the driven system (the B-oscillator) to the driver ($\psi_{O, R}$). Therefore, selection had clearly altered the phase-angle between the driven system and the driver, but not the phase relationship of the driver to the light. This demonstrates the reality of the two-oscillator system proposed by Pittendrigh and Bruce (1957, 1959), and makes abundantly clear the distinction between the light-sensitive driving *oscillation* (A) and the driven *rhythm* (B).

This important observation has been repeated for *Pectinophora gossypiella* (Pittendrigh and Minis, 1971). In this species selection resulted in an "early" strain emerging from its pupae about 5 hours before "late". Selection for eclosion time has also been achieved

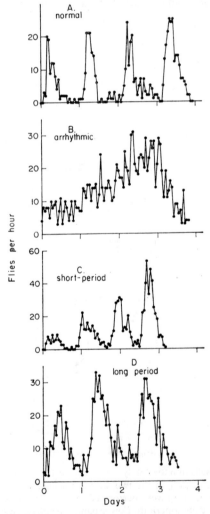

Fig. 3.25. The rhythms of pupal eclosion, in constant darkness, for populations of rhythmically normal and mutant *D. melanogaster*, previously exposed to LD *12* : 12 at 20°C. (From Konopka and Benzer, 1971.)

for *D. melanogaster* (Clayton and Paietta, 1972). These results lead to a concept of an adaptively adjustable coupling of driven systems to the common driving oscillation, and to the concept of the "circadian system" which will be examined in Chapter 4.

The treatment of *D. melanogaster* with the chemical mutagen, ethyl methane sulphonate (EMS) by Konopka and Benzer (1971) resulted in the isolation of three period or "clock" mutants. One of these was arrhythmic (per^0); the other two were rhythmic but had free-running periods (τ) of 19 hours (per^s) and 28 hours (per^1), respectively (Fig. 3.25). These altered periods were evident in both the eclosion rhythm and in a rhythm of adult locomotor activity. Since the latter was recorded using invidual insects, the arrhythmicity of per^0 was not due to a desynchronization of inviduals in the population, although it might have been due to a desynchronization of separate circadian subsystems within the individual. It was further shown that τ for each mutant was temperature-compensated; this suggests that the mutations were affecting the basic A-oscillator.

Recombination experiments with respect to morphological markers of known position enabled the three mutations to be located on the X chromosome. Indeed, the three mutations appear to represent changes in the *same* functional gene or cistron—a rather surprising fact, perhaps, since one might have expected a polygenic system to control such a complex event as an eclosion rhythm. Nevertheless, this was tested by constructing heterozygote females with different rhythm mutations on each of the two X chromosomes. Heterozygotes with an arrhythmic or long-period gene on one X chromosome and a "normal" gene on the other resulted in a rhythm of eclosion with a period close to normal. Heterozygotes with one short-period gene, however, showed a rhythm with a period intermediate between short and normal. These results demonstrate that per^0 and per^1 are recessive to normal, whereas per^s is only partially recessive. A heterozygote with one long and one short-period gene had a period close to normal, but when a short-period gene was opposed to an arrhythmic one, the rhythm displayed a short period. Similarly, the arrhythmic gene was overshadowed by a long-period one. These results suggest that the arrhythmic gene is simply inactive, and that the flies with per^0 have an inactive A-oscillator.

E. "Stopping" the Clock

1. *Low oxygen tension and cold torpor*

Kalmus (1935) and Pittendrigh (1954) showed that the clock controlling eclosion in *Drosophila* could be "stopped" by hypoxia. Pittendrigh (1954), for example, subjected cultures of *D. pseudoobscura* to nitrogen containing traces of oxygen for a 15-hour period. On returning to aerobic conditions the first peak of eclosion was found to be delayed by 15 hours, but the second and subsequent peaks by only 10. The similarities between this result and that for a temperature step suggest that the first effect was possibly due to an effect on the B-oscillator, whilst the maintained phase-shift was the result of "stopping" the basic A-oscillator.

Although τ for the eclosion rhythm in *D. pseudoobscura* is temperature-compensated, it is probable that cold torpor at temperatures approaching zero would effectively stop the motion of the oscillator. Apparently this effect has not been reported for the *D. pseudoobscura* case, but is well documented in other animals (Harker, 1958) and in plants.

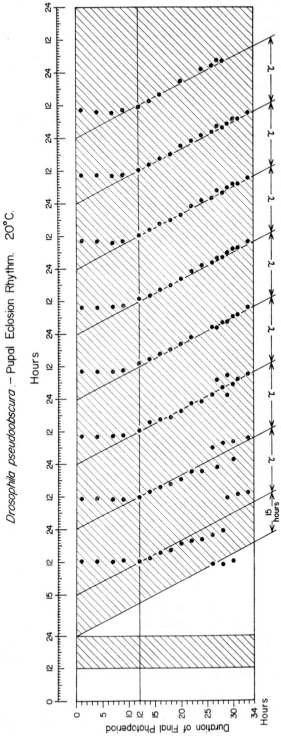

FIG. 3.26. The effect of varying the duration of the final photoperiod prior to releasing cultures of *D. pseudoobscura* into DD free-run. After a final photoperiod longer than 12 hours the rhythm obtains its principal "time-cue" from dusk, or lights-off; the inference is that the driving oscillation begins afresh from Ct 12 at the light/dark transition. (From Pittendrigh, 1966.)

2. Damping action of photoperiods in excess of about 12 hours

Reference has already been made to the fact that eclosion in *D. pseudoobscura* is arrhythmic in continuous light of sufficiently high intensity, but on transfer to DD a rhythm becomes apparent. These observations suggest that *LL* "damps out" the oscillation, and that the oscillation resumes its motion—with all individuals of the population in phase—on transfer to DD. The minimum length of light necessary to cause this damping out has been investigated by Pittendrigh (1960, 1966).

Pittendrigh (1966) raised 18 mixed-age populations of *D. pseudoobscura* in LD 12 : 12 and then exposed them to a last photoperiod of varying duration before release into DD (Fig. 3.26). The intensity of the light was about 100 f.c. Dawn of the last photoperiod came at its previously established phase, but dusk was either advanced or delayed, relative to the control, by various amounts in the several cultures. The results showed that after a terminal photoperiod of 12 hours or more the steady-state phase-points of the eclosion peaks occurred at fixed time intervals ($n\tau + 15$ hours) after the *LL*/DD transition, suggesting once again that the driving oscillation stops in protracted light. A systematic study of the $\Delta\phi$'s generated by 15-minute light pulses provided a phase response curve for the oscillation during the hours immediately following the *LL*/DD transition; this proved to be identical to that following *LD 12* : 12 (Fig. 3.27). The implication of this result is that photoperiods longer than 12 hours not only damp out the oscillation, but hold the oscillations (in each fly in the population) in the same fixed state which corresponds to that at Ct 12, so that, on entry or re-entry into the dark, all the oscillations in the population resume their motion at the same phase. A further implication, which

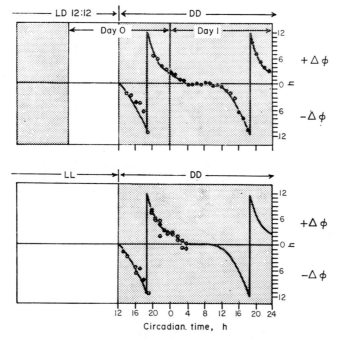

FIG. 3.27. Phase-response curves (new convention) for the *D. pseudoobscura* eclosion rhythm in DD free-run af terrelease from *LD 12* : 12 (top panel) or from *LL* (lower panel), confirming that the oscillation restarts at Ct 12 after light periods in excess of 12 hours. (From Pittendrigh, 1966.)

is especially important in the interpretation of some photoperiodic phenomena (Chapter 7), is that with photoperiods in excess of 12 hours, the principal (or only) time signal of significance is that associated with "dusk"; with shorter photoperiods, however, both dawn and dusk play a role in phase-control. The oscillation controlling adult eclosion in the flesh-fly, *Sarcophaga argyrostoma*, is similarly damped out by photoperiods in excess of 11 to 12 hours (Saunders, unpublished). In the silkmoth, *Antheraea pernyi*, however, the clock is stopped by as little as 30 minutes of light (Truman, 1971d).

3. The "singularity point"

Winfree (1970 a, b) showed that the oscillation controlling the eclosion rhythm in *D. pseudoobscura* could be "abolished" or placed in a completely "phase-less" state by a single short light signal provided that signal was applied precisely at the right moment. Populations of flies placed in this "phase-less" state showed an arrhythmic eclosion pattern as though the clock had been stopped. The critical signal necessary to stop the clock is referred to as the "singularity".

Mixed-age populations of *D. pseudoobscura* were raised in *LL* and then released into *DD* during intra-puparial development; this *LL*/*DD* transfer resulting, as we have seen earlier, in the generation of a clear rhythm of eclosion with peaks at circadian intervals. A variable time (T) after the *LL*/*DD* transfer a pulse (S) of dim blue light (10 μW/cm²) of less than 3 minutes duration was applied to the cultures, and the eclosion peaks then recorded 3 days later after any transients had subsided (Fig. 3.28). The values of T and S were varied systematically, and the interval between the end of the light perturba-

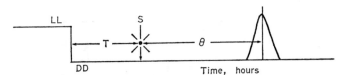

Fig. 3.28. *Drosophila pseudoobscura*. The format of Winfree's resetting experiments. *LL*-DD = the transition between constant light and constant darkness; T = the time interval between the *LL*/*DD* transition and the resetting signal; S = the duration of the resetting signal; θ = cophase, measured in hours from the end of the resetting signal and the centroids of the eclosion peaks. (From Winfree, 1970a.)

tion (S) and the centroid of the eclosion peaks (called the cophase, θ) plotted for each combination of T and S. When all the cophase data for a number of days were plotted in this manner the "cloud" of dots gave a helicoid or spiral ramp, described by Winfree as a "vertical corkscrew linking together tilted planes" (Fig. 3.29). This complex surface can also be depicted in two dimensions by plotting θ-contours above the $T \times S$ plane (Fig. 3.30). These contours represent the time after the stimulus (S) at which emergence peaks occur, plus multiples of 24 hours. The most interesting feature of this figure is that the cophase contours are confluent at a point (T^*S^*, or the singularity) at which the circadian rhythm is "abolished" or put into a completely "phase-less" state. In the spiral ramp depicted in Fig. 3.29 the symmetry axis corresponds to this singularity. Extensive experiments in which T and S were varied independently have shown the singularity to be a 50-second pulse placed 6.8 hours after the *LL*/*DD* transition. As the values of T and S approach this point there is a progressive broadening of the eclosion peaks; at the singularity itself eclosion becomes arrhythmic.

θ (T, S)

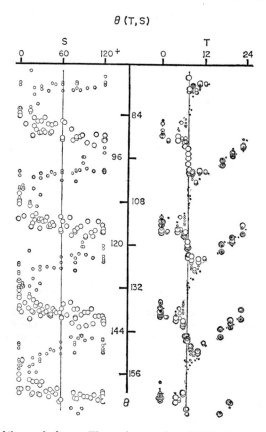

FIG. 3.29. *Drosophila pseudoobscura*. The cophase surface, θ (T,S), shown in two orthogonal projections. On the left the projection shows points $T < 7.3$ hours as larger circles and $T > 7.3$ as smaller circles. T from hour 15 to 24 is deleted, since this portion of the surface forms a tilted plane (visible in the other projection) which would obscure the spiral ramp between $T = 0$ and $T = 15$ hours. The right-hand projection shows data at $S = 0$ to 30 seconds as dots, 31 to 60 seconds as tiny circles, 61 to 90 seconds as larger ones, and the largest circles represent emergencies following perturbations longer than 90 seconds. (From Winfree, 1970b.)

These results are evidence for a critical annihilating light pulse T^*S^* which puts the clock into a non-oscillatory state, i.e. "stops the clock." It is of interest that this singularity point is close to the point of maximum phase shift (at Ct 18) where phase-inversion occurs (Pittendrigh, 1966). Two interpretations of the arrhythmicity generated by T^*S^* are therefore possible. In one, already mentioned, the motion of the oscillators is abolished; in the second the progressive broadening of the peaks and the final arrhythmicity represents nothing more than the mutual desynchronization of all the oscillators in the population. Winfree (1970b) recognized these alternatives and distinguished between them by giving a *second* perturbation of 120 seconds (pulse $S > S^*$) at different times after the singularity. If the arrhythmicity following perturbation at the singularity was due to a desynchronization of the constituent oscillators the resulting rhythm of eclosion would apparently be markedly bimodal, whereas a reinitiation of a population of oscillators "frozen" at the same phase would produce clearly unimodal eclosion peaks after the second perturbation. The experimental production of a unimodal distribution of phases suggested that reinitiation rather than resynchronization had occurred.

Extended resetting map $\theta\,(T,S)$

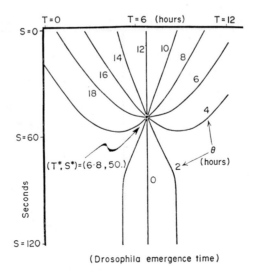

(Drosophila emergence time)

FIG. 3.30. *Drosophila pseudoobscura.* Contour map of the cophase surface, $\theta\,(T,\,S)$, showing the critical annihilating pulse (T^*S^*, or the singularity) which puts the clock into a non-oscillatory or phase-less state. (From Winfree, 1970a.)

4. The "fixed point"

In a mathematical analysis of the *Drosophila* eclosion rhythm Johnsson and Karlsson (1972b) have described a third way in which a light pulse might "stop" the driving oscillation. Using data from the phase response curve for 15-minute pulses (Fig. 3.11) the circadian time of the oscillation at the end of the pulse was plotted as a function of the circadian time of the oscillation at the beginning of the pulse, to give a "transformation curve" (Fig. 3.31). Thus a pulse starting at Ct 0 caused a phase advance of $+2.25$ hours,

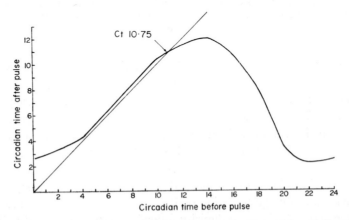

FIG. 3.31. *Drosophila pseudoobscura.* A plot of the circadian time of the eclosion rhythm after a 15-minute light pulse as a function of the circadian time before the pulse, according to the transformation method. The intersection between the transformation curve and the straight line drawn through the origin and with a slope of one, indicates the position of the "fixed point". A pulse of light commencing at the fixed point (Ct 10.75) theoretically leaves the oscillation at the *same* circadian time *after* the pulse, regardless of the duration of the light signal. (Redrawn from Johnsson and Karlsson, 1972b.)

so that at the end of the pulse the perturbed rhythm phase-led the control by 2.25 hours. Since the control was by then at Ct 0.25, the perturbed rhythm was at Ct $2.25+0.25=$ 2.50. Similarly, although a pulse commencing at Ct 8 caused no phase shift the duration of the pulse itself resulted in the circadian time at the end of the pulse becoming Ct $8.0+0.25 = 8.25$. A pulse at Ct 18, on the other hand, caused a phase-delay of -10.5 hours, so that the phase of the perturbed rhythm at the end of the pulse became Ct $18+0.25-10.5 = 7.75$ hours. The transformation curve is itself periodic, having a period (τ) equal to that for the *Drosophila* system.

Johnsson and Karlsson pointed out that at least one circadian time between Ct 0 and Ct 24 the phase of the driving oscillation should be the same at the end of the pulse as at its beginning. This can be graphically demonstrated by superimposing on Fig. 3.31 a straight line with a slope of one drawn through the origin. This line intersects the transformation curve at about Ct 10.75 and the intersection represents the position of the "fixed point". Evidently a pulse commencing at this time also leaves the oscillation at the end of the pulse at the *same* time (Ct 10.75). Similar fixed points were derived for longer pulses, the conclusion being that a light pulse given at Ct 10.75 should always leave it at Ct 10.75 regardless of its duration, as though the driving oscillation had been "frozen" during the time when the light was on. The fixed point has yet to be experimentally determined for *D. pseudoobscura*.

F. Spectral Sensitivity and Intensity Effects

Action spectra for light-induced phase shifts of the *D. pseudoobscura* oscillation have been determined by Frank and Zimmerman (1969). Larval cultures of mixed developmental age were raised in *LD* 12 : 12 and then transferred to DD as pupae. Pulses consisting of 15 minutes of monochromatic light were then applied to the free-running oscillation at two circadian times. In one series the pulse was placed at Ct 17 where white light generates a phase-delay $(-\varDelta\phi)$; in a second series it was placed at Ct 20 to generate a phase-advance $(+\varDelta\phi)$. Control groups experienced either white light pulses at Ct 17 and 20, or were allowed to free-run unperturbed in DD. The $\varDelta\phi$'s achieved with the monochromatic pulses were expressed as a percentage of the $\varDelta\phi$ caused by white light at that point. The intensities of the monochromatic pulse were systematically altered until a steady-state (7 days after the signal) equalled 50 per cent of the white light control.

The results showed that the direction of the phase shift (i.e. advance or delay) was not affected by wavelength, but the magnitude of the response increased with the intensity of the signal. The action spectra for both $+\varDelta\phi$ and $-\varDelta\phi$ were similar. The most effective wavelengths were between 420 and 480 nm (Fig. 3.32). Above 500 nm there was a sharp "cut-off" and no response could be achieved with an intense non-monochromatic light spanning a broad spectrum range from 600 nm into the infrared.

Working with the initiation of the egg-hatch rhythm in *P. gossypiella*, Bruce and Minis (1969) found a very similar spectral sensitivity. In this study populations of eggs were exposed to short light signals of monochromatic light after the midpoint of embryogenesis. The most effective wavelengths were between 390 and 480 nm; those above 520 nm were ineffective (Fig. 3.33).

Chandrashekaran and Loher (1969b) exposed populations of *D. pseudoobscura* to white light pulses of different intensities at Ct 15.5 $(-\varDelta\phi$ of ~ 5 hours) and at Ct 21.5

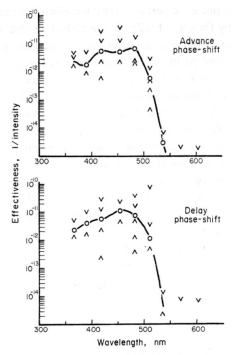

FIG. 3.32. Action spectrum for 50 per cent phase shifts of the *D. pseudoobscura* eclosion rhythm induced by 15-minute light signals: (*above*) advance phase-shifts; (*below*) delay phase-shifts. (From Frank and Zimmerman, 1969.) (Copyright 1969 by the American Association for the Advancement of Science.)

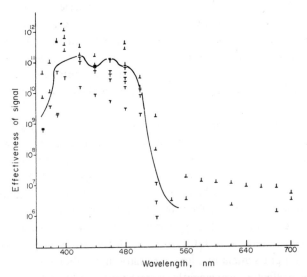

FIG. 3.33. Action spectrum for the initiation of the rhythm of egg-hatching in *Pectinophora gossypiella*. ⊥ signals judged to below the threshold; ⊤ signals above the threshold; ⊥ or ⊤ replicates giving the same results; ∓ replicates giving opposite results. (From Bruce and Minis, 1969.) (Copyright 1969 by the Americian Association for the Advancement of Science.)

($+\Delta\phi$ of ~ 5 hours). They found that the magnitude of the $\Delta\phi$'s—both advance and delay—was independent of intensity above about 10 lux; the same phase-shifts were achieved, for instance, with 10, 3000 and 10,000 lux. Pulses between 10 lux and 1.0 lux caused $\Delta\phi$'s of the correct sign but of a smaller magnitude, but below 1.0 lux only delays ($-\Delta\phi$) were discernible. Moreover, when the transients occurring between the day of treatment and the final steady state (days 4 or 5) were examined, low light intensities (0.1, 0.3 and 1.0 lux) were found to cause temporary "reversing transients" in a direction opposite to that finally achieved, or generated by higher intensities. These reversing transients were particularly well marked in populations exposed to pulses at Ct 21.5 ($+\Delta\phi$'s). Their significance, however, remains obscure.

The peak of sensitivity at 420 to 500 nm might suggest that a carotenoid pigment was involved in photoreception. However, Zimmerman and Goldsmith (1971) raised larvae of *D. melanogaster* on aseptic diets with and without β-carotene, and tested the adults for the photosensitivity of the compound eyes and the photosensitivity of the circadian rhythm controlling adult eclosion. Visual sensitivity was assayed by measuring the height in millivolts of the sustained corneal negative wave elicited in the dark-adapted eye by monochromatic light (454 nm) in pulses of approximately 1 second. The photosensitivity of the circadian rhythm was assayed by measuring the $-\Delta\phi$ generated by a 15-minute light signal of monochromatic light (454 or 458 nm) applied to a population of pupae at Ct 15.

Since carotenoids are only synthesized by plants, and carotenoid-deficiency in insect diets results in a loss of visual sensitivity, a similar impairment of the circadian oscillation might be expected if a carotenoid-derived chromophore was also involved in the latter system. The results, however, showed that although photosensitivity of the compound eyes in the carotenoid-depleted flies was about 3 log units lower, the responses of the circadian system in both groups were identical. Therefore, unless sufficient carotenoid is passed through the egg and is used preferentially for the circadian rhythm chromophore, these results suggest that carotenoids are not involved in the phase-shifting of the *D. melanogaster* oscillator. This conclusion is supported by the fact that the action spectrum for phase-shifts (Frank and Zimmerman, 1969) is quite different to that for the compound eyes, and also quite different from the absorption spectrum for common carotenoids.

Annotated Summary

1. "Once-in-a-lifetime" developmental events, such as hatching, moulting from one instar to another, and pupal eclosion, are frequently controlled by an endogenous circadian oscillation which only manifests itself in a mixed-age population of insects.

2. These rhythms are generally "damped out" by protracted light of high intensity, but may be "generated" by a single stepwise transfer from light to dark, a single exposure to a light pulse, or by the transfer from successive light/dark cycles into continuous darkness. The free-running rhythm so revealed exhibits its natural period (τ) which is close to 24 hours and is temperature-compensated.

3. In "population rhythms" the circadian oscillation *in each insect* dictates "allowed zones" or "gates" through which eclosion, hatching or hormone release may occur. In a mixed age culture this mechanism partitions the population into daily activity peaks. In continuous darkness (DD) the free-running oscillation shows successive gates occurring at intervals of modulo τ.

4. A free-running oscillation perturbed by light (or temperature) pulses shows steady-state phase-advances or phase-delays depending on the phase of the oscillation perturbed. Phase delays are generated by pulses falling in the early "subjective night"; phase advances by pulses in the late "subjective night". A plot of these phase changes against the time of the perturbation gives a phase response curve for the oscillation.

5. After a single unrepeated light perturbation the overt rhythm generally undergoes several transient cycles before reaching its new steady state. Pittendrigh's model to account for such phenomena envisages an A-oscillator (the "circadian pacemaker") which is immediately (i.e. within minutes) reset by the light pulse, and a driven element, the B-oscillator, which is coupled to A on the one hand, and to the physiological processes controlling eclosion, hatching, etc., on the other.

6. The free-running oscillator can be entrained by a variety of light cycles, including repeated pulses defining environmental cycles, and two or more pulses per cycle forming "skeleton" photoperiods. In each case the endogenous periodicity (τ) is adjusted to that of the environmental driving cycle (T) by discrete phase shifts caused by the light transitions. The phase shifts generated can be calculated from the phase response curve.

7. Both the light-on signal and the light-off signal of a pulse cause phase shifts, but the former is "dominant" in the late subjective night, whereas the latter is "dominant" in the early subjective night.

8. Circadian oscillations can also be entrained and phase-shifted by temperature pulses and temperature cycles, temperature steps-up generally simulating "lights-on" and steps-down simulating "lights-off".

9. The period of the oscillator is an inherited character, and in *Drosophila melanogaster* period- or clock-mutants have been isolated following treatment with a chemical mutagen. These mutations have been located on the X-chromosome and appear to represent changes in the same functional gene or cistron; they also appear to be mutations affecting the A-oscillator or circadian pacemaker. Other genetic experiments involving artificial selection for eclosion time have demonstrated the probable reality of the supposed A- and B-oscillators.

10. The driving oscillator (A) may be "stopped" by a variety of light treatments. In *Drosophila pseudoobscura*, for example, it is damped out by high light intensity lasting 12 hours or more, but resumes its motion (with all oscillators in the population at the same phase) on transfer to darkness. The A-oscillator also possesses a singularity, at which a pulse of light of a particular duration or energy can place the system in a "phase-less" state.

11. In *D. pseudoobscura* phase-shifts of the A-oscillator may be generated by light intensities down to less than 1 lux. Action spectra for both advance and delay phase-shifts are similar or show maximum sensitivity between 420 and 480 nm, and a sharp "cut-off" above 500 nm. Carotenoids, however, do not seem to be involved in the circadian photoreceptor.

CHAPTER 4

CIRCADIAN RHYTHMS AND PHYSIOLOGY: THE CIRCADIAN SYSTEM

THIS chapter contains a miscellany of rhythmic phenomena which, unlike the overt behavioural events described in Chapters 2 and 3, are at a more "physiological" level. It includes, for example, rhythms of cuticle deposition, which can be regarded as an activity of single epidermal cells; rhythms of colour change, bioluminescence, metabolism and pheromone activity; rhythms in endocrine and nervous tissue; and rhythmic sensitivity to drugs and insecticides. It also includes evidence that fundamental aspects of insect physiology such as longevity and growth rates are affected by changes in the periodicity of illumination, or lack of it, and that for "normal" functioning insects, like all other organisms, require a fluctuating and 24-hour environment. These phenomena are discussed in terms of a multioscillator concept for the total circadian system.

Many of these rhythmic phenomena, particularly, for example, colour change, rhythms of flashing in fireflies, and pheromone activity, have a clear functional significance. The advantage in daily growth layers possibly lies in strengthening the cuticle. The selective advantage or the biological significance of some of the other phenomena, however, is far from clear, although some of the rhythms in the endocrine and nervous systems may indicate an important role in the control of behavioural periodicities.

A. Rhythms of Cuticle Deposition

Although laminated cuticles had been known for some time, Neville (1963a) was the first to describe "daily growth layers". In a series of papers he showed that these growth layers were controlled by circadian oscillators which — unlike those controlling behavioural rhythms—leave behind a permanent morphological record.

In the desert locust *Schistocerca gregaria* the greater part of the endocuticle is deposited after ecdysis; in the adult it continues for 2 to 3 weeks before the insect is deemed "mature", i.e. non-teneral. Neville (1965) showed that in a normal fluctuating environment with a light-cycle (*LD 12* : 12) and a temperature cycle (days at 36°, nights at 26°) this endocuticle was laid down in alternate "day" and "night" layers, each pair approximately 10 μ thick (Fig. 4.1). The night layers showed the chitin organized into several lamellae arranged parallel to the cuticle surface; the day layers, however, although containing the same relative quantity of chitin, were non-lamellate. Removal of the protein by treatment with NaOH caused shrinkage of the cuticle but not the loss of its growth layers, thus confirming that this diurnal zonation was due to changes in the orientation of the chitin itself. Because the daily growth layers constitute a permanent record of the passage of time they can be used to determine the age structure of wild (and museum) populations of insects (Neville, 1963b).

Experimental manipulation of the light and temperature cycles has demonstrated the circadian nature of the controlling mechanism. For example, in constant dark (DD) the rhythm of cuticle deposition continued for at least 2 weeks (Neville, 1965) thus demonstrating its endogeneity. A rhythm of cuticle deposition was also generated by a transfer from LL into DD. In DD "free-run" the rhythm showed a natural period (τ) of about 23 hours, so that after 12 days the cuticle was about 180° out of phase with environmental time. The period of the cuticle rhythm was also shown to be temperature-compensated ($Q_{10} = 1.04$) although the quantity of material deposited was not ($Q_{10} = 2.0$). The difference between these two temperature coefficients resulted in the deposition of *thicker* layers at higher temperature, although with the same frequency. It was further shown that constant light (LL) stopped lamellogenesis and resulted in a wide non-lamellate zone; this, however, could be prevented by the use of a daily thermoperiod (12 hours at 36°, 12 hours at 26°). The similarities between this system and that controlling the rhythm of eclosion in *Drosophila pseudoobscura*, for example, are quite remarkable (Chapter 3).

In locust resilin, Neville (1963a) demonstrated daily growth layers by the use of ultraviolet light which causes the cross-linking amino acids dityrosine and trityrosine to autofluoresce. Under these conditions diurnal zonation was demonstrated in the prealar arm and the wing hinge ligaments. Confirmation that these layers were in the resilin itself was achieved by labelling the ligaments with tritiated tyrosine at known intervals; this later showed up in autoradiographs as discrete bands at predictable sites (Kristensen, 1966). Each ligament was shown to grow for 2 to 3 weeks after emergence. However, three pairs of layers were found to be deposited before emergence, and these "preimaginal" layers were clearly separated from those subsequently deposited (the "postimaginal" layers) by a pause in cuticle formation called the "emergence line". In contrast to the deposition of solid endocuticle the rhythmic formation of resilin persisted slightly in LL and seemed to be mainly influenced by the temperature cycle.

Daily growth layers have now been described in a large number of species from about nine insect orders (Neville, 1970). Their functional significance is far from clear, however, although Neville (1967) suspects that their advantage lies in strengthening the cuticle. In earlier papers (Neville, 1963 a, b) known examples were confined to the Exopterygota: adults of Lepidoptera, Diptera and Hymenoptera appeared to mature their cuticles so rapidly after emergence that only one endocuticular layer was formed. Since then, however, daily growth layers have been demonstrated in the honey bee (Menzel *et al.*, 1969) and in some of the thoracic apodemes of the flesh-fly *Sarcophaga falculata* ($=$ *argyrostoma*) (Schlein, 1972). Nevertheless, extensive endocuticular growth layers seem to be more characteristic of the hemimetabolous orders. Some of the similarities and differences between these insects will be noted here.

In the cockroach *Periplaneta americana*, cuticle deposition is controlled by a circadian oscillation, but it continues to oscillate in LL (Neville, 1965). In certain giant water bugs (Belostomatidae) daily growth layers are evident but the chitin crystallites change their orientation by 60° every 24 hours. Dingle *et al.* (1969) showed that the oscillation controlling cuticle deposition in the milkweed bug, *Oncopeltus fasciatus*, was initiated by the light-cycle experienced by the pharate adult. It was also temperature-compensated (Q_{10} about 1.0), but a smaller *number* of layers was deposited at higher temperature because the final thickness of the cuticle was the same at all temperatures. For example,

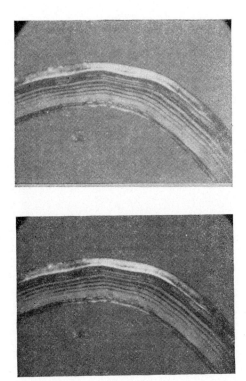

FIG. 4.1. Transverse section of the hind tibia of an adult locust, *Schistocerca gregaria*, photographed between crossed polaroids to show the daily growth layers in the endo-cuticle. (Courtesy of A. C. Neville, 1965.)

the final thickness of the cuticle in the mid-region of the metathoracic tibia was about 30 μ; at 19° this was reached after 10 to 11 days, whereas at 27° it took only 5 to 9 days.

Zelazny and Neville (1972) have recently shown that the deposition of the endocuticle in *Oryctes rhinoceros* and some other beetles, although rhythmic, is controlled by a non-circadian system. Firstly, there was no linear correlation between age and the number of layers deposited. In beetles maintained in DD since their adult ecdysis the number of layers increased steadily and then remained constant: in *O. rhinoceros*, for example, about twenty-two layers were evident in the pronotum after about 40 days. Secondly, the frequency of layer formation was higher at 34° than at 24°, and more endocuticle was deposited. However, the Q_{10} for the number of layers deposited (in 12 days) was 1.8, whereas that for the increase in thickness was only 1.46; consequently individual layers became *thinner* at the higher temperature. Thus, the period of the oscillation in *Oryctes* differs from 24 hours; the oscillation will not entrain to environmental cycles; and it is not temperature-compensated. Therefore, the system controlling cuticle deposition in these beetles is regarded as being non-circadian. In addition, since it lacks temperature-compensation it cannot, strictly speaking, be regarded as a "clock".

Neville (1967) reported that the rhythm of cuticle deposition in locusts continued when their head capsules were painted black, or when their eyes and ocelli were removed by cautery. Similarly, the rhythm continued in decapitated *Oryctes* (Zelazny and Neville, 1972). The site of photoreception, therefore, is not in the head, and a "dermal light sense" is suspected. Furthermore, cylinders of the hind tibiae cut from living specimens of *S. gregaria* continued to lay down growth layers when implanted into the haemocoels of recipient *gregaria* and *paranamense*, demonstrating that neither mechanical nor nervous connections were required. In *Oryctes*, different locations in the exoskeleton may grow with different frequencies suggesting that the oscillators are in individual epidermal cells, or in groups of neurosecretory cells with specific target innervation.

The action spectrum for the response in *S. gregaria* showed a peak of sensitivity close to 470 nm (Neville, 1967), similar to that described for the initiation and phase-shifting of overt behavioural rhythmicities (Bruce and Minis, 1969; Frank and Zimmerman, 1969), and for the photoperiodic induction of virginopara-production in the aphid *Megoura viciae* (Lees, 1968, 1971). This action spectrum suggested the involvement of a carotenoid-derived chromophore, particularly since β-carotene is abundant in locusts. However, the testing of locusts raised on carotenoid-depleted diets has not been performed.

B. Metabolic Rhythms

A number of metabolic functions in insects are known to show rhythmic diurnal changes. Some of these, such as the levels of oxygen consumption, CO_2-release, and levels of trehalose and glycogen probably reflect rhythms of activity. Others, such as luminescence and diurnal colour changes, are overt physiological rhythms with behavioural significance. All of these rhythms are undoubtedly as much "hands of the clock" as are the rhythms of activity and behaviour discussed in earlier chapters, and for this reason cannot be regarded, necessarily, as more "fundamental". Many of these diurnal changes have only been studied in light/dark (LD) cycles; consequently their endogeneity has not been established. Notable exceptions to this include the circadian rhythm of flashing in the fire-fly *Photinus pyralis* (Buck, 1937) and the rhythm of colour change in the stick insect *Carausius morosus* (Schliep, 1910, 1915). Most attention will be paid to those cases in which the endogenous nature of the rhythm has been established beyond doubt.

Respiratory rhythms have been investigated in a number of insect species. In *Tenebrio molitor*, for example, Michal (1931) and Campbell (1964) described a complex pattern of oxygen consumption with a maximum occurring during the night, but with several subsidiary peaks and troughs. Whether the rhythm persisted in constant light (*LL*) or constant dark (DD) was apparently not investigated. Similarly complex rhythms of oxygen consumption were described by Beck (1963, 1964) for the cockroach *Blatella germanica* and larvae of the European corn borer *Ostrinia nubilalis*. Both of these rhythms showed maxima just after dark and minima just after dawn, but there was also evidence of an 8-hour periodicity. In *B. germanica* the rhythm could be "entrained" by the light cycle in that the phase relationships of the peaks and the troughs altered with the length of the photoperiod. Beck (1968) stated that the rhythm in *O. nubilalis* persisted in DD and was therefore endogenous.

In *Drosophila melanogaster* the rhythms of oxygen consumption in larvae, puparia and adults appeared to be bimodal with peaks in the morning and evening (Rensing *et al.*, 1968). Belcher and Brett (1973) showed that the rhythms of oxygen consumption and of adult eclosion in *D. melanogaster* attained the same phase relationship to the light cycle, although the former was bimodal. In both LD *12* : 12 and LD *1* : 23 oxygen consumption was at its lowest at the time of the circadian cycle when eclosion was maximal. The respiratory rhythm, however, was not tested in DD; consequently its innateness is still open to question.

Nowosielski and Patton (1964) showed that the haemolymph titre of trehalose in the cricket *Acheta domestica* was at its peak about 3 hours before dawm, and Takahashi and Harwood (1964) showed that the highest levels of glycogen in the mosquito *Culex tarsalis* occurred towards the end of the light period. However, neither of these apparently rhythmic events were investigated in the absence of temporal cues, and glycogen levels, particularly, might be the result of increased storage during times when little activity occurs.

Among the most completely investigated physiological rhythms in insects are the daily changes in the colour of stick insects and the luminescence rhythm in the fire-fly. The stick insect *Carausius morosus* is normally dark coloured at night and light by day. Schleip (1910, 1915), however, showed that this rhythm persisted for several weeks in continuous darkness, but was absent from those insects kept in DD from the time of hatching. A reversal of the light cycle caused a reversal in the rhythm of colour change. This example, therefore, obeys at least some of the "rules" for an endogenous circadian rhythm. According to Kalmus (1938a) there is a hormone secreted in the head which controls the daily rhythm of pigmentation.

In the fire-fly *Photinus pyralis* there is a well-marked rhythm of luminescence (Buck, 1937). In constant dim light (*LL*) spontaneous flashes occur for a short time once every 24 hours; it is therefore fully endogenous. In addition, a change from darkness to dim light induced periodic flashing provided that the insects had been held in the dark for 24, 48, 72 or 96 hours, but not if they had been held in the dark for 12, 36, 60 or 84 hours. Since the free-running period (τ) of the rhythm is close to 24 hours, this experiment shows that flashing can only be induced at certain phases of an on-going oscillator (occurring at modulo τ), but not at points 180° out of phase (i.e. at modulo $\tau + \frac{1}{2}\tau$).

One of the most interesting physiological rhythms encountered in recent years is that of pheromone release and response in various Lepidoptera, such as the Noctuids *Autographa californica, Heliothis virescens, Spodoptera exigua* and *Trichoplusia ni* (Shorey and Gaston, 1965; Shorey, 1966). In *T. ni* the release of sex pheromone by the female and the reactivity of the male are governed by circadian clock mechanisms. Pheromone release reaches its maximum towards the end of the night in an *LD* cycle, and also persists with a periodicity close to 24 hours when transferred to constant dim light (0.3 lux) (Fig. 4.2)

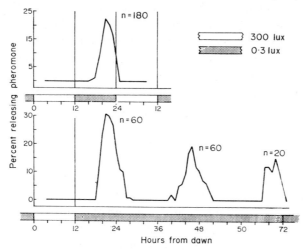

FIG. 4.2. Sex pheromone release by females of *Trichoplusia ni* in a light/dark cycle and in constant dim light (0.3 lux). (Redrawn, from Sower *et al.*, 1970.)

(Sower *et al.*, 1970). A circadian rhythm of pheromone activity has also been described for the tea tortrix moth *Adoxophyes fasciata* (Nagata *et al.*, 1972) and the mediterranean flour moth *Anagasta kühniella* (Traynier, 1970). The different timing of pheromone release in related sympatric species is an important mechanism for sexual isolation.

C. Rhythms in the Endocrine and Central Nervous Systems

Daily cycles of activity in the endocrine or nervous systems of insects have attracted attention because of the probable importance of these systems in the control of overt rhythms of physiology and behaviour. Many of these changes, however, have been detected and studied in cyclical rather than constant conditions so that their endogeneity remains doubtful. Definite links between histological evidence and overt rhythmicity are generally lacking, and these changes may only reflect the total rhythmicity in animal tissues. Brady (1967), for example, has pointed out: "the fact that synchronous changes take place in the nuclei of widely different tissues, varying from fat body to neurosecretory cells tends to suggest that one of the effects of the fundamental circadian organization of organisms may be to produce synchronous changes in all, or many, of their cells". Clearly, there is a need to recognize and distinguish cause from effect in these phenomena.

In the beetle *Carabus nemoralis*, Klug (1958) found diurnal changes in the relative numbers of two types of active neurosecretory cells in the brain, and also in the nuclear diameters of the cells in the corpora allata. Beck (1964) studied the diameters of the neurosecretory cells of diapausing larvae of the European corn borer, *Ostrinia nubilalis*, maintained in *LD 12*:12, and described a "well-defined" trimodal rhythm with maxima at 0, 8 and 16 hours after the onset of light. One peak coincided with dawn itself and was apparently phase-set by the light-cycle. However, the cell sizes varied widely, and Brady (1967) has drawn attention to the possible errors which may arise in measurements of this kind.

One of the most extensive investigations of this type has been carried out with *Drosophila melanogaster*. These studies revealed daily bimodal changes in the nuclear sizes of corpus allatum and pars intercerebralis cells in the adult (Rensing, 1964), and similar bimodal cycles of mean nuclear size in corpus allatum, fat body, prothoracic gland cells and brain neurosecretory cells in the larva (Rensing *et al.*, 1965). These cyclical changes were measured in *LD 12* : 12 and were considered to reflect metabolic activity. In the salivary glands these rhythms continued in *in vitro* preparations for 10 days, and also persisted in continuous light (*LL*) for at least 4 days (Rensing, 1969). The phases of the peaks could be shifted during larval development by a 6-hour photoperiod, or by the *in vitro* addition of ecdysone (Rensing, 1971).

Cymborowski and Dutkowski (1969, 1970) demonstrated daily cycles of RNA and protein synthesis in the neurosecretory cells of the brain and the suboesophageal ganglion of *Acheta domestica* by autoradiographic methods. In *LD 12* : 12, RNA synthesis in the brain neurosecretory cells was at its maximum during the period of least locomotor activity, 30 minutes after lights-on. Conversely, it was at its lowest when the crickets were most active; the two maxima were about 12 hours apart (Fig. 4.3). In the suboesophageal ganglion, on the other hand, intensified RNA synthesis occurred at two points with the first peak about 6 hours before maximum locomotor activity. The synthesis of protein and the accumulation of neurosecretory material were also cyclic (Cymborowski and Dutkowski, 1970) and a temporal and presumably causal sequence of RNA synthesis, protein synthesis and activity was posulated. These authors and their associates also found cyclical changes in acetylcholinesterase activity (Cymborowski *et al.*, 1970) and in the ultrastructure of the medial neurosecretory cells of the brain (Dutkowski *et al.*, 1971). These changes were not apparent in *LL* (which also suppresses locomotor activity in *Acheta*): the persistence of these rhythms in DD was not investigated.

A daily rhythm of the secretion of 5-hydroxytryptamine (serotonin) was reported in the brain and intestine of the harvestman *Leiobunum longipes* by Fowler and Goodnight (1966), and in the larvae, pupae and adults of *Drosophila melanogaster* (Fowler *et al.*, 1972). These rhythms were considered to be endogenous but were only examined in daily light cycles. Serotonin has also been recorded as a neurosecretory material in certain moths (*Noctua pronuba* and *Agrotis ipsilon*), and it is believed to have a role in circadian activity cycles (Hinks, 1967).

A circadian rhythm of nervous activity in a single neuron of the abdominal ganglion of the sea hare *Aplysia californica* has been extensively studied by Strumwasser (1965). These molluscs also show circadian rhythms in optic nerve impulses—even in isolated eyes—which persist in DD and constant dim light and can be phase-shifted by light pulses (Jacklet, 1969, 1971). Comparable studies have not been performed with insect material, although daily rhythms of the electroretinogram, persisting in DD, are known for some beetles (Jahn and Crescitelli, 1940; Jahn and Wulff, 1943), and in *Acheta domestica*, Tyschenko *et al.* (1972) have described "circadian neurones" in the brain which show increased impulse frequency just before the light is turned on and which maintain a high level of activity for 10 to 12 hours thereafter. Clearly, neurophysiological investigations of this kind may provide important insights into the mechanisms controlling overt rhythmicity.

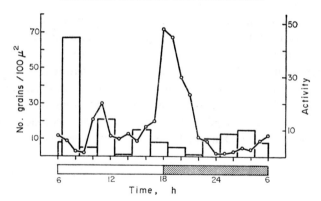

Fɪɢ. 4.3. Changes in the level of RNA synthesis in the neurosecretory cells of the pars intercerebralis of rhythmic house crickets *(Acheta domestica)* originating from LD *12* : 12 conditions. Open circles: locomotor activity expressed in number of movements per hour; histogram: the intensity of RNA synthesis. (Redrawn from Cymborowski and Dutkowski, 1969.)

D. Sensitivity to Drugs and Insecticides

Circadian rhythms of sensitivity to drugs were first recorded by Halberg and his associates (Halberg, 1960). At different times of the 24-hour cycle, mice were found to react differently to equivalent doses of *Escherichia coli* endotoxin, ethanol, ouabaine and the adrenocortical inhibitor, Su-4885. In the case of *E. coli* endotoxin, for example, highest mortality was recorded (in LD *12* : 12) towards the end of the light period and the lowest mortality close to midnight.

Similar examples of pharmacological sensitivity are now known in the insects. Beck (1963) maintained the cockroach *Blatella germanica* in LD *12* : 12 for 7 days and then injected standard doses of DDT, Dimetalan, Dichlorvos and KCN into the haemocoel at different times of the 24-hour cycle. The insects were then held in *LL* to determine the resulting mortalities. Sensitivity to cyanide (2 μg/insect) was highest shortly after dark and lowest soon after dawn. Dimetalan (2.5 μg/insect) was most toxic in the middle of the night. The rhythm of sensitivity to Dichlorvos (0.15 μg/insect) was bimodal with peaks during the late subjective day and the late subjective night; that for DDT was unimodal but very shallow.

Adults of the two-spotted spider mite *Tetranychus urticae* showed rhythms of sensitivity to a variety of insecticides and pharmacological agents. In LD *14* : 10 maximum sensitivity to Dichlorvos occurred about 2 hours after dawn (Polcik *et al.*, 1964), whereas maximum sensitivity to ether, chloroform and carbon tetrachloride occurred around dusk (Nowosielski *et al.*, 1964). The house cricket *Acheta domestica* showed maximum sensitivity to these agents during the first part of the night when the insects were most active. Apparently circadian cycles of the formation of various hydrolytic products and metabolites following the injection of radiophosphorus-labelled Disulphoton into the larvae of *Heliothis zea* have also been recorded by Bull and Lindquist (1965), and Fernandez and Randolph (1966) showed that larvae of *Musca domestica* were more susceptible to DDT, dieldrin and aldrin when reared in LD *14* : 10 than in LD *10* : 14 or in *LL*.

The importance of these observations is that the differential sensitivity to these drugs and insecticides at different circadian times must reflect biochemical oscillations in the organisms which are entrained and phase-controlled by the light-cycle in a manner comparable to those overt rhythms of physiology and behaviour described earlier. Together with the complex reactions to the periodicity of the environment discussed in the next section, these results indicate the general and pervasive nature of oscillatory processes in insect physiology and biochemistry. Most observations, however, have been made in LD cycles, although test insects have often been removed to constant light after treatment. It would be instructive, therefore, if rhythms of sensitivity to pharmacological agents were also studied after transfer to DD in which the oscillators, if endogenous, would be expected to free-run.

E. Environmental Periodicity and Fundamental Aspects of Physiology

Organisms maintained in constant conditions of light and temperature, or in environmental cycles with a periodicity far from that which matches the Earth's rotation, often show impaired growth and survival, or deleterious changes affecting general "fitness". Many plants, for example, become abnormal and necrotic when maintained in LL and constant temperature (Arthur and Harwill, 1937; Hillman, 1956), or show suboptimal growth rates in light cycles longer or shorter than 24 hours (Highkin and Hanson, 1954; Went, 1959).

The first clear example of this phenomenon in the insects was described by Pittendrigh and Minis (1972) for the survival of adults of *Drosophila melanogaster* in different light-cycles. These insects were exposed to LD *12* : 12 ($T = 24$ hours), LD *10.5* : 10.5 ($T = 21$ hours), LD *13.5* : 13.5 ($T = 27$ hours) and to LL throughout adult life at a constant temperature of 25°. In both males and females of a wild-type, and females of a tumorous strain (tu^g), survival was significantly greater when they were "driven" at $T = 24$ hours than in the other regimens (Fig. 4.4). The *amount* of light received was not the factor involved because it constituted 50 per cent of the cycle in each case. A similar effect of T and the periodicity of the light regime on adult longevity has also been observed in the blowfly *Phormia terraenovae* (Aschoff *et al.*, 1971).

Saunders (1972) raised larval populations of the flesh-fly *Sarcophaga argyrostoma* in a wide range of light cycles of different length ($T = 21$ to $T = 72$ hours) with 12, 14 or 16 hours of light in each. The results showed that the rate of larval development (from the deposition of the first instar larva to puparium formation) was a function of both T and photoperiod. With a 12-hour photoperiod larval development was most protracted at $T = 24$ (LD *12* : 12) and at $T = 48$ (LD *12* : 36) and most rapid at $T = 36$ (LD *12* : 24) and at about $T = 60$ (LD *12* : 48) (Fig. 4.5). The difference in mean developmental time between larvae maintained at $T = 36$ hours and at $T = 48$ hours was about 3 days (or 20 per cent of the total). In the case of a 14-hour photoperiod the length of larval development at $T = 42$ hours was about 4 days shorter than that at $T = 48$. These differences are clearly independent of the amount of light received since larvae reared at $T = 42$ (LD *14* : 28) completed development in about 14 days whereas those at $T = 32$ (LD *14* : 18), although receiving a greater *total* illumination, showed a mean developmental time of about 18 days. The cyclical nature of the results was interpreted as a temporal interaction between the light pulses and innate physiological oscillations within the insects.

Figure 4.5 also shows an additional effect of photoperiod on developmental rates. Thus, at $T = 24$ hours, a short photoperiod of LD *12* : 12 caused a more protracted larval

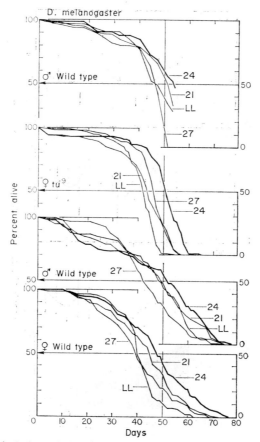

FIG. 4.4. The survival of populations of *Drosophila melanogaster* in light/dark cycles whose period (*T*) is 24, 21, or 27 hours, and in *LL*. In all four experiments the flies on a 24-hour day lived significantly longer than the flies in the other environments. (From Pittendrigh and Minis, 1972.)

development than a long daylength (*LD 14* : 10 or *LD 16* : 8). This and other photoperiodic effects on growth rate will be dealt with more fully in Chapter 5. In Fig. 4.6 the results are shown in the form of an "extended circadian topography" of the type predicted by Pittendrigh (1972), in which the contours connect points of equal larval duration. The surface of this topography clearly demonstrates the three points of protracted development and, at *T* = 24 hours, the acceleration in long days.

The results obtained for larval development in *S. argyrostoma* and for the survival of *D. melanogaster* (Pittendrigh and Minis, 1972) suggest that organisms, having evolved an innate periodicity in their metabolic functions close to the natural cycle of the Earth's rotation around its axis, perform "more efficiently"—or, at least, "differently"—when driven by light cycles close to τ (their natural circadian periodicity), or modulo τ, than when driven at modulo $(\tau + \frac{1}{2}\tau)$. These observations have a bearing on the interpretation of the nature of the circadian system in multicellular organisms (Section 4F).

In Chapter 3 it was shown that pupal eclosion in *Drosophila pseudoobscura* becomes arrhythmic in *LL* but rhythmicity may be restored, or arrhythmicity prevented, by a 24-hour temperature cycle. Clearly, continuous light disorganizes the circadian system in some way, but a temperature cycle can maintain it. A similar effect of thermoperiod has been observed in the maintenance of "normal" form and function in plants and

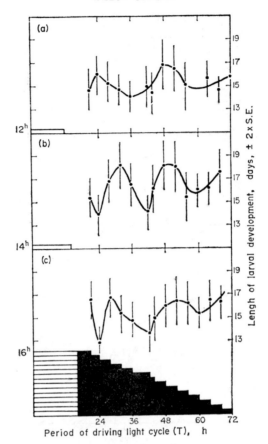

FIG. 4.5. The length of larval development of *Sarcophaga argyrostoma* as a function of the period (*T*) of the driving light-cycle: (a) 12-hour photoperiod; (b) 14-hour photoperiod; (c) 16-hour photoperiod. Each point represents the mean number of days to puparium formation of 200 to 900 larvae. Vertical lines indicate $\pm 2 \times$ S.E. The white and black bars at the bottom illustrate the experimental design. (From Saunders, 1972.)

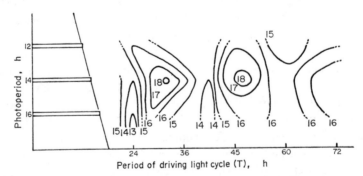

FIG. 4.6. Data from Fig. 4.5 redrawn as a "circadian topography". The contours connect points of equal larval developmental times (in days). Note the three points of protracted development, and (at *T* = 24) the accelerated development with a 14- and a 16-hour photoperiod. (From Saunders, 1972.)

animals. Hillman (1956), for example, showed that the deleterious effects of *LL* on tomatoes could be avoided if a 24-hour temperature cycle was imposed on the plants. Few carefully controlled experiments of this kind have been performed with insects, although several observations support the conclusion that a temperature cycle is beneficial. Messenger (1964) found that a rhythmically fluctuating environment (temperature, humidity and illumination) produced the highest rate of survival, growth and reproduction in the aphid *Therioaphis maculata*. More recently, Hollingsworth (1969) maintained females of *D. melanogaster* in daily temperature cycles (25/30°C) as part of a test of Pearl's (1928) "rate of living" theory of ageing, which maintains that if poikilotherms are kept for half their lives at one temperature and half at another, their expected lifespan would be the harmonic mean of the expectations of life at the two constant temperatures. The results showed that the females lived significantly longer than predicted from this hypothesis when maintained in fluctuating temperatures, and those kept for 3 hours at 30°C and 21 hours at 25°C lived *as long as* those kept at the lower temperature (25°C) throughout. Although the light regime was not specified, these results may be interpreted as support for a hypothesis that a cyclical environment is less deleterious than a static one. Nayer (1972), working with adults of the mosquito *Aëdes taeniorhynchus*, showed that females kept in a fluctuating temperature (12 hours at 22°C, 12 hours at 27°C; mean = 24.5°C) lived slightly longer (42.4 days) than those kept at 22°C throughout (41.2 days), and significantly longer than the age-span predicted for a constant temperature of 24.5° (~36 days). These experiments were apparently performed in a light cycle (*LD 12* : 12) but may constitute additional evidence that a daily temperature cycle is more favourable than a constant temperature.

Another form of "metabolic stress" arising from abnormal environmental cycles might be expected from large or repeated changes in the phase of the driving light-cycle, or from the exposure of an organism to a temperature cycle out-of-phase with the light. The deleterious nature of these procedures has been demonstrated for adult survival in *Phormia terraenovae* (Aschoff *et al.*, 1971). Harker (1958) also reported that suboesophageal ganglia removed from cockroaches *(Periplaneta americana)* and implanted daily for 4 days into the haemocoels of recipients *12 hours out-of-phase* led to the appearance of transplantable tumours in the midgut. Nishiitsutsuji-Uwo and Pittendrigh (1967) were later unable to reproduce these results, however, and the validity of these observations remains open to question.

F. Temporal Organization and the Circadian System

The foregoing chapters of this book illustrate the fact that an insect, like other eukaryotic organisms, displays a wide array of rhythms of activity, behaviour and physiology which have evolved a close match to the period of the Earth's rotation around its axis. Many of these rhythms have, or are presumed to have, an adaptive significance in that in their entrained steady-state they attain a particular phase relationship to the environmental cycle so that the insects perform certain functions at particular times of the day. One of the fundamental questions about this temporal organization is the nature of the circadian system, particularly if there is a single "master clock" or whether organisms are a "population" of clocks.

The idea of a single driving oscillation (a "master clock"), which on one hand is entrained by the light-cycle and on the other is coupled to a number of driven rhythms controlling overt activity, appears to be acceptable only in the case of the Protista (Sweeney, 1969). Most authors agree that the circadian system in higher organisms consists of a larger number—even a "population"—of oscillators, each associated with a number of driven rhythms controlling behavioural or physiological phenomena.

These oscillations may be loosely coupled one to another, or independently coupled to the environmental light-cycle, perhaps by different pigments. Some may be phase-set by the whole photoperiod, others only by dawn or by dusk.

The evidence for such a multioscillator circadian system comes from a number of sources. In the moth *Pectinophora gossypiella*, for example, three overt behavioural rhythms have been studied, namely: egg hatch, pupal eclosion and oviposition (Pittendrigh and Minis, 1971). In steady-state entrainment to the light cycle all three achieve a different phase relationship to the *Zeitgeber*, egg hatch and eclosion occurring during the early part of the day and oviposition after dusk. When transferred into DD, however, the two "adult" rhythms (eclosion and oviposition) reveal a free-running period (τ) of about 22.5 hours, whereas τ for the egg-hatch rhythm is close to 24 hours. Furthermore, red light fails to entrain either of the "adult" rhythms, but perceptibly *shortens* τ for one of them (eclosion). These observations can be interpreted as evidence for more than one oscillator in the circadian system.

Photoperiodic time measurement in the parasitic wasp *Nasonia vitripennis* (Chapter 7) is a function of the circadian system (Saunders, 1970), and separate "dawn" and "dusk" oscillators are evident (Saunders, 1974). In organisms other than insects, evidence for more than one oscillator has been obtained from the activity cycles of rodents (Pittendrigh, 1960, 1967), the tree shrew *Tupaia glis* (Hoffmann, 1969) and man (Aschoff, 1969) which, in constant conditions, may split spontaneously into two components with distinctly different frequencies. Such findings are evidence that several circadian oscillators control overt rhythms of activity and physiology.

Since single-celled organisms such as the marine dinoflagellate *Gonyaulax polyedra* (Sweeney, 1969) show an array of circadian rhythms it must be assumed that a multicellular organism is a "population" of oscillators in the same sense that it is a "population" of cells. By an extension of the same argument a population of organisms such as a culture of fruit-flies—is also a population of oscillators. When such a "population" is entrained by an environmental light-cycle all the constituent oscillators adopt the same frequency and are thus synchronized both to the *Zeitgeber* and to each other. There is then a high degree of internal temporal organization in the circadian system. The arrhythmicity which may occur in *LL* is probably due to an internal desynchronization of the constituent subsystems. Furthermore, the multioscillator hypothesis enables one to interpret the deleterious effects on survival and growth observed in organisms kept in *LL* or in abnormal cycles. In other words, it is likely that the population of oscillators in the organism is synchronized when T is close to τ or modulo τ, but becomes desynchronized when T is far from τ, or in *LL*. Significant changes in the internal temporal organization is bound to have far-reaching effects if, for example, optimal concentrations of enzymes and of substrates occur at random times of the 24-hour cycle.

The fact that individual cells of a metazoan organism make up a population of oscillators is illustrated by the cuticle deposition rhythm in *Schistocerca gregaria* in which each epidermal cell appears to constitute its own "clock". In normal *LD* cycles these constituent clocks are synchronized and act together in the construction of laminated cuticle (Neville, 1965). In the beetle *Oryctes rhinoceros*, however, the epidermal cells contain endogenous oscillators which cannot entrain to the light-cycle. In this situation, therefore, individual cells are not coupled to the environment but rely on a system of coupling between neighbouring cells; this results in the deposition of layers with different frequencies in different areas of the body (Zelazny and Neville, 1972).

Complex multicellular animals, such as the insects, are of course not merely aggregations of cells. Just as morphological differentiation has occurred to produce tissues and organs with different functions, differentiation has also occurred to produce organs with a specific time-keeping or time-measuring function. Thus, although each cell still constitutes its own "clock", an insect will have an organ, usually the brain or another part of the central nervous system, which is primarily concerned with the control of overt rhythms of activity and behaviour. The site of such clocks and their output is the subject of a later chapter (Chapter 10).

One aspect of temporal organization has practical significance but is commonly overlooked. Experimental physiologists often take elaborate precautions to ensure that their experimental organisms are kept in conditions which remain as constant as is physically possible. Whilst the reasons for this are obvious and understandable it is possible that internal desynchronization takes place and abnormal effects become manifest. On the other hand, when circadian organization *is* maintained, the experimenter is not always aware that treatments may have profoundly different effects at different times of the day or night.

Annotated Summary

1. Circadian oscillations are known to control a wide variety of physiological and metabolic rhythms in insects, including colour change, respiration, bioluminescence, pheromone activity, and cuticle growth.

2. In many hemimetabolous insects, and a few holometabola, the endocuticle is layed down in discrete "day" and "night" layers in which the organization of the chitin lamellae differs. In *Schistocerca gregaria* these layers are controlled by a circadian "clock": they "free-run" in DD with a period (τ) of about 23 hours, are temperature-compensated, and are damped out in *LL*.

3. Rhythms in the CNS and endocrine systems may be important in relation to the control of overt physiological and behavioural rhythms. In some cases these rhythms have been sufficiently studied to establish their circadian nature.

4. Insects, like other organisms, show differential sensitivity to drugs and insecticides at different times of the day and night. This is thought to reflect internal biochemical oscillations.

5. Insects appear to perform "more normally" or "better" when maintained in natural 24-hour light or temperature cycles; but when maintained in cycles with an unnatural period or in constant light, constant dark, or in constant temperature, they may show reduced longevity or a protracted rate of growth.

6. The insect, like other multicellular organisms, is regarded as a population of independent or semi-independent oscillators each, perhaps, driving a series of driven elements or rhythms which, in turn, control physiological or behavioural events. This oscillatory temporal organization is called the circadian system.

CHAPTER 5

PHOTOPERIODISM AND SEASONAL CYCLES OF DEVELOPMENT

PHOTOPERIODISM comprises a miscellany of clock phenomena in which organisms distinguish the long days (or short nights) of summer from the short days (or long nights) of autumn and winter, and thereby obtain "information" on *calendar* time from the environment. A wide range of organisms, principally "higher" plants and animals living in a terrestrial environment, use this "noise-free" information to control various seasonally appropriate switches in metabolism, most of which have a clear functional significance or survival value.

Although a number of botanists in the last decades of the nineteenth and the early part of this century were aware of some daylength effects on flowering plants (see Cumming, 1971), the phenomenon of photoperiodism was first adequately described by Garner and Allard (1920) for a variety of plants including tobacco, soybean, radish, carrot and lettuce. They found that many plants could only flower and fruit when daylength fell between certain limits; some plants responded to long days, others to short. The Maryland Narrowleaf variety of tobacco *(Nicotiana tabacum)*, for example, grew by vegetative means to an "extraordinary" height at daylengths down to 12 hours per day, but produced flowers and seeds at *LD 7: 17*. The Biloxi variety of soybean *(Soja max)* was also a short-day plant. Garner and Allard found that the type of growth could be manipulated experimentally by supplementing or curtailing the natural daylength.

These observations were soon followed by Marcovitch's (1923, 1924) demonstration that the appearance of seasonal morphs in several species of aphids was similarly controlled. The strawberry root aphid, *Aphis forbesi*, for example, produced sexual forms when the natural daylength was curtailed to 7 hours per day, even at the height of summer. Conversely, long exposure to artificial light in September inhibited the sexual forms and induced viviparous reproduction. Rowan (1926) later demonstrated that daylength could have important effects on reproductive behaviour and physiology of birds. Kogure (1933) and Sabrosky *et al.* (1933) were the earliest workers concerned with the photoperiodic control of insect diapause. Sabrosky and his collaborators showed that diapause in the grasshopper *Acrydium arenosum* could be averted by exposure to continuous light; Kogure's very extensive investigation showed that the commercial silkmoth *Bombyx mori* was a short-day insect with a winter diapause in the egg, and included observations on light-intensity thresholds and the spectral sensitivity of the response. Shortly afterwards, Baker (1935) described the photoperiodic termination of diapause in a number of overwintering tree-hole mosquitoes and midges.

Since these early papers many of the seasonal activities of animals, including insects, crustacea (e.g. Stross and Hill, 1968), acarina (e.g. Lees, 1953a; Belozerov, 1964), birds, mammals and reptiles, have proved to be under a similar type of daylength control. Amongst the insects, the induction and termination of diapause, and the control of seasonal morphs are probably the most widespread phenomena, and certainly the most extensively studied. Seasonal polymorphism as well as a number of other aspects of physiology and behaviour—such as the adoption of winter coloration, deposition of fat, protracted development, and migration to hibernation sites—are often connected with the induction of the diapause state, and are therefore brought about by the same environmental factors.

Since the discovery of photoperiodism just half a century ago a very large number of scientific papers on the subject has appeared. This is particularly true for the control of diapause in the northern temperate zones (approximately 30° to 60°N), and because of the obvious importance of seasonal cycles in the study of agricultural and other pest species. Danilevskii (1965), for example, was able to list over 100 insect species with such a clock, and Beck (1968) raised this number to about 150. The number now is very much higher and only a fraction of these will be reviewed here. This chapter will describe the principal photoperiodic phenomena observed in the insects; the properties of the clock and the nature of time measurement will be dealt with in subsequent chapters.

A. Dormancy: Quiescence and Diapause

Unless insects have evolved some specialized habit or habitat-choice, periods of unfavourable climate demand from them a state of reduced metabolism, or dormancy, to enable them to overwinter or withstand a dry season. During the favourable season the insect may produce several successive generations (a multivoltine life cycle) or a single generation per year (univoltine); other species may have a life-cycle occupying several years with a number of dormancy periods. In tropical areas insects often show bursts of activity when the rainy season begins and become dormant with the start of the dry season. In temperate latitudes insects characteristically commence activity in the spring and become dormant as the winter approaches. In certain subtropical areas, where the summers are hot and dry and the winters cold or cool, two periods—spring and summer—may be favourable for growth and activity, and dormancy may intervene in both winter and summer.

Ecologically it is possible to distinguish two types of dormancy: hibernation and aestivation. The physiological mechanisms controlling dormancy, however, may be quite diverse; quiescence and diapause are the two most important mechanisms involved (Lees, 1955). Since only the latter appears to be controlled by a photoperiodic clock, quiescence will be dealt with very briefly.

The chief characteristic of quiescence, as opposed to diapause, is that the state of dormancy is *directly* imposed by the adverse conditions, and recovery occurs soon after these restrictions are removed. For example, insects in a state of "cold torpor" resume activity when the temperature rises, and dehydrated larvae continue development when water is supplied.

Dehydration occurs most frequently as a mechanism for aestivation. The Chironomid, *Polypedilum vanderplanki*, for example, breeds in pools of water on exposed rocks in parts of west and east Africa. During the dry season these pools dry up and the larvae become almost totally dehydrated (Hinton, 1951). They can tolerate repeated periods of dehydration and hydration and have remained viable in the dry state for many years.

Laboratory experiments (Hinton, 1960) have demonstrated a remarkable resistance to extremes of temperature whilst in this condition. Dehydrated larvae, which may contain as little as 8 per cent of moisture, can withstand total immersion in liquid helium, and exposure to 102 to 104°C for short periods without preventing recovery and subsequent development when returned to water. Dried larvae have also recovered after immersion in absolute ethanol for 24 hours, and up to 7 days in pure glycerol. It is probable that many other insects which live and breed in ephemeral aquatic habitats can aestivate in a similar dehydrated condition.

Diapause differs from quiescence in two fundamental ways. Firstly, it is an "actively induced" state most frequently involving the cessation of neuro-endocrine activity, usually at species-specific points in the insect's life-cycle. Secondly, the onset of diapause is brought about by environmental factors which, although signalling the approach of "unfavourable" conditions, are not, in themselves, adverse. These factors have been called "token stimuli" (Lees, 1955), the most important of which is photoperiod. Since local populations of an insect respond to daylength in the same manner, and enter diapause at the same developmental stage, this mechanism also serves to synchronize the seasonal activities of the particular species.

Müller (1970) has recognized three types of diapause. (1) *Parapause* is an obligatory diapause observed in univoltine species. There is a clearly defined phase of induction and the diapause supervenes in every generation in a species-specific instar. The onset of this type of diapause is genetically determined and appears to be independent of the environment. (2) *Eudiapause* is a facultative cessation of development with species-specific sensitive periods (inductive phases) and diapausing instars. In favourable conditions development proceeds unchecked; as unfavourable seasons approach diapause supervenes. This type of diapause is usually induced by photoperiod, but terminated by a period of chilling (overwinter), or by a change in the level of the temperature. (3) *Oligopause* is also a facultative arrest, often with induction and termination of diapause under photoperiodic control. Mansingh (1971) and Thiele (1973) have also contributed to, or commented on, this classification.

Insects entering winter diapause generally react to short daylength, decreasing daylength, or to a change from long to short daylength. By analogy with plants these insects are called long-day species because they are active during the long days of summer and become dormant in the autumn. Those insects which aestivate, or are winter-active, usually show an opposite, or short-day response. One notable exception to this is the commercial silkmoth *Bombyx mori*, which is a short-day insect but overwinters as an embryo within the egg because the stages *sensitive* to daylength (the eggs and young larvae of the maternal generation) occur in the height of summer when the days are long (Kogure, 1933). These reactions to daylength will be considered in detail in Chapter 6.

In some insects such as *Dendrolimus pini* (Geispits, 1965), larvae may enter diapause during one or more of the larval stadia, and some long-lived insects, such as the dragonfly *Tetragoneuria cynosura* (Lutz and Jenner, 1964), may diapause as early nymphs in one winter and mature nymphs in the next. Other species, such as *Mamestra brassicae* (Masaki and Sakai, 1965), *Hyphantria cunea* (Umeya and Masaki, 1969) and certain carabid beetles (Thiele, 1969), may hibernate *and* aestivate, sometimes in different instars. The majority of insects, however, diapause at a single species-specific point in their life cycle. The silkworm *Bombyx mori* (Kogure, 1933), the mosquito *Aëdes togoi* (Vinogradova, 1960) and the green vetch aphid *Megoura viciae* (Lees, 1959), for example, diapause as eggs. A vast assemblage of species, including the pink bollworm *Pectinophora gossypiella* (Adkisson, Bell and Wellso, 1963), *Ostrinia nubilalis* (Beck and Hanec, 1960), *Nasonia vitripennis* (Saunders, 1965 a, b), *Gryllus campestris* (Fuzeau-Braesch, 1966) and the green blowfly *Lucilia caesar* (Ring, 1967), enter diapause as larvae or nymphae. Pupal diapause is particularly common in the *Lepidoptera*, such as *Acronycta rumicis* (Danilevskii, 1965), *Pieris rapae* (Barker *et al.*, 1963) and *Antheraea pernyi* (Tanaka, 1950), and in the Diptera, such as *Erioischia brassicae* (Hughes, 1960), *Lyperosia irritans* (Depner, 1962) and *Sarcophaga argyrostoma* (Fraenkel and Hsiao, 1968). Adult or reproductive diapause is also widespread: examples include *Musca autumnalis* (Stoffolano and Matthysse, 1967), *Leptinotarsa decemlineata* (De Wilde and De Boer, 1961) and various coccinellid beetles (Hodek, and Cerkasov, 1958). An updated but certainly incomplete list of those species in which a photoperiodically induced or terminated diapause has been recorded is included in an appendix.

Larval, pupal and nymphal diapause is generally regarded as an inactivation of the brain-prothoracic gland system (Williams, 1952; Novak, 1966). Under the appropriate photoperiodic influence, the neurosecretory calls of the brain fail to release the brain or the prothoracotrophic hormone (PTTH), thus the prothoracic glands remain inactive and, in the absence of ecdysone, growth and development stop (Fig. 5.1). This form of endocrine control of larval and pupal diapause has been established or postulated for *Hyalophora cecropia* (Williams, 1946, 1952), *Cephus cinctus* (Church, 1955), *Leuhdorphia japonica* (Ichikawa and Nishiitsutsuji-Uwo, 1955), *Mimas tiliae* (Highnam, 1958), *Ostrinia nubilalis* (Cloutier *et al.*, 1962) and *Lucilia caesar* (Fraser, 1960). It is frequently accompanied by physiological changes such as reduced metabolic rate, reduced water content, increased fat, and by changes in behaviour, such as the spinning of silken hibernacula in some Lepidoptera.

In *Diatraea grandiosella*, however, the diapausing larva apparently retains a functional endocrine system throughout its period of dormancy. Chippendale and Yin (1973) and Yin and Chippendale (1973) showed that diapause larvae undergo "stationary moults" in a diapause-maintaining photoperiod of LD 12: 12. All larvae raised in LD 12: 12 moult from a "spotted" (non-diapause) form to an "immaculate" (diapause) form. About half of them then undergo a second immaculate–immaculate moult, and about 14 per cent undergo a third. At each ecdysis the head capsule remains the same size and the larvae show physiological characters of the diapause state: they do not feed, they have a low respiratory quotient, partial dehydration, increased cold hardiness, and accumulation of fat. Injection of ecdysone (20-hydroxyecdysone) into diapausing larvae caused further immaculate–immaculate moults. Neck ligation caused a premature termination of diapause and pupation of the thoraco-abdominal compartment, showing that a cephalic factor was necessary for the maintenance of the diapause state. Neck ligation and subsequent injection of ecdysone into the body caused premature pupation of that part behind the ligature, showing that ecdysone only causes diapause termination in the absence of this cephalic factor. Topical application of juvenile hormone (JH) caused non-diapause larvae to become immaculate and enter diapause, and repeated application of JH prolonged diapause and increased the number of stationary ecdyses. The cephalic factor which maintains diapause in *D. grandiosella* was therefore identified as juvenile hormone: diapause in this insect, therefore, is quite unlike that in any other so far described.

The immediate cause of imaginal or reproductive diapause is the inactivation of the

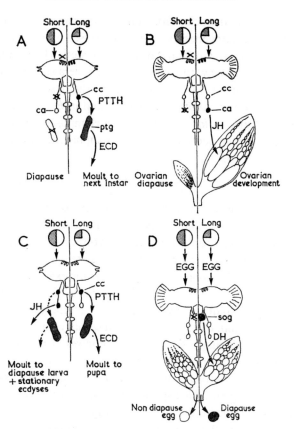

FIG. 5.1. Four types of diapause in insects (schematic). A. Larval-pupal diapause in which short days inactivate the neurosecretory cells of the brain. This results in inactive prothoracic glands, no ecdysone, and diapause supervening before the next moult. Long days allow brain hormone and ecdysone production, and moulting to the next instar. (After Williams, 1952.) B. Adult (ovarian) diapause in which short days inactivate the brain neurosecretory cells and hence the corpora allata. The consequent lack of juvenile hormone causes yolk synthesis in the oocytes to cease. Long days allow ovarian development. In ovarian diapause, metabolites are frequently redirected to the fat body as the insect enters diapause. (After de Wilde, 1959.) C. Larval diapause in *Diatraea grandiosella* in which a reduced titre of juvenile hormone in short days causes moulting to the diapause larva. The corpus cardiacum-prothoracic gland system may remain active during diapause causing stationary ecdyses. Long days cause a greater fall in the titre of juvenile hormone and the larva pupates. (After Yin and Chippendale, 1973.) D. Egg diapause in *Bombyx mori*. Long days perceived by the eggs and young larvae of the maternal generation result in the secretion of a diapause hormone by the suboesophageal ganglion of the pupa which enters the ovarian egg and imposes the diapause state. (After Fukuda, 1963.) Active glands shown in black or shaded, inactive glands shown with a cross. cc — corpora cardiaca; ca — corpora allata; ptg — prothoracic gland; sog — suboesophageal ganglion. PTTH — prothoracotrophic (brain) hormone; ECD — ecdysone; JH — juvenile hormone; DH — diapause hormone.

brain neurosecretory cells controlling the corpora allata; inactive corpora allata result in an absence of juvenile hormone (JH) and suppression of the ovaries, often at the point when yolk deposition should begin. This type of endocrine control has been demonstrated for *Dytiscus marginalis* (Joly, 1945), *Leptinotarsa decemlineata* (De Wilde *et al.*, 1959; De Wilde and de Boer, 1961), *Pyrrhocoris apterus* (Slama, 1964), *Anacridium aegyptium* (Geldiay, 1967) and *Galeruca tanaceti* (Siew, 1965 a, b, c). It is usually accompanied by physiological changes such as the deposition of fat in the fat body rather than in the ovaries (gonotrophic dissociation). Male insects can also overwinter in a repro-ductive diapause and spermatogenesis is often suppressed or arrested during dormancy.

Egg diapause can occur at any stage of embryogenesis (Lees, 1955); some species diapause as fully formed larvae within the egg shell, others such as *Bombyx mori* are at an earlier stage in development. In many such cases, the onset of egg diapause is determined by the photoperiod experienced maternally, and in *B. mori* (Fukuda, 1951, 1963; Hasegawa, 1951) and *Orgyia antiqua* (Kind, 1965) it is imposed on the embryo by a "diapause hormone" produced in the maternal sub-oesophageal ganglion and passed into the ovarian egg.

B. Seasonal Morphs

Season-bound morphological forms have been described in the Lepidoptera, Orthoptera, Homoptera, Heteroptera and in the Thysanoptera. In most of these instances photoperiod is the decisive environmental factor involved. At certain points in development, daylength appears to operate a genetic switch thereby opening or closing alternate pathways of development. The long days of summer, for example, may favour the production of summer forms of butterflies, short-winged bugs, or viviparous parthenogenetic aphids which reproduce rapidly to avail themselves of the favourable food supplies. Short days, on the other hand, often favour winged sexual forms which become dormant or, in the case of aphids, lay diapausing eggs. The control of polymorphism is therefore frequently associated with the onset of dormancy.

The Nymphalid *Araschnia levana* provides one of the best-known examples of seasonal dimorphism in the Lepidoptera. This species exists in two forms—*levana* and *prorsa*—which at one time were regarded as distinct species (Fig.5.2). Danilevskii (1948) and Müller (1955), however, showed that the short days of autumn induced a diapausing pupa from which the "typical" *levana* form (orange-brown with a pattern of black spots) emerged in the spring. The long days of summer, on the other hand, led to the production of the black and white *prorsa* which was not associated with a pupal diapause. In the southern part of the Soviet Union the green silver-lined moth *Hylophila prasinana* also occurs in a spring and summer form. The spring form, which is the typical *prasinana*, emerges from a diapause pupa, whereas the summer form (formerly thought to be a distinct species, "*H. hongarica*") does not include a diapause stage in its life cycle. Further north in the Soviet Union, this species is univoltine and only the *prasinana* form is known (Danilevskii, 1965). Sakai and Masaki (1965) have similarly shown that short photoperiods (less than 13 hours) induce the spring and autumn form of *Lycaena phlaeas daimio*, which shows a ground colour of bright coppery-orange with small black spots and a narrow marginal band, whereas long photoperiods (more than 14 hours) induce the more darkly pigmented summer form. This species also possesses a photoperiodically determined diapause (in the larva) but the induction of diapause appears to be independent of wing coloration. Most larvae at high temperature, for example, even in short daylength, avert diapause, but still give rise to the appropriate wing colour. Similar seasonal effects are known in the great southern white butterfly, *Ascia monuste* (Pease, 1962), which produces a summer melanic form in *LD 16 : 8* but a white form in short days (*LD 8 : 16*), in the Nymphalids *Polygonia c-aureum* (Aida and Sakagami, 1962) and *Kaniska canace no-japonicum* (Aida, 1963), and in the Pierids *Eurema hecabe mandarina* (Aida, 1963) and *Colias eurytheme* (Watt, 1969; Hoffmann, 1973).

In most cases the selective advantage of these seasonal colour polymorphisms in

Lepidoptera are far from clearly understood. In *Colias eurytheme*, however, Watt (1968, 1969) has shown that the dark hind undersides of the spring and autumn (short day) broods increase the efficiency of the absorption of solar energy, thus promoting activity and reproductive success during the cooler periods of the year. The light yellow or orange hind underwings of the summer generation, on the other hand, minimizes overheating in the warm season. How widespread this phenomenon is cannot be commented upon, but in the case of *Araschnia levana*, *Lycaena phlaeas* and *Ascia monuste*, the long-day summer forms are darker than the short-day broods of spring and autumn.

Seasonal forms in the Orthoptera, Heteroptera and Thysanoptera often involve alary dimorphism. In the ground cricket *Nemobius yezoensis*, for instance, experimental exposure to long days induces some of the nymphs to develop without a diapause and to become macropteres (Masaki and Oyama, 1963). Normal development in this species, however, is univoltine, and the insects are short winged and diapause as a nymph. Working in southern Finland, Vepsäläinen (1971 a, b) found that *Gerris odontogaster* occurred in short- and long-winged forms. The overwintering population was almost entirely macropterous. Their offspring, however, were dimorphic, micropteres emerging up to the middle of July and macropteres thereafter. These long-winged individuals leave the pond for overwintering sites and hibernate in a state of reproductive diapause. Experimental manipulation of the photoperiod showed that long days (over 18 hours per day) plus incremental changes in daylength during the early nymphal instars were required to produce adult micropteres. In the Thysanoptera, Köppä (1970) demonstrated that micropterous specimens of *Anaphothrips obscurus* were produced by short-day illumination.

It is probably amongst the Homoptera that seasonal polymorphism is most widespread, particularly in the leaf hoppers (Cicadellidae), white flies (Aleyrodidae), jumping plant

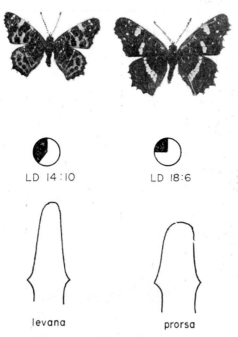

LD 14:10   LD 18:6

levana   prorsa

FIG. 5.2. *Araschnia levana*. Coloration and form of the penis in the spring or short-day form (*levana*) which emerges from a diapausing pupa, and the summer or long-day form (*prorsa*). (After Reinhardt, 1969.)

lice (Psyllidae) and in the aphids. Müller (1954, 1957), for example, showed that the leafhoppers *Euscelis plebejus* and *E. lineolatus* each occurred in a number of seasonal forms, many of which had previously been given separate specific status (Fig. 5.3). These forms differ in a number of characters including size, colour and the shape of the aedeagus. *E. plebejus* was found to be sensitive to daylength in the middle nymphal instars. In short-day conditions (4 to 15 hours light per day) only the spring form *(incisus)* was obtained, but in long days (17 to 18 hours) nearly all became the summer or *plebejus* form. At intermediate and ultrashort daylengths, or in DD, leafhoppers with an intermediate type of aedeagus were produced. The cicadellids *Nephotettix apicalis* and *N. cincticeps* (Kisimoto, 1959), the delphacids *Stenocranus minutus* (Müller, 1954) and *Delphacodes striatella* (Kisimoto, 1956), and the psyllid *Psylla pyri* (Bonnemaison and Missonnier, 1955; Oldfield, 1970) all show photoperiodic control of size, wing dimorphism and diapause. The whitefly *Aleurochiton complanatus* produces non-diapause and unpigmented "puparia" during long days, but diapausing "puparia" with thick sclerotized cuticles during the autumn (Müller, 1962 a, b).

The most complex seasonal cycles are undoubtedly found in the aphids. Most temperate species reproduce during the summer months as a series of viviparous, parthenogenetic females usually referred to as virginoparae—which may be alate (alatae) or wingless (apterae)—but produce sexual forms (males and oviparae) as the days become shorter in the autumn (Fig. 5.4). The fertilized oviparae generally deposit diapausing eggs which overwinter on the primary host. This alternation of sexual and asexual forms is referred to as holocycly. The rapid multiplication, viviparity and parthenogenetic reproduction of the virginoparae allows maximum utilization of available resources

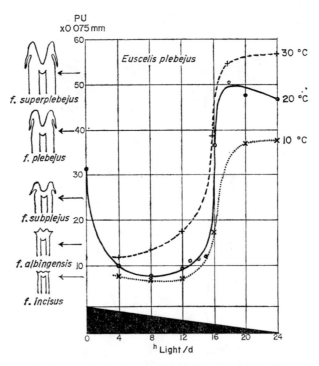

FIG. 5.3. Photoperiodic control of polymorphism in *Euscelis plebejus*. The aedeagus in the various forms is shown on the left; the ordinate is based on measurements of the aedeagus outline. (After Müller, 1960.)

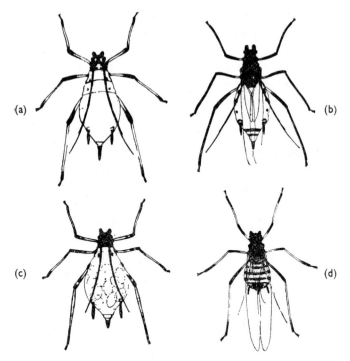

FIG. 5.4. Seasonal and sexual forms in the aphid *Megoura viciae*. a — apterous virginopara formed under long days; b — alate virginopara formed under long days and crowded conditions; c — ovipara formed in short days; d — male. (After Lees, 1959.)

during the summer months, whereas sexual reproduction allows genetic recombination and the production of the hibernating stage.

Some aphids such as *Aphis forbesi* (Marcovitch, 1924), *Brevicoryne brassicae* (Bonnemaison, 1951) and *Megoura viciae* (Lees, 1959) are monoecious and spend their active life on a single plant species. Others such as *Aphis fabae* (Davidson, 1929), *Myzus persicae* (Bonnemaison, 1951) and *Dysaphis plantaginea* (Bonnemaison, 1958, 1965) are heteroecious and change their host plant during the season, often from a primary host (a tree, for example) to a variety of herbaceous plants during the summer months. In different aphid species the determination of sex, and the determination of virginoparae and oviparae, are under photoperiodic control; the production of alate forms, however, is more often dependent on population density. A detailed account of these morphs and their induction is given by Lees (1966).

C. Growth Rates

In a number of insect species the rate of development is controlled by photoperiod and, as with other such phenomena, long- and short-day responses may be observed. Amongst the cutworms, for example, larvae of *Agrotis occulta* develop more rapidly in long days, whereas those of *A. triangulum* develop more rapidly in short (Danilevskii, 1965). Vinogradova (1967) showed that larvae of the mosquito *Aëdes triseriatus* developed more rapidly in LD *20* : 4 than in LD *10* : 14, the latter causing a pronounced delay during the fourth instar. The sod webworm, *Crambus tutillus*, however, is a short-day species and develops more rapidly in LD *12* : 12 than in LD *16* : 8. In addition, larvae of this species exposed to daylengths *decreasing* at a rate of 2 minutes per day from an initial 16-hour photoperiod grew more quickly than those exposed to LD *16* : 8 throughout (Kamm, 1972).

Atwal (1955) showed that larvae of the diamond back moth *Plutella maculipennis* completed their development in about 18 days at LD *9* : 15 but took only about 15.5 days at LD *15* : 9. Furthermore, adult females of this species raised in LD *15* : 9 laid more eggs more than those raised in LD *9*: 15, even though those reared at the shorter daylength had developed more slowly and might be expected to be heavier and to give rise to more fecund adults. This effect of photoperiod on fecundity was confirmed by Harcourt and Cass (1966) who showed that females of *P. maculipennis* raised from the larval stage in LD *12* : 12 laid about 37 eggs whereas those raised in LD *16* : 8 laid about 74. According to these authors there is no "true" diapause in *P. maculipennis* although the species overwinters most frequently as an adult, and short-day females deposited more fat and developed fewer eggs than those raised in long days. This species, therefore, may possess a "weak" or incipient form of ovarian diapause. Short-day promotion of ovarian development is indicated in the results of Meudec (1966). These showed that the ovaries of 24-hour-old females of *Acrolepia assectella* contained more eggs when the larvae had been raised in LD *9* : 15 (8.3 eggs per female) than when raised in DD (4.1 eggs per female) or in LL (2.9 per female). Minis and Pittendrigh (1968) showed that the embryonic development of *Pectinophora gossypiella* was more rapid in LL than in DD. This so-called "light-growth effect" is probably another example of the same phenomenon. It also seems clear that these photoperiodic effects on growth and development are similar to those effects of environmental periodicity on growth and survival discussed in Chapter 4. In the flesh-fly *Sarcophaga argyrostoma*, for example, larval growth rates were affected by photoperiod *and* the length of the driving light cycle (Fig. 4.5) (Saunders, 1972).

In many instances photoperiodic effects on larval growth rate appear to be associated in some way with the diapause response. In the viceroy butterfly *Limenitis archippus*, for example, larvae maintained in long days show rapid development to the adult instar without any arrest. In short-day conditions, however, larval development is much slower and the larvae spin silken hibernacula in the third instar within which they become dormant (Clark and Platt, 1969). Similarly, in the flesh-flies (*Sarcophaga* spp.), short days induce slow development *and* a subsequent pupal diapause whereas long days induce rapid development and non-diapausing pupae (Denlinger, 1972; Saunders, 1972). The exact relationship between these two phenomena remains to be clarified, however. Some authors, including Denlinger (1972), favour the view that short-day larvae develop more slowly because they are "diapause-committed". His evidence for this is that larvae of *Sarcophaga* exposed to short days as embryos within the maternal uterus show a slower rate of development than those similarly exposed to long days, even when both groups were subsequently raised in identical conditions (short photoperiod and 26°). The present writer, however, favours an alternative hypothesis in which protracted development in short days is a result of photoperiod, and is also one of the variables which raises the incidence of pupal diapause because more short-day cycles are "seen" by the larvae before the end of the sensitive period. This hypothesis will be developed later (Chapter 8).

D. Migration

The role of photoperiod in the control of insect migration has been little studied despite the fact that a large number of species display seasonal movements, often over quite long distances. Southwood (1962) considered that migration and diapause were two alternatives open to insects in order to counteract changes in their environment. Insects may diapause *in situ*, or move to more favourable areas. Other insects, such as the pentatomid *Eurygaster integriceps*, the pond skater *Gerris odontogaster*, the milkweed butterfly *Danaus plexippus* and various coccinellid beetles, to name but a few, migrate to hibernation sites and return to breeding and feeding areas in the spring. Aphids also migrate from summer to winter hosts. It is clear that many of these migrations are closely associated with the induction of a diapause condition, and are probably controlled by daylength.

The coccinellid beetles, *Semiadalia undecimnotata* and *Coccinella septempunctata*, for example, enter a reproductive diapause in short daylengths (*LD 8 : 16* and *LD 12 : 12*) but not at long daylength (*LD 16 : 8* and *LD 19 : 5*) (Hodek and Cerkasov, 1960; Hodek, 1967). Once diapause has been induced both males and females migrate to winter hibernation sites, the females with undeveloped ovaries and large reservoirs of fat. In Czechoslovakia, *C. septempunctata* moves to winter refuges amonst cultivated areas such as pastures and the edge of woods, preferably in elevated areas. *S. undecimnotata* migrates to hibernation sites on or near the summits of hills, and overwinters in large aggregations in rock crevices or at the base of plants (Hodek, 1960). In the spring the reactivated insects disperse in any direction.

E. Miscellaneous Photoperiodic Phenomena

Apart from the more frequently occurring photoperiodic phenomena—such as the induction of diapause and seasonal morphs, or the effects of photoperiod on growth rates—there is a variety of other events so controlled, only a few of which will be discussed here. Some of them occur in association with the diapause state and others appear to be independently controlled. Many of them have not been studied systematically in a wide range of photoperiods, however, and in the absence of a clear-cut critical daylength cannot be regarded, necessarily, as examples of "true" photoperiodism.

In the fruit-fly *Drosophila melanogaster*, which has no known "classical" photoperiodic response, Pittendrigh (1961) showed that recovery of the adults from a heat stress (40°C for 12 minutes) was more rapid in long (e.g. *LD 18 : 6*) than in short daylengths (e.g. *LD 6 : 18*). In the house-fly *Musca domestica*, which is also considered to be "day-neutral" (Danilevskii, 1965), adults raised as larvae in *LD 14 : 10* were more susceptible to DDT, dieldrin and aldrin than those raised in *LD 10 : 14* or in *LL* (Fernandez and Randolph, 1966). These two examples show that photoperiod can exert an effect on quite fundamental aspects of physiology.

Working with a north Italian strain of *Locusta migratoria*, Perez *et al.* (1971) demonstrated that male sexual behaviour was a function of daylength. Males were raised either in short daylength (*LD 12 : 12*) or in long daylength (*LD 16 : 8*). Those males raised in short days spent about 70 per cent of their time in vigorous sexual behaviour for the first 5 or 6 weeks of their adult life. On the other hand, those raised in long daylength exhibited only slight sexual behaviour, or none, during the first 2 or 3 weeks, and only reached their maximum (about 60 per cent of their time) after about 10 weeks. In this species there is no clear-cut diapause, although females maintained in short days produce many more egg pods, and more rapidly, than those kept in long days.

Lumme *et al.* (1972) have recently demonstrated a complex effect of photoperiod on the testis pterin content of a non-diapausing strain of *Drosophila littoralis*. Both long days (*LD 18* : 6) and very short days (*LD 6* : 18) caused a marked decrease in pterin content as compared with either *LL*, *LD 12* : 12 or DD. This work is of interest because the amount of pterin in the testes is an indication of the amount of that pigment in the whole insect, and pterins have an absorption which concides with the action spectra for the phase-setting of circadian oscillations (Chapter 3).

In the ichneumon *Compoletus perdistinctus* the sex ratio is reported to be affected by photoperiod, the greatest proportion of females being produced at *LD 12* : 12 (Hoelscher and Vinson, 1971).

In certain other cases photoperiodic phenomena occur in association with the induction of the diapause state. Many insects which overwinter as adults, for example, also lay down fat instead of developing their ovaries. One such example is the mosquito *Culex tarsalis* which enlarges its fat body under the influence of short days as part of its "preparation" for winter diapause (Harwood and Halfhill, 1964). Harwood and Takata (1965) have also demonstrated that overwintering females of this species (i.e. those raised at short daylengths) lay down a greater proportion of their fat in the form of unsaturated fatty acids.

McLeod (1967) showed that the assumption of the winter colour (pale green, yellow or brown) in adults of the lacewing *Chrysopa carnea* was associated with a reproductive diapause brought about by exposure to short days. Long days induced a full summer coloration (bright green) and no diapause. Tauber *et al.* (1970) further demonstrated that the *full* winter coloration (waxy green or brown with dark reddish-brown markings on the dorsum) and a more intense diapause were brought about by a transfer from *LD 16* : 8 in the immature stages to *LD 12* : 12 as adults.

Annotated Summary

1. In "classical" photoperiodism insects are able to distinguish between the long days (or short nights) of summer and the short days (or long nights) of autumn, and respond with a seasonally appropriate switch in metabolism. Photoperiodic switches control diapause, seasonal morphs, growth rates and a variety of associated physiological states.

2. Diapause involves the temporary inactivation or alteration of the endocrine system, triggered by the appropriate photoperiodic stimulus acting on the brain. Larval-pupal diapause generally involves the inactivation of the brain-prothoracic gland system with a resulting lack of ecdysone; adult, or reproductive, diapause the inactivation of the brain-corpus allatum system, a lack of juvenile hormone, and gonotrophic dissociation. In a few insects, however, the diapause state is imposed by a "diapause hormone" or by a particular hormone titre.

3. Quiescence differs from diapause in that it is the *direct* response to adverse environmental conditions, e.g. dehydration or cold torpor. It does not occur in response to photoperiod, and is quickly lifted with the advent of a more favourable environment.

4. The stage at which diapause occurs is nearly always species-specific, but even closely related species may diapause in a different instar. Some long-lived insects may diapause in successive winters. Others living in subtropical areas may hibernate in winter and aestivate in summer, both in the diapause state.

CHAPTER 6

THE PHOTOPERIODIC RESPONSE

A. Types of Response

Most experimental work on the induction of insect diapause has been carried out using stationary photoperiods. Groups of insects are usually exposed to a series of daylengths (all at $T = 24$), and to DD and LL, throughout their development or "sensitive period", and the proportion of the "population" entering diapause plotted as a function of daylength. The curves obtained are called photoperiodic response curves.

Since the majority of insects are summer-active the most frequent type of curve is the long-day response. In this type insects grow, develop or reproduce in long days but become dormant in short days. This response is particularly common, therefore, in multivoltine species with a facultative winter diapause. Only a few examples of this type will be given here, the choice generally being restricted to those species which have been investigated most intensively (Fig. 6.1). The most important feature of the response is the so-called critical daylength which separates the long photoperiods resulting in non-diapause development from the short photoperiods which ultimately lead to the dormant state. The critical photoperiod is frequently very abrupt and, in a sense, is a "measure" of the accuracy of the clock. In some species, for example, a change of as little as 10 or 15 minutes in the length of the daily light period may result in a significant change in the proportion of the population entering diapause, and a change of 1 hour usually converts all of the individuals from one developmental pathway to the other. The steepness of the curve at the critical point, however, may be a product of selection (accidental or otherwise) occurring during the course of laboratory colonization and subsequent experimentation. The photoperiodic response curve is also, of course, a "population response": individual insects presumably have their own threshold, but the response can only be analysed in a group of sufficiently large size.

Danilevskii (1965) has pointed out that a photoperiodic response curve includes responses at both natural and unnatural photoperiods. Those towards the right-hand side of the curve, particularly on either side of the critical point, represent those daylengths which occur naturally during that part of the year when the temperature and other climatic factors are suitable for insect development. This part of the curve, therefore, has an adaptive significance and is a product of natural selection (Fig. 6.2). Photoperiods outside this range are never met with in natural conditions or they occur in the depth of winter when insect morphogenesis is at a standstill. Nevertheless, the responses of insects to ultrashort daylengths, and to DD and LL, have a physiological significance; they must be "explained", for instance, in attempts to determine the mechanism of time measurement.

INSECT CLOCKS

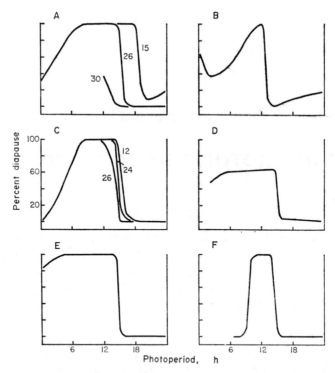

FIG. 6.1. A selection of long-day photoperiodic response curves. A — *Acronycta rumicis* (after Danilevskii, 1965); B — *Pectinophora gossypiella* (after Pittendrigh and Minis, 1971); C — *Pieris brassicae* (after Danilevskii, 1965); D — *Nasonia vitripennis* at 15°C (after Saunders, 1966a); E — *Megoura viciae* at 15°C (after Lees, 1965); F — *Ostrinia nubilalis* at 30°C. (After Beck, 1962.) Figures on the curves indicate temperatures in °C.

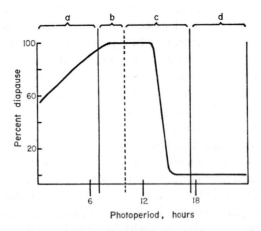

FIG. 6.2. A long-day photoperiodic response curve showing its properties (schematic). The solid vertical lines indicate the range of natural photoperiods at 55°N. Regions a and d are therefore never experienced in nature. Region b only occurs during the winter when the temperature is probably below the minimum for development (and the insect is in diapause). Only region c is of ecological importance; note that this region is dominated by the critical daylength which operates the seasonal switch in metabolism.

Photoperiodic response curves may differ in a number of ways. In some species, such as *Leptinotarsa decemlineata*, the response in DD is the same (about 100 per cent diapause) as in "strong" short daylengths (de Wilde, 1958). In others, such as *Pectinophora gossypiella* (Pittendrigh and Minis, 1971), *Ostrinia nubilalis* (Beck and Hanec, 1960), *Pieris brassicae* (Danilevskii, 1965) and *Sarcophaga argyrostoma* (Saunders, 1971), the proportion entering diapause falls off in ultrashort daylengths. In conditions of constant darkness the response may vary from zero as in *P. brassicae* to 100 per cent as in *L. decemlineata*. It usually varies widely with temperature. In *P. gossypiella* the proportion entering diapause is apparently greater in DD than in photoperiods of 2 to 6 hours (Pittendrigh and Minis, 1971). Similarly, in very long photoperiods and in *LL*, the incidence of diapause may be higher than in natural long daylengths and, once again, be more variable than at points just longer than the critical (Williams and Adkisson, 1964; Pittendrigh and Minis, 1971). These unstable responses at the extremes presumably reflect the absence of any selective pressure.

In a number of species such as *Leucoma salicis*, *Euproctis similis* (Geispits, 1953) and *Leptinotarsa decemlineata* (de Wilde, 1958) diapause is induced in all photoperiods except in a narrow range of long days (Fig. 6.3). These species clearly show a tendency towards a univoltine life cycle which, in an extreme example, would show an obligate

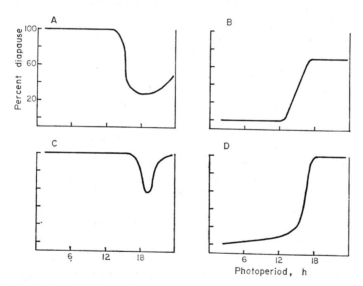

FIG. 6.3. A selection of "intermediate" (A and C) and short-day (B and D) photoperiodic response curves. A — *Leptinotarsa decemlineata* at 25°C (after de Wilde, 1958); B—*Bombyx mori* at 15°C (after Kogure, 1933); C — *Leucoma salicis* (after Danilevskii, 1965); D — *Stenocranus minutus* (after Müller, 1958).

diapause in every individual regardless of the photoperiod. At the opposite extreme, insects such as the house fly *Musca domestica* and the fruit-fly *Drosophila melanogaster* appear to have no diapause and to be "day-neutral". The immature stages of the house fly, for example, pass the winter in a state of quiescence (cold torpor), developing slowly when the conditions allow. The overwintering generation then emerges in the spring and is augmented by migration from the warmer regions (Sacca, 1964; Somme, 1961).

The opposite, or short-day response, is shown by a smaller number of species, particularly those which are spring-, autumn- or winter-active, and pass the summer in an aestival diapause (Fig. 6.3). The geometrid *Abraxas miranda*, for example, spends the summer (June to August) in a pupal diapause but the adults emerge in September and October and the larvae of the next generation actively feed and grow during the short days of winter (Masaki, 1957). The commercial silkmoth *Bombyx mori*, on the other hand, has a winter diapause in the egg, but a short-day response because the stages sensitive to photoperiod (the eggs and young larvae of the maternal generation) occur during the preceding summer. Thus long days perceived during the summer will give rise to moths laying diapausing eggs, whereas short days (i.e. in the spring) will give rise to moths laying non-diapause summer eggs (Kogure, 1933).

A number of species living in southerly latitudes, where the summers are hot and dry but the winters cold, are found to enter diapause in the summer and in the winter, and become active at two seasons, spring and autumn. These species may show both long- and short-day responses according to the season. In Japan, southern races of the cabbage moth, *Mamestra brassicae*, for example, have a pupal diapause in the winter and the summer. Both the hibernal and aestival diapauses are induced by photoperiod, and are of the long- and short-day types respectively (Masaki, 1956, 1968). In the moth *Hyphantria cunea* larvae exposed to short days gave rise to pupae with a winter diapause (Masaki *et al.*, 1968). Under long-day conditions, however, some pupae undergo a brief summer arrest (Umeya and Masaki, 1969) so that the adults tend to emerge in two distinct peaks. Even in more northerly latitudes long-lived insects may show two dormant periods. Thiele (1969) showed that the beetles *Nebria brevicollis* and *Patrobus atrorufus* hibernate as larvae. The young adults which emerge in the spring and early summer then undergo an aestivation diapause before maturation and reproduction. Aestivating adults of *P. atrorufus* require first long days and then short days for full gonadal development.

In the examples discussed above the reactions to stationary photoperiods, either above or below a critical value, were considered. Such conditions, of course, do not exist in nature: natural photoperiods are constantly changing, the rate of change being a function of latitude. Except for the summer and winter solstices each daylength occurs twice in a year, once when the days are increasing and once when they are decreasing. For this reason it is natural that responses to the *direction* of such changes have been sought amongst the insects, particularly amongst those which are long-lived. It should be remembered, however, that in many insects the problem of "deciding" whether days are increasing or decreasing probably does not arise. In some species, for instance, the stages sensitive to photoperiod may only be present at certain seasons, e.g. in the autumn but not in the spring. In others, such as those inhabiting the higher latitudes, cold spring weather delaying the resumption of activity may extend well into the season, far beyond the critical daylength.

Some insects respond to the direction of change in daylength by determining a sequence of photoperiods, either long to short, or short to long, at different stages of development. Norris (1959, 1962, 1965), for example, showed that the red locust *Nomadacris septemfasciata* entered an intensive reproductive diapause if the hoppers experienced a long day regime ($\sim$13 hours light per day) but the adults a short day ($\sim$12 hours). Diapause-free development followed a transfer from short to long days. In its natural environment (parts of tropical Africa) this type of response ensures that the species reproduces during the "summer" rains but becomes dormant during the "winter" drought. A similar response has been reported by Wellso and Adkisson (1966) and Adkisson and Roach (1971) for the bollworm *Heliothis zea*. The most effective

treatment for inducing diapause in this species was found to be when the adults and eggs
were exposed to longer daylengths than the larvae. Amongst the Carabidae, the univoltine
spring breeders *Pterostichus nigrita*, *P. augustatus*, *P. oblongopunctatus*, *P. cupreus*,
P. coerulescens (Thiele, 1966, 1968, 1971; Krehan, 1970) and the Staphylinids *Tachyporus*
spp., *Tachinus* spp. and *Philonthus fuscipennis* (Lipkow, 1966; Eghtedar, 1970) require
long days following short days for the full development of the ovaries. The autumn
breeders *Patrobus atrorufus* and *Nebria brevicollis* (Thiele, 1969, 1971), on the other hand,
have an adult reproductive aestivation and require a period of long days and then short
days for full maturation. More recently, Tauber and Tauber (1970) have provided another example. In the lacewing *Chrysopa carnea* stationary long days (*LD* 16: 8) promote
continuous development and reproduction, whereas short days (*LD* 12: 12) produce
a relatively short (~34 days) imaginal diapause. The critical daylength for stationary
photoperiods was about *LD 13*: 11. A considerably more intense and durable diapause
(~95 days), however, was brought about by a transfer from *LD 16*: 8 to *LD 12*: 12
during development. More importantly, a transfer from *LD 16*: 8 to *LD 14*: 10 (both
above the critical daylength) induced a proportion of diapause (29 per cent) amongst
the resulting adults, and a change from *LD 8*: 16 to *LD* 12: 12 (both *below* the critical)
gave 0 per cent diapause. Diapause termination was also effected by such a mechanism:
transfer of diapausing adults from *LD 8*: 16 to *LD 12*: 12 caused oviposition to commence in about 16 days, although in stationary photoperiods of 8 and 12 hours the pre-
oviposition periods were about 75 and 34 days respectively. These results show that insects can respond to changes in daylength even when such changes do not cross the critical value as defined by stationary or "absolute" photoperiods.

Several investigators have examined the effects of photoperiods increasing or decreasing at a natural rate. In this type of experiment, care has to be taken to ensure that
the changes occur either entirely above or entirely below the established critical point,
otherwise interpretation becomes difficult. Corbet (1955, 1956) found that the nymphs
of *Anax imperator* developed without interruption if they entered their last instar during
the spring, but remained in diapause for about 8 months if the last moult occurred after
the summer solstice. He pointed out that the same absolute daylengths occurred at
both seasons, and attributed the induction of diapause later in the year to decreasing
daylength. Delayed development in another long-lived insect, *Anthrenus verbasci*, has
also been ascribed to decreasing as opposed to stationary photoperiods (Blake, 1960,
1963) (see also Chapter 9). In *Acronycta rumicis*, shortening the daylength *above* the
critical photoperiod did not induce diapause, but lengthening the day *below* the critical
appeared to reduce its incidence (Danilevskii, 1965). Other responses to changing daylengths have been demonstrated, or claimed, for *Gerris odontogaster* (Vepsäläinen, 1971),
Crambus tutillus (Kamm, 1972), and for the stone-fly *Capnia bifrons* (Khoo, 1968a).
Tauber and Tauber (1973) have recently shown that diapausing females of *Chrysopa
carnea* remain responsive to natural illumination throughout diapause. As the autumn
days became shorter (between October 25th and the winter solstice) the rate of diapause
development slowed and therefore diapause duration increased. After the winter solstice
the time to reactivation got steadily shorter as the season progressed. These results
hardly suggest a simple reaction to absolute stationary daylengths, but rather a continuous sensitivity to photoperiodic changes.

B. The Sensitive and Responsive Stages

The period of the life-cycle which is sensitive to photoperiod never extends to the whole of development. In some species the "sensitive period" occurs in the same instar as the resulting diapause, in many others it precedes it. Very frequently, young larvae are photoperiodically sensitive and diapause occurs in the last larval instar. This is observed, for example, in *Grapholitha molesta* (Dickson, 1949) and *Ostrinia nubilalis* (Beck and Hanec, 1960). Often the larvae are sensitive and diapause supervenes in the pupal instar; this is seen in *Antheraea pernyi* (Tanaka, 1950) and in *Diataraxia oleracea* (Way and Hopkins, 1950). These "delayed photoperiodic responses" led De Wilde (1962) to distinguish (a) photoperiodic induction, a reversible, partly photodynamic process from (b) photoperiodic determination, the induced state of the overall controlling centre of growth and reproduction, and (c) the photoperiodic response, or the reaction of the effector system.

In *A. pernyi* (Williams and Adkisson, 1964), *O. nubilalis* (McLeod and Beck, 1963), *P. gossypiella* (Bell and Adkisson, 1964), the pitcher plant midge *Metriocnemus knabi* (Paris and Jenner, 1959), *Chironomus tentans* (Engelmann and Shappirio, 1965) and *Chaoborus americanus* (Bradshaw, 1969), to name but a few, the sensitive period extends to cover the diapause stage, so that diapause *termination*, as well as induction, is governed by photoperiod. These species therefore enter diapause in the autumn when days shorten below the critical value and resume development in the spring when the critical daylength is exceeded, provided that the temperature is above the threshold for morphogenesis. In *Antheraea pernyi* the critical daylengths for induction and termination are "mirror images" of each other, indicating that the same clock process is involved in both (Fig. 6.4). This type of control is regarded by Müller (1970) as a photoperiodically induced quiescence, or "oligopause".

As indicated above, the sensitive period frequently comes to an end *before* the diapausing instar. In *Sarcophaga* spp., for example, the larvae are particularly sensitive as

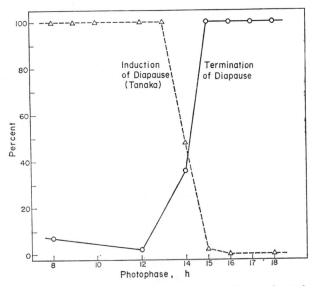

FIG. 6.4. Photoperiodic induction and termination of pupal diapause in *Antheraea pernyi*. Note that the critical daylength for termination is a "mirror image" of that for induction. (From Williams and Adkisson, 1964.)

embryos with the maternal uterus (Denlinger, 1972). The larvae then remain sensitive throughout their development (Saunders, 1971) but become insensitive at the time of puparium formation. The diapause stage is the pupa. In *Araschnia levana* the second to fourth larval instars are the most sensitive, but the pupa becomes dormant (Müller, 1955). A more extreme example is provided by the vine leaf roller, *Polychrosis botrana* (Komorova, 1949): in this species the pupal diapause is induced by photoperiods experienced by the eggs and early larval instars. Since the sensitive period comes to an end before the diapause stage, reactivation of diapause is controlled by non-photoperiodic mechanisms, most usually a prolonged period of exposure to low temperature. In Müllers' (1970) terminology this constitutes "true" diapause or "eudiapause". However, although diapause cannot be reversed in this species by photoperiodic treatment during the dormant stage, it can be reversed by long days applied during the sensitive period.

A selection of insect life cycles showing the sensitive and diapausing stages is illustrated in Fig. 6.5.

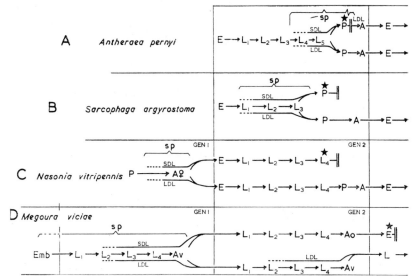

FIG. 6.5. A selection of insect life-cycles showing the sensitive and diapausing stages. L_1-L_5 — larval instars; P — pupa; A — adult; Av — adult virginopara; Ao — adult ovipara; Emb — embryo; sp — "sensitive period"; SDL — short daylength; LDL — long daylength. The asterisk marks the diapausing instar, and each "box" represents one generation. A. In *Antheraea pernyi* short daylength experienced by the larvae induces pupal diapause, but sensitivity to daylength is retained throughout pupal development so that diapause termination is also photoperiodically controlled. B. In *Sarcophaga argyrostoma* the larval instars constitute the sensitive period which finishes at puparium formation. Diapause termination is not controlled by photoperiod. C. In *Nasonia vitripennis* the adult female parents are sensitive to daylength and diapause does not supervene until the progeny reach the fourth larval instar. D. In *Megoura viciae* morph determination depends on the photoperiod received by the parent female aphid during its development. Short-day-produced oviparae lay diapausing eggs.

C. Maternal Induction of Diapause and Seasonal Forms

The most extreme examples of a delayed photoperiodic response are to be seen when the sensitive stage occurs in one generation and the arrest of development in the next. Such maternal influences are particularly interesting because they raise a number of important questions about the nature of photoperiodic determination (Chapter 8).

The classical example of a maternal influence on diapause induction is, of course, *Bombyx mori*, in which photoperiodic signals experienced by the eggs and young larvae of one generation determine the diapause or non-diapause status of the eggs in the next (Kogure, 1933). It is known that this mechanism involves the production of a diapause hormone by the suboesophageal ganglion of the adult female which enters the ovarian egg (Fukuda, 1963). In a growing number of species, however, photoperiods experienced maternally are known to induce diapause in the *larvae* of the next generation. Since this form of diapause is most probably of the normal larval-pupal type involving a cessation of activity in the brain neurosecretory cells, this kind of control differs from that in *B. mori*. A few examples of insects with a maternally operating photoperiod will be described here.

Females of the parasitic wasp *Nasonia vitripennis* deposit their eggs within the puparia of "higher" flies. Under conditions of long days ($>15\frac{1}{4}$ hours per 24) these eggs hatch to give rise to non-diapause larvae which pupate and produce the next generation of adults without interruption. In short-day conditions ($<15\frac{1}{4}$ hours per 24), however, females "switch-over" from the production of non-diapausing progeny to those which enter a larval or pre-pupal diapause late in the fourth instar, just before defaecation and the pupal ecdysis (Saunders, 1965b, 1966a). The type of progeny produced can be reversed during the maternal sensitive period by an appropriate manipulation of the light cycle (Fig. 6.6), but once the eggs have been deposited within the host puparium the type of development (i.e. diapause or non-diapause) is fully determined: it cannot be

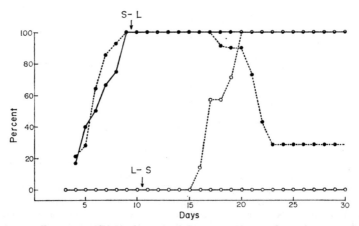

FIG. 6.6. The effect of transferring females of *Nasonia vitripennis* from short daylength to long daylength (●------●)or from long daylength to short daylength (o-----o) at the arrows, on the production diapausing offspring. Ordinate shows the proportion of females producing diapause larvae. This demonstrates the reversibility of the inductive process within the sensitive period. ●——● short days throughout; o——o long days throughout. (From Saunders, 1965b.)

reversed by exposing the immature instars to different daylengths, or to temperature changes within the ecological range (Saunders, 1966a). The diapause larvae are also insensitive to photoperiod and require a prolonged period of cold for diapause development and reactivation (Schneiderman and Horwitz, 1958). A topical application of ecdysone will also promote pupation (H. Holmes, personal communication).

A similar type of control has been described for the braconid *Coeloides brunneri* (Ryan, 1965) and for the blowflies *Lucilia caesar* and *L. sericata* (Ring, 1967). Pupal diapause in the horn-fly, *Lyperosia irritans*, is also dependent on a maternally acting photoperiod (Depner, 1962), as is the egg diapause in a variety of Aëdine mosquitoes (Anderson, 1968; McHaffey and Harwood, 1970). Vinogradova and Zinovjeva (1972b) have recently demonstrated that larval diapause in *Calliphora vicina* is controlled, below 15°C, by the maternal light regime. Flies exposed to photoperiods in excess of 16 hours per day produce offspring which showed no arrest in development, whereas those maintained at short daylength (less than 15 hours per day) produced most of their offspring as diapause larvae. As with *N. vitripennis* an inversion of the photoperiod during the maternal sensitive period resulted in a reversal of the response. It is probable that this type of control is much more widespread than previously realized. For example, the onset of pupal diapause in the cabbage root maggot, *Erioischia brassicae*, which was formerly attributed to indirect photoperiodic influences transmitted through the tissues of the host plant (Hughes, 1960), is now known to depend, at least in part, on the photoperiods "seen" by the female progenitor (Read, 1969).

The most complex cases of maternal control are to be observed in the aphids. Their extremely rapid reproduction during the summer months is a consequence of their viviparous, parthenogenetic habit and the "telescoping" of successive generations. A reproducing virginopara of *Megoura viciae*, for example, may contain the early embryonic stages of her grandchildren. In this species the type of progeny produced (virginoparae or oviparae) is determined by the photoperiod acting on the parent virginopara (Lees, 1959). The photoperiodic centre in the brain (Chapter 10) begins to function about 2 to 3 days before the birth of the parent but the first embryos do not become responsive until the parent has developed to the second instar. These embryos are only sensitive when they are in in the second and third positions in the embryonic chain; before this they cannot respond, and after this they are fully determined as either virginoparae or oviparae. During the summer months when days are longer than about $14\frac{1}{2}$ hours all the progeny are produced as a further generation of virginoparae. When the photoperiod falls below the critical point, however, oviparae (and males) are born, and the former proceed to deposit diapausing eggs. Close to the critical daylength parent aphids switch from the production of daughter virginoparae to oviparae and back again, in a "flip-flop" fashion, indicating an unstable equilibrium in the maternal controlling centres. Because of the "telescoped" development the interval between the sensitive period and the resulting egg diapause can be regarded as spanning three generations.

D. Factors which Modify the Photoperiodic Response

1. Temperature

Temperature can affect the expression of the photoperiodic clock and the induction of diapause in a number of ways. Constant temperatures frequently modify the *degree* of the response, or alter the position of the critical daylength. Low- or high-temperature pulses may completely reverse the photoperiodic response, from short to long or long to short, depending on the time in the light: dark cycle at which the pulse is applied. Finally a daily temperature cycle, or thermoperiod, may simulate most or all of the effects of a light cycle, by acting as a *Zeitgeber* in its own right.

The results of a large number of investigations with long-day species have shown that a high constant temperature and long daylength act together to avert diapause, whereas low temperature and short daylength act together to induce it. Similarly high temperatures may facilitate diapause termination in those species reactivated by long daylength. In short-day species, however, an opposite effect has been noted: in *Bombyx mori*, for example, low temperatures reduce the diapause-inducing effect of long days, whereas high temperatures enhance it (Kogure, 1933). High temperatures also promote diapause in *Abraxas miranda* (Masaki, 1958).

Figure 6.7 shows the modifying effects of temperature on the photoperiodic response in a number of insect species. In the cabbage white butterfly *Pieris brassicae* a short day

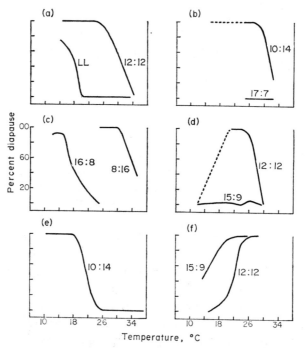

FIG. 6.7. The effects of temperature on the incidence of diapause at short and long daylength. A — *Acronycta rumicis* (after Danilevskii, 1965); B — *Antheraea pernyi* (after Mansingh and Smallman, 1971); C — *Diataraxia oleracea* (after Way and Hopkins, 1950); D — *Grapholitha molesta* (after Dickson, 1949); E — *Sarcophaga argyrostoma* (after Saunders, 1971); F — *Bombyx mori* (after Kogure, 1933).

(*LD 12*: 12) is fully inductive (100 per cent diapause) up to about 25°C. At higher temperatures the proportion of larvae entering diapause drops sharply so that at about 30°C all of the insects develop without arrest (Danilevskii, 1965). Pupal diapause in the tomato moth *Diataraxia oleracea* is fully expressed in short days at temperatures below 30°C; above this incidence drops (Way and Hopkins, 1950). Similarly in the flesh-fly *Sarcophaga argyrostoma*, practically all of the larvae become diapausing pupae in short daylength (*LD 10*:14) at 15° to 18°, there is a drop at 20°, and at 25°C and above the short-day response is eliminated (Fig. 6.8) (Saunders, 1971). In the silkmoth *Antheraea pernyi* the photoperiodic response was at one time thought to be virtually independent of temperature (Tanaka, 1944). The more recent investigation of this species by Mansingh and Smallman (1971), however, shows that the same "rule" applies: at temperatures of 24° to 26°C short days induced about 100 per cent of diapause in the pupae, at 28° to

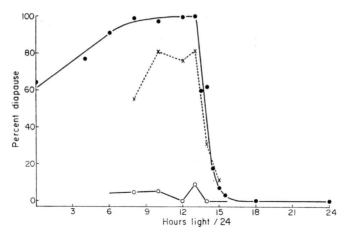

FIG. 6.8. The effect of temperature on the photoperiodic response in *Sarcophaga argyrostoma*.
●——● 15°C; ×——× 18°C; ○——○ 25°C. (From Saunders, 1971.)

30°C it was 94 per cent, but at 32° the proportion of dormant pupae dropped to 32 per cent. In some long-day species, normally diapause-averting (long) photoperiods will induce diapause if the temperature is sufficiently low. A model to account for these temperature effects is developed in Chapter 8.

The oriental fruit moth *Grapholitha molesta* shows a more complex relationship between temperature and diapause induction. In this species short-day induction only occurs within the range 20° to 27°C; diapause is avoided at both higher (30°C) and at lower (12°C) temperature (Dickson, 1949).

Constant temperatures are also known to affect the critical daylength in a number of long-day species (Fig. 6.1) In some insects, such as *Pieris brassicae*, temperature up to about 26°C has little effect on the critical photoperiod (Danilevskii, 1965). This species, therefore, is regarded as having a temperature-compensated response over practically the entire range of ecologically important temperatures. The critical daylength in the aphid *Megoura viciae* shifts to shorter values by about 15 minutes for every 5°C rise in temperature, but at 23°C and over, short-day induction of oviparae fails completely (Lees, 1963). The European corn borer, *Ostrinia nubilalis* (Beck and Hanec, 1960), similarly shows a rather slight effect of temperature on the critical point: at 19°C it is only about 30 minutes longer than at 29°C.

In other species the critical photoperiod decreases steadily as the temperature rises. In *Acronycta rumicis*, for example, it moves to shorter values by about 1.5 hours with every 5°C increment (Goryshin, 1955). In the spinach-leaf miner *Pegomyia hyosciami* and the cabbage-root maggot *Erioischia brassicae* the critical daylength shortens by 3 and 4.5 hours, respectively, with a 7°C rise in temperature (Zabirov, 1961).

The question of temperature-compensation of the critical daylength is one of importance since it must affect the date at which natural populations of insects begin to enter diapause in the autumn. At a physiological level, however, the manner in which temperature affects the response is far from clear. It could, for example, affect the "clock" mechanism itself, or it could modify the response at a more superficial level. Figure 6.9 demonstrates that an overall reduction of the diapause response caused by an elevated temperature could result in a shortened critical value without necessarily affecting the mechanism of time measurement itself. Conversely a lowered temperature might cause

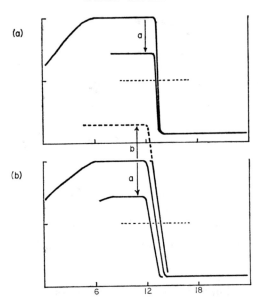

Fig. 6.9. The effect of temperature on the photoperiodic response: theoretical considerations. a — raising temperature; b — lowering temperature. In (a), a species with a steep critical daylength, raising the temperature may reduce the amount of diapause at short daylength, but has little effect of the position of the critical photoperiod. In (b), a species with a less steep critical daylength, altering the temperature may have a more pronounced effect on the position of the critical daylength. These effects may be confused with those shown in Fig. 6.1, A and C.

an increase in the critical daylength. Figure 6.9 also shows that the degree of the change might be a function of the slope of the curve at the critical point. Such effects have not been demonstrated with certainty, but data on the critical daylength in *Pectinophora gossypiella* after selection for "early" and "late" eclosion strains (p. 140) indicates its possibility (Pittendrigh and Minis, 1971).

Short "pulses" of high or low temperature may have spectacular effects on the diapause response; in some insects complete reversals have been reported. Working with the larvae of *Acronycta rumicis*, for example, Danilevskii (1965) showed that chilling at 5°C for 3 hours daily at the beginning or at the end of the light period in *LD 17 : 7* converted the response from a long to a short day. However, chilling the middle of a 17-hour light period, or similar periods of chilling in the dark component of the cycle, had no such reversing effects. Danilevskii concluded that low temperature was equivalent to darkness in that the insects were unable to "see" the light at 5°C. In plants, reversals of the photoperiodic control of flowering in the short-day plant *Xanthium pensylvanicum* by chilling have been reported by de Zeeuw (1957) and by Nitsch and Went (1959). Schwemmle (1960) also reported reversals by chilling in *Hyoscyamus niger* and *Perilla ocymoides*. These authors, like Danilevskii, considered that chilling in the light had a similar effect to that of darkness. Danilevskii (1965) also showed that brief periods of heating applied at the end of the night in an *LD 13 : 11* cycle produced a drop in the diapause response provided that the pulse was above about 38°C.

In the parasitic wasp *Nasonia vitripennis* photoperiodic reversals have been achieved by chilling and by heating at different phases of the light : dark cycle. In a cycle of *LD 14 : 10*, which is just short of the critical daylength ($15\frac{1}{4}$ hours per 24), a daily period

TABLE 6.1. THE REVERSAL OF PHOTOPERIODIC EFFECT IN *Nasonia vitripennis* BY A DAILY PERIOD OF CHILLING

Temperature (°C)	Chilling applied	Females (no.)	Mean days to "switch" ($\pm$ S.E.)	Delay (+) or acceleration (−) in "switch" (days)
		LD cycle *14*: 10		
18	No chilling (control)	58	9.7±0.35	
2	4 hours daily after lights-on	40	8.3±0.37	−1.4
2	4 hours daily in middle of light	18	8.6±0.50	−1.1
2	4 hours daily after lights-off	39	24.0*	+14.3
2	4 hours daily in middle of dark	36	23.0*	+13.3
		LD cycle *16*: 8		
18	No chilling (control)	36	23.0*	
2	4 hours daily after lights-on	16	10.6±0.70	−12.4
2	4 hours daily in middle of light	19	12.7±0.72	−10.3
2	4 hours daily after lights-off	20	23.0*	0.0
		LD cycle *8*: 16		
18	No chilling (control)	36	10.6±0.69	
2	4 hours daily after lights-on	16	12.2±1.09	+1.6
2	4 hours daily after lights-off	19	7.6±0.65	−3.0
2	4 hours daily in middle of dark	18	9.5±0.80	−1.1

* None of the females had "switched" after 23 to 24 cycles. (From Saunders, 1967.)

of 4 hours at 2°C applied at the beginning or in the middle of the long night converted the response from that of diapause-inducing to diapause-averting (Saunders, 1967, 1968, 1969) (Table 6.1). Chilling during the light component of this cycle merely had a "strengthening" effect on the short-day response. In the converse experiment at LD 16 : 8, chilling in the dark had no effect, but chilling in the light reversed the response from a long-day (diapause-averting) to that of a short-day (diapause-inducing). In a very short-day cycle of LD 8 : 16—which is well short of the critical daylength—no such reversals were observed. These results differ from those obtained by Danilevskii (1965) for *A. rumicis* in that both "night" and "day" were sensitive to a period at low temperature, so that, following his argument, the period at low temperature could be "seen" either as dark or light depending on the phase-point at which it is experienced. An alternative explanation is that the clock mechanism is stopped or phase shifted by the low-temperature pulse so that long days and long nights are effectively shortened. Subsequent experiments with 3-hour daily periods of chilling (2°C) or heating (35°C) at all points of the "clock" have shown that both forms of treatment have a similar effect on the response of

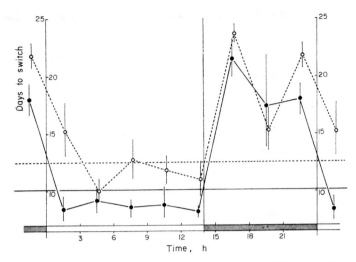

FIG. 6.10. The effects of daily periods of chilling (3 hours at 2°C, ●——●) or heating
(3 hours at 35°C, ○——○) on the photoperiodic response of adult females of *Nasonia
vitripennis*, otherwise kept at 18°C and short daylength (*LD 14* : 10). The plotted points
show the mean number of days to the "switch" to diapause larva production; the vertical
lines show 2×SE of the mean. The horizontal dotted line shows the unheated control, the
solid line the unchilled control. Note that both chilling and heating have maximum diapause-averting (long-day) effects at the beginning and end of the night.

N. vitripennis in *LD 14* : 10 (Saunders, unpublished) (Fig. 6.10). In particular it was observed that maximum photoperiodic reversal occurred when the low- or high-temperature
pulse was placed at the beginning or at the end of the 10-hour "night"; chilling or heating
in the middle of the night gave a smaller response.

The effects of 24-hour temperature cycles, or thermoperiods, have been studied in
continuous dark (DD) and with concomitant light cycles. The latter will be examined
first. Goryshin (1964) showed that the proportion of *Acronycta rumicis*, *Leucoma salicis*
and *Pieris brassicae* entering diapause was a function of "night" rather than "day"
temperature. With *A. rumicis*, for example, diapause incidence in short-day cycles
(*LD 12* : 12 to *LD 16* : 8) was almost as high in regimes with days at 30°C and nights at
17°C as in the same photoperiods at a constant temperature of 17°C. If the days were at
17°C and the nights at 30°C, however, the incidence of diapause was very much lower
although still above that for 30°C throughout. Lees (1953b) also demonstrated a reduction in the proportion of winter females if the mite *Panonychus ulmi* was maintained in
regimes with a warm night. In the corn borer *Ostrinia nubilalis* maintained at *LD 15* : 9,
Beck (1962a) showed that a day temperature of 31°C and a night temperature of 21°C
resulted in as much diapause (96 per cent) as 21°C throughout, whereas a day at 21°C
and a night at 31°C gave merely 15 per cent diapause, the same as at 31°C throughout
the 24 hours. In the examples given above it is clear that night temperature is more important than day temperature. This is perhaps hardly surprising—at a superficial level—
because insects have evolved in an environment in which the nights are colder than the
days. It may also demonstrate, however, that night-length measurement is "more important" than day-length measurement (Chapter 7).

Using the pink bollworm *Pectinophora gossypiella*, Pittendrigh and Minis (1971) have
examined the effects of a light-cycle (*LD 8* : 16) and a concurrent sinusoidal temperature
cycle (20° to 29°C daily) in which the phase angle between the two cycles was varied

systematically. They too found "long-day" effects when the low point of the temperature cycle fell during the day, and "short-day" effects when it fell at night. The experiments, however, were designed as a specific test of the "external coincidence" model (p. 137) and will be discussed more fully in Chapter 7.

A daily thermoperiod is also known to have diapause-inducing properties in the *absence* of a light cycle. Beck (1962b), for example, maintained larvae of *O. nubilalis* in continuous darkness but in a temperature cycle consisting of 11 hours at 31°C, 11 hours at 10°C, and the remaining 2 hours in the warming and cooling phases. Such a regime induced nearly all of the larvae to enter diapause, whereas constant temperatures of 31°, 26° and 21°C and DD caused very few larvae to become dormant. Beck (1968) also quotes unpublished work by D. G. R. McLeod in which "long-day" thermoperiods (in which the warm phase lasted for 16 hours per day) were found "not as effective as long-day photoperiods in the avoidance of diapause". Menaker and Gross (1965) raised the larvae of *Pectinophora gossypiella* in continuous darkness and fluctuating temperatures. They showed that a thermoperiod of 12 hours at 31°C and 12 hours at 21°C induced a higher proportion of diapause larvae than a regime consisting of DD and a constant temperature of 26°C (the mean of 31° and 21°C). In a more extensive investigation using *Acronycta rumicis*, *Pieris brassicae* and *Spilosoma menthastri*, Goryshin and Kozlova (1967) demonstrated that "short-day" temperature cycles (12 hours at 26° to 32°C per day) produced a somewhat higher proportion of diapause than "long-day" temperature cycles (18 hours at 25° to 29°C per day).

The most clear-cut dependence of diapause induction on thermoperiod has been demonstrated for the parasitic wasp *Nasonia vitripennis* (Saunders, 1973a). Wasps were raised from the egg stage in the complete absence of light and then subjected as adults to a variety of "square-wave" temperature cycles obtained by transferring the insects from

TABLE 6.2. THE THERMOPERIODIC CONTROL OF DIAPAUSE INDUCTION IN *Nasonia vitripennis* WITH A 13° TO 23°C TEMPERATURE CYCLE

The experimental insects were raised from the egg stage in the total absence of light

HL ratio*	Females (no.)	Progeny produced on days 15–17			
		Developing	Diapause	None†	Percent diapause
Control					
13°	118	32	25	61	43.9
18°	56	40	1	15	2.4
Experimental					
6:18	110	0	83	27	100.0
8:16	59	1	43	15	97.7
10:14	40	0	22	18	100.0
12:12	95	10	26	59	72.2
14:10	11	8	0	3	0.0
16:8	31	17	0	14	0.0
18:6	43	24	0	19	0.0

* HL ratio (in hours) of high temperature (23°C) to low temperature (13°C) in thermoperiodic cycle.
† The number of experimental females of *N. vitripennis* that failed to lay eggs during the test period (day 15 to 17 of adult life); hence the empty host puparia. (From Saunders, 1973a.)

one incubator (at 23°C) to another (at 13°C). The type of progeny produced (diapause or non-diapause larvae) was then examined during a 2-day "test period" between the 15th and 17th days of adult life. In short-day thermoperiods (between 6 and 10 hours at 23° per day) almost all of the wasps produced diapausing larvae. On the other hand, diapause was completely absent when the females experienced the warmer conditions for 14 or more hours per day (Table 6.2). The proportions of females producing diapausing progenies at constant temperatures of 13°C and 18°C (the mean of 12 hours at 13°C, and 12 hours at 23°C) were about 44 and 2 per cent, respectively. The "critical thermoperiod" was observed to be about 13 hours at 23°C per day. This is shorter than the critical photoperiod observed earlier for this species ($15\frac{1}{2}$ hours light per 24) (Saunders, 1966a), but the difference was assumed to reflect the slow rate of cooling (and perhaps heating) consequent upon the manner of transfer from one incubator to the other. More abrupt temperature transitions would undoubtedly produce a critical thermoperiod closer to that for light. These experiments demonstrate that a daily temperature cycle can simulate the diapause-inducing or averting effects of photoperiod in the complete absence of light, although temperature may be a "weaker" *Zeitgeber* than photoperiod. Since these important observations have a relevance in considering the nature of the clock mechanism, thermoperiodic effects on diapause induction will be discussed again in another section (Chapter 7).

2. Diet

Insect diets rarely remain the same throughout the year; they frequently change with the season, both in quality and quantity. Plant-eating insects such as Lepidopterous larvae or aphids, for example, may have access to young green leaves in the spring and early summer, but only to yellowing foliage in the autumn. Those feeding in fruits and seeds are subjected to chemical changes associated with maturation and ripening. Predaceous insects are subjected to seasonal changes in both the number and type of prey available. In a number of insects these qualitative and quantitative changes in nutrition are known to affect diapause induction and the expression of the photoperiodic mechanism. As with temperature, diet usually modifies the degree of the photoperiodic response or occasionally the position of the critical daylength. In a few insects diet forms a major influence in the seasonal cycle of activity.

One of the best-known examples of nutritional quality affecting the photoperiodic response is afforded by the pink bollworm, *Pectinophora gossypiella* (Adkisson, 1961; Bull and Adkisson, 1960, 1962; Adkisson et al., 1963). Larvae of this species have been raised on artificial diets containing different quantities of fats and oils. Those raised at 27°C and short daylength (*LD 10 : 14* to *LD 12 : 12*) produced only about 15 per cent diapause on diets containing 0.25 per cent of wheat-germ oil, but almost 80 per cent diapause on diets containing 5 per cent of cotton-seed oil. This difference is thought to reflect natural increases in oil content as the bolls ripen. In this example the degree of the response to short daylength is clearly modified by diet. In another cotton pest, *Chloridea obsoleta*, however, a diet of cotton leaves—as opposed to cotton bolls—caused a shift in the critical daylength from about $13\frac{1}{2}$ to about $14\frac{1}{2}$ hours (Danilevskii, 1965). As with the rather similar effects of temperature, modifications of the degree of the response and the photoperiodic threshold may be aspects of the same phenomenon.

Lees (1953a) showed that females of the red spider mite *Panonychus ulmi* laid mainɪy non-diapause eggs in long days when reared on young apple foliage. Those transferred to yellowing leaves, however, produced about 68 per cent of winter eggs. The same effect was observed when the mites were supplied with foliage which had been "bronzed" by previous and heavy infestations of *P. ulmi*. The white-fly *Aleurochiton complanatus* also showed a tendency to produce an increased proportion of diapause stages (winter "puparia") when fed on yellowing foliage (Müller, 1962b). In a few instances the *species* of plant is known to have an effect: Danilevskii (1965), for example, found that the larvae of the beet webworm *Loxostage sticticalis* developed without arrest on pigweed *(Chenopodium album)*, but showed a high incidence of diapause on wormwood *(Artemesia incana)*.

Several recent publications have demonstrated the importance of the *presence* of food, or the *quantity* of food. Termination of larval diapause in the phantom midge *Chaoborus americanus*, for example, requires the action of long days and the presence of a source of food such as mosquito larvae (Bradshaw, 1969, 1970). When brought in from the field, diapausing larvae of this species never developed to pupation if starved and kept under short daylength, although short-day insects supplied with prey showed a variable degree (2 to 49 per cent) of reactivation. Under conditions of long daylength, however, 4 to 50 per cent of the larvae pupated if starved, but nearly all of them (92 to 98 per cent) terminated diapause if provided with an abundance of prey. The synergistic action of long days and food was found to occur only when the two components were present simultaneously. Clay and Venard (1972) found that the incidence of larval diapause in *Aëdes triseriatus* increased when short-day larvae were provided with an inadequate diet (less than an optimum quantity of pulverized chow) or were kept at low temperature. In this case both inadequate food and low temperature exerted their effect by slowing larval development (see Chapter 8).

Tauber and Tauber (1973a) have recently described an effect of diet on the incidence of diapause in the lacewing *Chrysopa mohave*, which feeds on insects such as aphids during its larval and adult instars. In short-day conditions ($<$ 14 hours light per 24) the adults were found to enter a reproductive diapause which was terminated by exposure to photoperiods in excess of 16 hours. However, withholding prey also induced the diapause syndrome—even in adults maintained in long days—and supplying prey terminated it. Natural populations of this insect in California were found to reproduce during April, May and June when the days were long and prey abundant, but some insects were found to enter a "food-mediated" summer diapause when food was in short supply. In October, November and December the whole population entered a short-day-induced reproductive diapause which enabled them to overwinter.

In the parasitic wasp *Nasonia vitripennis* the withholding of host puparia—on which the adult wasps feed as well as deposit their eggs—is known to modify the response to short daylength (Saunders, 1966b). Female wasps maintained in *LD 12* : 12 at 18°C and supplied each day with two host puparia *(Sarcophaga argyrostoma)*, produced about 73 per cent of their progeny as diapausing larvae. Those deprived of hosts for the first 3, 5 and 7 days of adult life, however, produced about 86, 91 and 99 per cent diapausing progeny, respectively (Table 6.3).

In a number of early studies it was suggested that the seasonal cycles of phytophagous insects were influenced, or even controlled, by chemical changes associated with the photoperiodic responses of the plant host. This view was particularly prevalent with root-feeding insects which were not exposed directly to daily cycles of illumination. In his

TABLE 6.3. THE EFFECT OF HOST DEPRIVATION ON THE PRODUCTION OF DIAPAUSE LARVAE BY FEMALES OF *N. vitripennis* AT 18°C

	Deprived of hosts (days)	Females (no.)	Mean adult life-span (days±S.E.)	Mean age at "switch" (days±S.E.)	Mean number of offspring per female ±S.E.	Offspring in diapause (%)
Short daylength (*LD 12:12*)	0	19	32 0±1.95	9 1±0.44	632.8±40.74	72.6
	3	20	30.0±2.22	7.7±0.38	537.5±39.67	86.5
	5	18	34.3±1.62	8.2±0.41	552.6±30.72	91.1
	7	31	35 6±1.15	8.5±0.15	532.0±18.21	99.0
Long daylength (*LD 18:6*)	0	19	25.9±1.81	22.7 (9)*	590.1±36.29	4.6
	7	10	28.7±3.08	27.0 (3)*	434.7±81.44	5.9

* At long daylength only a small number of females survive long enough to show a "switch" to the production of diapause larvae. (From Saunders, 1966b.)

work with the strawberry root aphid, for example, Marcovitch (1924) was inclined to believe that the observed photoperiodic response was mediated through the tissues of the plant on which the insects were feeding. In more recent years this has been suggested for the cabbage-root maggot *Erioischia brassicae* (Hughes, 1960), and even for the aphid *Megoura viciae* (Von Dehn, 1967) which feeds above ground. Lees (1968), however, has pointed out that the strawberry-root aphid tends to feed at or near ground level where access to light might occur, and he has countered von Dehn's suggestion with the observation that a prenatal sensitivity to photoperiod occurs in *M. viciae* (Lees, 1967), and that differences in the manner in which the *parents* of the experimental aphids were kept could therefore account for von Dehn's results. In the case of *E. brassicae*, subsequent investigations by Read (1969) have also demonstrated a maternal sensitivity to daylength. Furthermore, several authors, including Way and Hopkins (1950) with *Diataraxia oleracea* and Lees (1953a) with *Panonychus ulmi*, have made unequivocal demonstrations that arthropod photoperiodic reactions are independent from those of their host plants.

Interactions between the seasonal cycles of insect parasites and that of their hosts do seem to occur, however. An extensive literature has built up on this subject mainly concerned with the problem of whether diapause is independently induced by environmental factors, or whether it is more or less influenced by the diapause state, or other physiological characteristics of the host species. Masslennikova (1968) has demonstrated that examples of both may occur.

Rearing the braconids *Apanteles glomeratus* and *A. spuria* in a variety of Lepidopterous hosts, Geispitz and Kyao (1953) showed that photoperiod acted directly and independently upon the parasites: in short daylength the parasites entered diapause within the body of the host whereas under long daylength they emerged and pupated within their cocoons. This independent sensitivity to daylength in *A. glomeratus* was confirmed by Masslennikova (1958). Nevertheless she was able to demonstrate a modifying influence of the host species. When this braconid parasitizes the univoltine black-veined white butterfly *Aporia crataegi* it hibernated within the body of its host as a first instar larva and adopted the host's univoltine life cycle. In the multivoltine species *Pieris brassicae*, however, the parasite was also multivoltine, the correspondence in the number of generations being due to a similarity between the independent photoperiodic reactions of the two species.

Vinogradova and Zinovjeva (1972a) have recently found similar effects on the braconids *Aphaereta minuta* and *Alysia manducator*. These species attack the larvae of various Calliphorid flies but complete their development within the puparia. Larvae of *Parasarcophaga similis* were raised in long-day conditions, parasitized by *A. manducator*, and then exposed to either long-day (*LD 20* : 4) or short-day (*LD 12* : 12) photoperiods. In the long-day group about 15 per cent entered diapause, whereas in the short-day group about 82 per cent became dormant; this demonstrated that the induction of diapause in the parasite was to a large extent independent of the host. On the other hand, when the experiment was repeated with host larvae raised in short-days (i.e. with "diapause-committed" hosts), 43 per cent of the parasites entered diapause after subsequent long-day treatment, whereas nearly 80 per cent did so after short days. This experiment demonstrated an influence of the host's diapause condition upon that of the parasite. With the gregarious parasitoid *Aphaereta minuta* Vinogradova and Zinovjeva (1972a) showed that host *species* modified the parasites' reaction to photoperiod. This braconid was raised in a variety of Calliphorid larvae subjected to different daylengths. In long daylength (*LD 20* : 4) and 18°C almost all of the parasites had emerged from their hosts

TABLE 6.4. THE PRODUCTION OF DIAPAUSE LARVAE BY FEMALES OF *Nasonia vitripennis* WHEN SUPPLIED WITH PUPARIA OF *S. argyrostoma*, *C. erythrocephala*, *P. terrae novae*, AND DIFFERENT HOST SPECIES ON ALTERNATE DAYS

Host	No. of females	Mean adult life-span (days±S.E.)	Mean age at "switch" (days±S.E.)	Mean no. of offspring/female (+S.E.)	Offspring in diapause (%)
Sarcophaga argyrostoma	19	24.2 ± 1.34	8.7 ± 0.57	364.2 ± 23.0	72.2
Calliphora erythrocephala	20	54.4† ± 2.31	12.4† ± 0.47	482.2† ± 21.9	46.5
Phormia terrae nova	19	57.9† ± 2.73	12.0† ± 0.61	518.9† ± 24.9	48.4
Sarcophaga/ Calliphora on alternate days	19	42.0† ± 2 19	11.9† ± 0.57	533.6† ± 24.7	54.8
Calliphora/ Sarcophaga on alternate days	19	42.9† ± 2.50	10.5* ± 0.55	497.8† ± 27.9	57.2
Phormia/ Sarcophaga on alternate days	20	42.3† ± 2.76	10.4 ± 0.59	526.5† ± 29·0	66.0

* P < 0.05.

† P < 0.01. Difference between mean marked with asterisks daggers and corresponding mean for the *Sarcophaga* group. (From Saunders *et al.*, 1970.)

by the 60th day, regardless of the host species involved. But under a short-day regime (LD 12 : 12) the wasps emerged quickly from *Parasarcophaga similis*, less quickly from *P.* (= *Sarcophaga*) *argyrostoma* and *Calliphora vicina*, but very slowly from *Bellieria melanura*.

In the pteromalid *Nasonia vitripennis* larval diapause is controlled by the photoperiod experienced maternally (Saunders, 1966a). Female wasps maintained at short daylength ($<15\frac{1}{4}$ hours light per 24) switch to the production of diapausing larvae after a few short day cycles, but those exposed to long daylength produce nearly all their progeny as non-diapause broods. Furthermore, once the eggs are deposited within the puparia of the host (Cyclorrhaphous flies) subsequent development cannot be modified by exposure to an opposing photoperiod, or to changes in temperature. The species of host used does modify the response to short daylength, however (Saunders *et al.*, 1970). With *Sarcophaga argyrostoma*, the life-span of the adult wasps was relatively short (21 to 27 days), but the switch to the production of diapausing progeny occurred fairly early in adult life (8 to 11 days) so that the proportion of diapause larvae produced was high (65 to 75 per cent). With puparia of *Calliphora erythrocephala* (= *vicina*) or *Phormia terrae novae*, however, the wasps had a longer life (34 to 58 days), but the switch was delayed so that the final proportion of diapause larvae was lower (30 to 48 per cent, Table 6.4). This difference was attributed, in part, to qualitative differences associated with the haemo-lymph on which the adult wasps had fed. For example, wasps supplied with different host species on alternate days showed an intermediate rate of switching to the production of diapause larvae as if the production of diapause progeny was being modified by a mixed diet consisting of the haemolymph of the two species. If the nutritional differen-ces were affecting the larvae directly, a different incidence of diapause would have been found on alternate days.

3. Geographical factors

The length of the day and its rate of change are functions of latitude. After the spring equinox, for example, days are longer at higher latitudes and daylengths increase more rapidly than in areas further south. With an increase in latitude, however, the climate is usually colder, thus restricting the length of the breeding season: in higher latitudes the short summer comes to an end earlier and whilst the days are still long. Even at the same latitude the climate is generally harsher at an increased altitude, or in areas well removed from the influence of an ocean. Insects have responded to these geographical variations in climate by evolving appropriate modifications to their photoperiodic responses.

The study of geographical variation has been dominated by Russian entomologists, possibly because their national boundaries encompass such a vast area and such extremes of climate. Danilevskii (1965) and his associates, for example, have investigated the photoperiodic responses of a number of species, particularly Lepidoptera, from the south (40°N.) to the north (60°N.), and from the warm climate of the Black Sea to Siberia.

An increase in latitude has been shown to be associated with an increase in the critical daylength (Fig. 6.11). In *Acronycta rumicis*, for example, a population from the Black Sea coast (Sukhumi, 43°N.) showed a critical value of about $14\frac{1}{2}$ hours per 24, whereas populations from Belgorod (50°N.), Vitebsk (55°N.) and Leningrad (60°N.) showed criti-cal photoperiods of $16\frac{1}{2}$, 18 and $19\frac{1}{2}$ hours, respectively. These data show that the critical

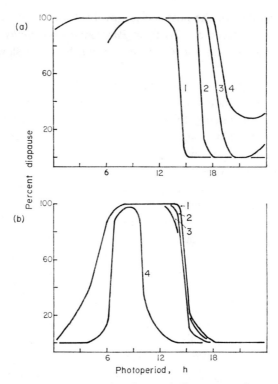

FIG. 6.11. The effects of latitude on the photoperiodic response, for two insects in the
Soviet Union. (a) *Acronycta rumicis* at (1) Sukhumi 43°N., (2) Belgorod 50°N., (3) Vitebsk
55°N., (4) Leningrad 60°C. (b) *Pieris brassicae* at (1) Leningrad 60°N., (2) Brest-Litovsk
52°N., (3) Belgorod 50°N., (4) Sukhumi 43°N. (From Danilevskii, 1965.)

daylength changes by about $1\frac{1}{2}$ hours with every 5° of latitude. The selective advantage of this is clear: the insects at higher latitude compensate for the longer summer
daylengths and the earlier onset of winter with a higher diapause threshold. Southerly
populations, on the other hand, are able to exploit the longer growing period by delaying
the onset of diapause with a shorter critical photoperiod. The ecological realities of
these differences were demonstrated by transferring populations from one latitude to
another. Southern races transferred to Leningrad, for example, failed to diapause and
died with the first frosts. Northern races transferred to the Black Sea, on the other hand,
entered diapause after only one generation, even though the environment would have
supported a second.

Associated with a change in the critical daylength is also a change in voltinism. Southern populations are usually bivoltine or multivoltine with a facultative winter diapause.
The Leningrad population of *A. rumicis*, on the other hand, is univoltine and the photoperiodic response curve shows a distinct tendency towards an obligate form of diapause
since a proportion of the pupae enter diapause even at daylengths in excess of 20 hours.

Similar data are known for a number of other species. The cabbage moth *Mamestra
brassicae*, for example, shows a critical daylength of about $14\frac{1}{2}$ hours on the Black Sea
coast, but about $18\frac{1}{2}$ hours near Leningrad (Danilevskii, 1965). In Japan, this species is
bivoltine throughout its distribution, but its phenology is complicated by the appearance
of a summer diapause in southern latitudes (Masaki, 1956, 1968). Thus, in Hokkaido
(41° to 45°N.) the moth completes two generations during the relatively short breeding

season and overwinters in pupal diapause. In northern Honshu (40°N.) some pupae have a brief period of summer aestivation. This summer diapause increases in intensity and duration in more southerly latitudes until in the Ryukyu islands (28°N.) the insect, although still bivoltine, is active only in the spring and the autumn, and spends the winter (November to March) and the summer (May to September) in diapause.

In eastern Europe both *A. rumicis* and *M. brassicae* form a continuous series, or cline, from south to north with no clearly defined geographical races. In *Pieris brassicae*, on the other hand, the critical daylength is the same (about 15 hours per 24) in Leningrad (60°N.), Brest-Litovsk (52°N.) and Belgorod (50°N.). On the Black Sea coast (43°N.), however, a distinct geographical race occurs with a critical photoperiod of about 10 hours per 24 (Danilevskii, 1965). These races appear to be genetically distinct.

Examples of geographical races or clines with respect to the photoperiodic reaction are now known in a large number of insect species. Beck (1968), for instance, was able to list 34, and Danilevskii *et al.* (1970) many more. Recent examples include *Nasonia vitripennis*, with a critical daylength of $15\frac{1}{4}$ hours in a population from Cambridge, England (52°N.) and $13\frac{1}{2}$ hours from Woods Hole, Massachusetts (42°N.) (Saunders, 1966a), and the lacewing *Chrysopa carnea* with a critical photoperiod of 13.5 to 14 at Ithaca, New York (42°27′N.), but 12.5 to 13 in Arizona (33°19′N.) (Tauber and Tauber, 1972b).

Of particular interest is the phenological adaptation of recently introduced species to local conditions. The European corn borer *Ostrinia nubilalis*, for example, was introduced into North America and first recorded in New Hampshire in 1921. Beck and Apple (1961) have shown that populations in different areas now show different critical photoperiods, different numbers of generations per year, and different rates of larval growth. The pink bollworm *Pectinophora gossypiella*, on the other hand, has retained remarkably constant photoperiodic characters since its introduction into the Americas early in the present century. Ankersmit and Adkisson (1967) studied populations of this insect from Texas (32° and 28°N.), the Virgin Islands (18°N.), Venezuela (10°N.), Colombia (3°N.) and Argentina (27°S.). In all of these the critical daylength was about the same (12 to 13 hours/24), although the intensity of the response was more pronounced in the most northerly and most southerly populations. In the tropical populations (Venezuela and Colombia) the photoperiodic response was weak and almost eliminated at high temperature. The Colombian strain produced a high incidence of diapause at LD *10*: 14 but this photoperiod is never encountered at that latitude. The photoperiodic responses of the tropical populations, therefore, were rarely operating, and the insect was showing a strong tendency towards a homodynamic type of development.

Geographical differences in the photoperiodic response at different localities at the same latitude have been investigated in a number of species (Fig. 6.12). The small cabbage white butterfly *Pieris rapae*, for example, accomplishes six generations a year on the Black Sea coast and shows a critical daylength of less than 12 hours/24. However, at Vladivostok—which is almost at the same latitude but is subjected to a much longer and more severe winter—the same species only manages three generations a year and enters diapause at a critical daylength of $14\frac{1}{2}$ hours/24 (Danilevskii, 1965). Altitude puts similar restraints on the number of generations achieved because of the retarded development at lower temperature. The white satin moth *Leucoma salicis* achieves one generation per year above 2000 m but two generations on the plains. In the Caucasus, the green-veined white *Pieris napi* occurs in two distinct forms associated with altitude. Up to about 1600 m there is a genetically uniform race (ssp. *meridionalis*) with a 13-hour

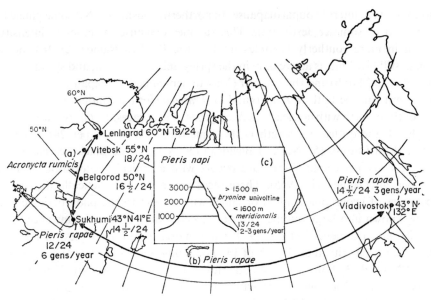

FIG. 6.12. The effects of latitude (a), longitude (b), and altitude (c) on the diapause response (critical daylength and number of generations per year) in *Acronycta rumicis*, *Pieris rapae* and *P. napi* in the Soviet Union. (Data from Danilevskii, 1965.)

critical photoperiod and several generations a year. Above about 1500 m, however, there is a univoltine and phenologically separate subspecies *(bryoniae)* which has only a very rudimentary photoperiodic response at high temperature.

E. The Genetics of the Photoperiodic Response

The foregoing sections clearly demonstrate that the photoperiodic response and the state of diapause it controls are genetically "programmed" in the development of the organism. Thus, in a particular species diapause occurs at a specific stage in the life-cycle and the photoperiodic response, particularly the length of the critical photoperiod, is an evolved character matching the local conditions of photoperiod and temperature in the area from which the population was drawn.

Studies on the genetics of photoperiodism fall into two main categories. The first depends on the intrinsic genetic variability of natural or laboratory populations and involves the selection for either diapause or diapause-free lines. The second approach involves the cross-mating of individuals from different geographical populations with different diapause and photoperiodic characteristics. A few examples of each will be given here.

The spruce budworm *Choristoneura fumiferana* is a univoltine species which diapauses as a second-instar larva within a silken hibernaculum. It is typical of all insects, however, in that its genome contains a considerable "store" of genetic variability. Thus Harvey (1957) was able to select an almost diapause-free line in only six generations by breeding from the few insects which failed to diapause, and thereby to isolate a strain with a facultative rather than an obligate form of arrest. Similarly, Barry and Adkisson (1966) maintained the pink bollworm *Pectinophora gossypiella* at short daylength (*LD 12*: 12) and 28°C, and bred from the few non-diapausing insects; after twenty-three generations an almost diapause-free strain was obtained. Tanaka (1951), on the other hand, bred only from the diapausing individuals of *Antheraea pernyi* and developed a strain which showed univoltine characters. In some cases selection has taken place inadvertently: House (1967), for example, recorded a gradual loss of diapause through successive laboratory-reared generations of the parasitoid *Pseudosarcophaga affinis*. Finally, and of particular interest, is the selection of a short-day strain of *A. pernyi* from an originally long-day population by breeding from individuals which completed only one generation a year (Chetverikov, quoted in Danilevskii, 1965). In this case a complete inversion in the photoperiodic response was achieved which shows that the long- and short-day responses probably only differ in the "clock-dependent" processes linking time measurement with the endocrine control mechanisms.

Pittendrigh and Minis (1971) maintained *P. gossypiella* in *LD 14*: 10 and selected those moths which emerged from their pupae either earlier or later than the "stock". After quite a small number of generations (four to eight) the diapause response was also found to have altered, the "late" strain apparently having a shorter critical photoperion and a reduced response in the formerly "strong" short daylengths (*LD 12*: 12). This result indicates that the genes involved in photoperiodism are linked in some way with those controlling other aspects of the insect's reaction to the light cycle.

The cross-mating of different geographical strains was performed by Danilevskii (1965) for *Acronycta rumicis*, using races isolated in Leningrad (60°N.) and in Sukhumi (43°N.). At 23°C the Leningrad strain showed a critical daylength of about 19 hours/24 and the Sukhumi strain a critical daylength of about 15 hours/24. First generation crosses (F₁) between these two races gave rise to larvae which showed an intermediate re-

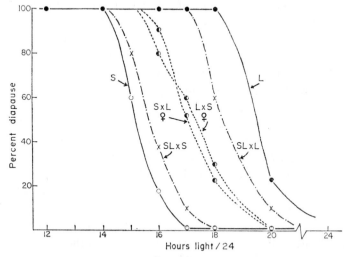

FIG. 6.13. The results of genetic crossing experiments between strains of the moth *Acronycta rumicis* from two different latitudes. The critical daylength for the strain from Sukhumi (S) on the Black Sea coast (43°N.) was about 15 hours; that for a strain from Leningrad (L) (60°N.) was about 19 hours. The critical daylengths for the females of both F₁ crosses (S×L and L×S) were about 17 hours, and the back-crosses between these hybrids and moths of the original strains (SL×S and SL×L) were similarly intermediate in respect of their critical daylengths. Not all the crosses are shown, and only the females of the F₁ are included. The males resulting from these particular crosses showed a critical daylength closer to the maternal type. These results show both the genetic nature of the photoperiodic response and that inheritance is multifactorial. (From Danilevskii, 1965.)

sponse of about 17 hours/24. The inbred progeny of these individuals (the F_2) also showed a 17-hour critical photoperiod. Danilevskii pointed out that this threshold was equivalent to that for a population of *A. rumicis* drawn from a latitude of about 50°N., and considered that these results indicated a continuous gradient of gene frequencies from south to north with continuous hybridization.

When the F_1 progeny were analysed according to sex, it was found that the male offspring of either cross (Leningrad female × Sukhumi male, or Sukhumi female × Leningrad male) showed a critical photoperiod closer to that of the *female* parent (Fig. 6.13). Female offspring, on the other hand, showed critical values strictly intermediate (about 17 hours) between the parental types. The reaction of the male hybrids therefore clearly inclines towards the maternal characteristics.

Although all crosses between these widely separated geographical strains were achieved without difficulty and the progeny developed quite normally, there was clear evidence of "hybrid vigour", or heterosis. In the parental strains the duration of larval development was about 20 to 21 days in short daylength and about 19 days in long. The F_1 hybrids, however, developed more rapidly: in short daylength larval development was completed in about 18 days and in long days in about 16 days. Larval weights also showed evidence of heterosis. Sukhumi larvae were usually larger than those from Leningrad, female larvae of the former weighing about 310 mg as opposed to about 274 mg. Female larvae derived from crossing the two races, however, weighed about 320 mg.

The photoperiodic responses of larvae resulting from back-crosses to the parental strains were also investigated. In this type of cross the response was identical for either sex and showed a critical daylength intermediate between the two parents. Thus, a cross between a female of the Sukhumi×Leningrad strain (with a critical daylength of 17 hours) and a male of the Leningrad strain (about 19 hours) gave progeny with a critical photoperiod close to 18 hours. Similarly, a cross between a Sukhumi×Leningrad female (17 hours) and a Sukhumi male (15 hours) gave progeny with a 16-hour threshold. The absence of segregation in these back-crosses clearly indicates that the genetic control of the photoperiodic response is polygenic.

Crosses between geographical races have now been performed in a number of other insect species. Russian entomologists, for instance (see Danilevskii, 1965), have demonstrated a similar intermediate photoperiodic response in F_1 hybrids of *Spilosoma menthastri*, *Pieris brassicae* and *Leucoma salicis*. In the parasitic wasp *Nasonia vitripennis* the rate of switching to the production of diapause larvae has been investigated in crosses between strains from Cambridge, England (52°N.), and Woods Hole, Massachusetts (42°N.) (Saunders, 1965b). In conditions of constant darkness females of the English strain switched to all-diapause broods more rapidly than those from Massachusetts. When females of one strain were mated with males of the other the pattern of diapause production in the larvae so produced was *identical* to the maternal type, confirming that diapause induction in this species is determined in the adult female. Crosses between these offspring, however, revealed the F_1 type with an intermediate rate of switching to diapause. In this species, therefore, the males only contribute towards the diapause characteristics of their grandchildren and subsequent generations.

F. The Spectral Sensitivity and the Intensity Threshold of the Photoperiodic Response

In principle, an experimentally observed action spectrum should correspond directly with the absorption spectrum of the involved pigment molecule. Action spectra were thus used to recognize and to isolate the pigment phytochrome (Borthwick *et al.*, 1952). Proper action spectra should also plot the energy (ideally the reciprocal number of quanta in the wavelength interval) required to obtain a particular constant biological response (e.g. 50 per cent diapause) against wavelength. This procedure has been carried out with very few insects, the best documented example being the aphid *Megoura viciae* (Lees, 1966, 1971). Nevertheless, a number of workers have investigated the spectral sensitivity and the light intensity thresholds of the photoperiodic response in insects, at least partially, with a view to the characterization of the pigment(s) involved. These studies show that most species are optimally sensitive to light in the blue-green region of the spectrum and largely insensitive to red. In some species, however, sensitivity extends into the red end of the spectrum. There is evidently, therefore, a considerable diversity in the pigments involved.

Species which are not red-sensitive include *Bombyx mori* (Kogure, 1933), *Grapholitha molesta* (Dickson, 1949), *Panonychus ulmi* (Lees, 1953a), *Dendrolimus pini* (Geispitz,.

TABLE 6.5. THE SPECTRAL SENSITIVITY OF THE PHOTOPERIODIC RESPONSE

Species	Effective wave-lengths in nm	Non-effective wave-lengths in nm	References
1. Red-insensitive species			
Bombyx mori	350–510	>600	Kogure (1933)
Grapholitha molesta	430–580	<430; >600	Dickson (1949)
Panonychus ulmi	365–540	>600	Lees (1953a)
Dendrolimus pini	violet–green	red	Geispitz (1957)
Pieris brassicae	violet–green	red	Geispitz (1957)
	400–520	>580	Claret (1972)
Antheraea pernyi	398–508	>580	Williams *et al.* (1965)
	400–500		Hayes (1971)
Euscelis plebejus	365–550	>550	Müller (1964)
Anthonomus grandis	blue–orange	red	Harris *et al.* (1969)
Carpocapsa pomonella	400–500	>600	Norris *et al.* (1969)
	peak at 450		Hayes (1971)
Chaoborus americanus	peak at 550		Bradshaw (1969)
Megoura viciae	peak at 450–470	>550	Lees (1966, 1971)
2. Red-sensitive species			
Acronycta rumicis	407–655		Geispitz (1957)
Leptinotarsa decemlineata	423–675		De Wilde and Bonga (1958)
Pectinophora gossypiella	480–680		Pittendrigh *et al.* (1970)
Nasonia vitripennis	480–640	>650	Saunders (1975a)
Pimpla instigator	386–700		Claret (1973)

1957), *Pieris brassicae* (Geispitz, 1957; Claret, 1972), *Antheraea pernyi* (Williams *et al.*, 1965; Hayes, 1971), *Megoura viciae* (Lees, 1966, 1971), *Euscelis plebejus* (Müller, 1964), *Anthonomus grandis* (Harris *et al.*, 1969), *Carpocapsa pomonella* (Norris *et al.*, 1969; Hayes, 1971) and possibly *Chaoborus americanus* (Bradshaw, 1969). These species generally show a maximum sensitivity between 400 and 500 nm and a sharp cut-off at longer wavelengths (Table 6.5).

In his study of *M. viciae*, Lees (1966, 1971) used narrow-band filters (average bandwidth 7 nm at half-peak transmission) in conjunction with a series of neutral density filters. Test virginoparae were exposed to a 1-hour pulse of monochromatic light placed $1\frac{1}{2}$ hours after dusk (i.e. *LD* $13\frac{1}{2}$: $1\frac{1}{2}$: 1: 8) or to a 30-minute pulse $7\frac{1}{2}$ hours after dusk (i.e. *LD* $13\frac{1}{2}$: $7\frac{1}{2}$: $\frac{1}{2}$: $2\frac{1}{2}$), the two positions corresponding to an early and a late light break (Chapter 7), either of which induces the production of daughter virginoparae. Unbroken nights of $10\frac{1}{2}$ hours (*LD* *13.5*: 10.5), or nights pulsed with an ineffective or subthreshold stimulus, on the other hand, give rise to oviparae. The intensity of the light at each wavelength was adjusted in different experiments until a 50 per cent response (50 per cent virginopara-producers) was obtained. The results for the early night interruption showed that maximum sensitivity was in the blue (450 to 470 nm) with a threshold intensity at that wavelength of about 0.2 μW cm^{-2} (Fig. 6.14). Fifty times that

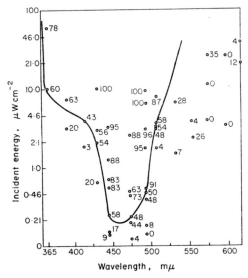

FIG. 6.14. Action spectrum for the photoperiodic control of progeny type in the aphid *Megoura viciae*. Semilogarithmic plot showing the response to 1-hour interruptions of monochromatic light introduced into a 10.5-hour dark period 1.5 hours after its inception. The "main" photoperiod of 13.5 hours was of white light (100 ft-candles). The intensity of the response is given by the percentage of aphids that produced viviparous progeny. These values are also shown by their respective points and the curve is drawn at approximately the 50 per cent response level. (From Lees, 1966.)

energy was required to give a 50 per cent response at 365 nm, or in the green at 550 nm; there was no sensitivity in the red. For the late night interruption, maximum effectiveness was again in the blue, but sensitivity extended some way into the red.

Williams *et al.* (1965) studied the termination of pupal diapause in *A. pernyi* by providing 8 hours of white light supplemented with 8 hours of filtered light. Reactivation was found to be rapid with 398, 434 and 508 nm, but longer wavelengths (580 and 640 nm) were inactive. Hayes (1971) similarly showed that this species was optimally sensitive to wavelengths between 400 and 500 nm. In *E. plebejus*, Müller (1964) provided filtered light as a short supplementary photoperiod and adjusted the distance from the light source to produce an irradiance of 160 μW cm^{-2}, well above the threshold. Maximum sensitivity was found to be between 365 and 550 nm resulting in the production of the long-day or *plebejus* form. Wavelengths above 550 nm were not "seen" and the *incisus* form was produced.

A pronounced red sensitivity has been found in five insect species. Larvae of *Acronycta rumicis* were found to be responsive to all three spectral regions (407, 530 and 655 nm) at the energy levels used (Geispitz, 1957). De Wilde and Bonga (1958) exposed the Colorado potato beetle *Leptinotarsa decemlineata* to a constant energy level of 2.0 μW cm^{-2} transmitted through narrow-band interference filters. The beetles responded to all wavelengths between 423 and 675 nm, but were unaffected by those greater than 675. Larvae of the pink bollworm *Pectinophora gossypiella* could also "see" red light of low intensity. Pittendrigh *et al.* (1970) demonstrated that this insect could distinguish in a photoperiodic sense between 14 hours of red light per day (640 nm, 0.59 ergs/cm²/sec) and 12 hours of red light per day (620 nm, 0.57 ergs/cm²/sec). The former was "read" as a long day and gave rise to only 5 per cent diapause, whereas the latter was "read" as a short day and gave rise to 98 per cent diapause. Exposure of the larvae to 14 hours of 680 nm at 0.14 ergs/cm²/sec induced 17 per cent diapause in contrast to the 73 per cent in DD; evidently the spectral sensitivity in *P. gossypiella* extends at least as far as 680 nm. These studies were of particular interest because they suggested that the photoperiodic clock is distinct from that controlling some of the overt behavioural rhythms studied in this species (Chapter 7); the action spectrum for the initiation of the egg-hatch rhythm, for example, showed a maximum effectiveness between 390 and 480 nm and a sharp cut-off above 520 nm (Bruce and Minis, 1969). The Ichneumonid *Pimpla instigator* also responds to wavelengths as long as 700 nm (Claret, 1973).

The parasitic wasp *Nasonia vitripennis* can distinguish between 12 and 18 hours of red light per day (>600 nm) and produce 89.3 and 1.8 per cent diapausing progeny, respectively (Saunders, 1973b). They can also distinguish between 12 hours of white light per day (100 per cent diapause) and 12 hours of white light supplemented with 6 hours of red (>640 nm) (5.7 per cent diapause). Preliminary observations on the action spectrum for this species have been carried out supplying filtered light as a 3-hour supplement following a 13-hour white photoperiod (Saunders, 1975a). Light in this position is regarded as the lights-off signal which entrains and phase-sets the "dusk" oscillator (Chapter 7). Maximum sensitivity occurs between about 554 and 586 nm with a sharp cut-off above about 650 nm. At 586 nm the wasps respond to less than 1 μW cm^{-2}.

The data obtained from spectral sensitivity studies indicate that the intensity threshold for the photoperiodic response may be quite low. The threshold for *M. viciae* at 450 to 470 nm, for example, is about 0.2 μW cm^{-2} (Lees, 1966, 1971). Other observations, often with white light, support this. Table 6.6. shows that there is a low threshold value (between 0.025 and 25 lux, depending on the species) above which the receptors appear to be saturated so that they respond in an identical fashion to low light intensities and to full sunlight. In most cases the threshold is above that of the full moon (0.7 lux, 0.65 μW cm^{-2}) so that moonlight probably does not affect the photoperiodic response. The thresholds, however, are generally *below* that of the available light at dawn and dusk so that the "natural photoperiod" is probably closer to the time which includes the two periods of civil twilight (de Wilde, 1962).

The question of intensity thresholds is complicated by the fact that the photoperiodic receptors appear to be in the brain (Chapter 10) and that light reaching the brain has to pass through the cuticle of the head capsule; this must reduce the intensity and probably alter the spectral quality of the light where broad-band filters are used. In addition, many insects live in "cryptic" situations (e.g. within fruit) which must also modify the quantity

TABLE 6.6. APPROXIMATE INTENSITY THRESHOLDS FOR THE PHOTOPERIODIC REACTION

Species	Intensity, in lux	References
Bombyx mori, eggs and larvae	0.1–0.8	Kogure (1933)
Acronycta rumicis larvae	5	Danilevskii (1948)
Diataraxia oleracea larvae	10	Way and Hopkins (1950)
Panonychus ulmi	10–20	Lees (1953a)
Leptinotarsa decemlineata adults	0.1	De Wilde and Bonga (1958)
Metriocnemus knabi larvae	0.025	Paris and Jenner (1959)
Antheraea pernyi, brain	<10.8	Williams *et al.* (1965)
Grapholitha molesta, larvae in young apples	10–30	Dickson (1949)

and quality of the light eventually reaching the receptors. Larvae of the midge *Metriocnemus knabi*, for instance, respond to light energies down to 0.025 lux but normally live in the water collecting within the pitcher plant *Sarracenia purpurea* (Paris and Jenner, 1959). Conversely, the larvae of the oriental fruit moth *Grapholitha molesta* show a threshold of about 10 to 30 lux whilst burrowing in young apples (Dickson, 1949); the light actually reaching the larvae must be much less than this. For example, the threshold for the reactivation of diapausing larvae of the codling moth *Carpocapsa pomonella*—now outside the fruit—was found to be less than 0.1 μW cm^{-2} at 450 nm (Hayes, 1971).

The problem of the intensity of light actually reaching the brain has been investigated in *Antheraea pernyi* by Williams *et al.* (1965). Diapausing pupae of this species are enclosed in an apparently opaque cocoon; the pupal cuticle is also heavily pigmented apart from a small transparent "window" overlying the brain itself. The transmission of light through the cocoon was measured with a spectrophotometer for a range of wavelengths. At 400 nm it was found to be only 0.000009 per cent of the incident light, and only 0.014 per cent at 700 nm. The transmission of blue light (460 nm) by the transparent facial cuticle was found to be between 2 and 5 times that of the pigmented cuticle elsewhere on the pupa, which in turn was about 5000 times as transparent as the cocoon. A combination of these measurements on the cocoon and the facial cuticle suggested that as little as 0.0000003 per cent of the blue light (460 nm) falling on the outside of the cocoon could reach the brain by simple transmission.

The geometry of the cocoon in *A. pernyi*, however, apparently makes it an ideal "light-integrating sphere". It collects scattered light, especially in the spectral range 440 to 510 nm, within it as a blue haze. In the effective range (440 to 490 nm) more than 0.14 per cent of the light energy reaches the brain which is fully saturated by less than 10.8 lux of this blue light.

The transparent facial window referred to earlier seems to play a minor role in the transmission of light to the brain. Shakhbazov (1961) reported that the photoperiodic response of *A. pernyi* could be eliminated by coating the window with an opaque material. Williams (1969a), however, failed to repeat this observation: the photoperiodic responses of pupae with their windows painted black *and* returned to their almost opaque cocoons were fully preserved. Williams concluded that the transparent facial cuticle was merely a "safety device for cocoons in shady situations".

The pronounced blue-sensitivity of many insect species has suggested that the photoperiodic pigment is either pink (Williams *et al.*, 1965) or orange (Lees, 1966). However, both in *M. viciae* and in *A. pernyi* the brain appears colourless to visual inspection, and this suggests that the pigment is in very low concentration. These authors, and others, have suggested that the chromophore might contain a carotenoid. Pteridine pigments have also been proposed as a possible candidate (L'Helias, 1962). The concordance between available action spectra and absorption spectra for these pigments is insufficiently similar, however, and one is reminded of the fact that carotenoids are apparently *not* involved in the entrainment of the rhythm of pupal eclosion in *Drosophila pseudoobscura* (Zimmerman and Goldsmith, 1971) which has a similar action spectrum (Chapter 3). Hayes (1971) claimed to find small orange spots distributed in a random fashion on the brain of *A. pernyi*. The absorption spectrum for these orange areas showed similarities to the measured action spectrum, but their relationship to the photoperiodic response appears to be doubtful.

Annotated Summary

1. Insects may show long- or short-day responses. Others show responses to changes from long to short days, from short to long days, or to naturally changing daylengths. Some species, on the other hand, are day-neutral. In the long-day response curve the most important physiological and ecological feature is the sharp discontinuity at the critical daylength which separates the diapause-inducing short photoperiods from the long days which promote uninterrupted development.

2. Insects are usually sensitive to photoperiod during a restricted part of their development. In some the sensitive period occurs in the same instar as the resulting diapause; in many more it precedes it. In the former case, diapause reactivation is frequently controlled by a photoperiodic mechanism, whereas in the latter it is more usually in response to a period of chilling.

3. In a number of species the sensitive period occurs in the maternal generation, and the eggs may be irreversibly "determined" for diapause or non-diapause development before they are deposited.

4. The diapause response is variously modified by temperature. In most long-day species constant high temperatures act in accordance with long-days to avert diapause and low temperatures act in the same "direction" as short days. Low or high temperature *pulses* may completely reverse the photoperiodic response, and a daily temperature *cycle* (or thermoperiod) may act as a *Zeitgeber* in its own right and simulate most if not all of the effects of a light cycle.

5. Both the quality and the quantity of the diet may modify the photoperiodic response. Senescent vegetation or ripening fruit, for example, may act to intensify the short-day response in the autumn. In parasitic species the species of host may modify the response.

6. The photoperiodic response curve is a product of natural selection and differs in populations from different latitudes and geographical areas. At higher latitudes the longer summer days, the greater rate of change in daylength, and the shorter summer season are reflected in a longer critical daylength than in areas further south. More northerly populations frequently achieve fewer generations per year than

those in the south. Altitude and the proximity to the warming influence of oceans also affect the photo-periodic response and voltinism.

7. The important characters of the photoperiodic response are genetically controlled and therefore inherited, in all cases by a polygenic system. The diapause response is therefore open to natural and artificial selection, and to reciprocal crosses between different geographical populations.

8. Action spectra for the photoperiodic response indicate a variety of photopigments. Some insects show a sensitivity not unlike that for the phase-shifting of circadian oscillations with a sharp cut-off above 500 nm; others are distinctly red-sensitive. Intensity effects are generally related to habitat, and sensitivity may be as low as 0.025 lux.

CHAPTER 7

THE PHOTOPERIODIC CLOCK

THE photoperiodic response curve describes the reaction of a population of a particular insect to different static photoperiods. Ecologically and physiologically the most important part of this curve is the sharp discontinuity between the long-days which allow continuous or diapause-free development and the short-days which induce diapause. This "critical photoperiod" infers that the insect possesses some kind of inherited and intrinsic clock which enables it to "measure" the length of the day or the night (or perhaps both) and respond with the appropriate seasonal switch in metabolism. As shown by the steepness of the curve at the critical point this form of time measurement is often performed with considerable accuracy. Some species, for instance, are able to detect differences of as little as 10 or 15 minutes in daylength, and a difference of 1 hour usually converts the response from one metabolic pathway to the other.

In earlier chapters we have discussed some of the seasonal events controlled by this mechanism, and the way in which the response is modified by certain environmental factors such as temperature, diet and latitude. In this chapter we are concerned with the physiological mechanisms by which this time measurement is achieved. To date, practically nothing is known about the concrete physiological processes involved, but we have a considerable knowledge of the formal properties of the system. Attention will be directed almost entirely at the long-day response because this is the most common type in the insects, and practically all of the published work is concerned with this form of the response.

Photoperiodic induction is a two-stage process. The first involves the measurement of the duration of the dark or the light phases of the diel cycle by the clock. The second involves the summation of successive days information on daylength to a point at which induction occurs, or is irreversible; this process is accomplished by the "photoperiodic counter", and will be the subject of Chapter 8.

Early workers were apparently not much concerned with the mechanism of time measurement in photoperiodism. Very often, however, the *implication* was that day- or night-length measurement (or sometimes both) was accomplished by means of an "hour-glass" type of mechanism which was set in motion at either dawn or dusk and proceeded at a (presumably) temperature-compensated rate. If—in the case of a dawn hour-glass—dusk intervened before the process had reached a certain stage the photoperiod was "read" as a short-day; if, on the other hand, the process reached completion before dusk it was "read" as a long-day. Dickson (1949), in his very comprehensive investigation of diapause induction in the oriental fruit moth *Grapholitha molesta*, found that induction was only fully expressed in cycles consisting of between 11 and 15 hours of dark and 8 to 15 hours of light: he discussed both "light" and "dark reactions".

Way and Hopkins (1950) found that diapause in the tomato moth *Diataraxia oleracea* was averted in photoperiods of more than 15 hours, and supposed that "prolonged stimulation of the last instar larval brain by light falling on the photoreceptors induces formation of the adult growth hormone". Both of these theories imply an hour-glass mechanism.

A totally different explanation, however, had been put forward in 1936 by the German botanist Erwin Bünning to explain similar phenomena in plants. His very original proposition was that photoperiodic time measurement was dependent on the "endogenous diurnal rhythm" then known to provide temporal organization in plants, such as the daily up-and-down leaf movements of *Phaseolus*. Bünning (1936) proposed that the 24-hour period comprised two half-cycles differing in their sensitivity to light. The first 12 hours constituted a photophil or "light-requiring" half-cycle and the second 12 hours

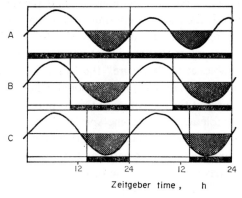

Zeitgeber time, h

FIG. 7.1. Bünning's (1936) model for the execution of the photoperiodic time measurement by a circadian oscillation. A — the free-running clock in DD or *LL*. B — under short days the light does not extend into the second or scotophil half-cycle, whereas in C (long days) it does. (After Bünning, 1960.) (From Pittendrigh and Minis, *Am. Nat.* **93**, 261, fig. 3 (1964), University of Chicago Press).

a scotophil or "dark-requiring" half-cycle. Short-day effects were thus produced when the light was restricted to the photophil, but long-day effects when it extended into the scotophil (Fig. 7.1). This model for the clock, therefore, was an oscillator as opposed to an hour-glass, and focused attention on the many similarities between photoperiodism and the overt diurnal rhythms of behaviour and physiology.

Although this model received some attention, particularly in the German botanical literature of the 1950s, it was not considered as a serious alternative to the hour-glass in insect photoperiodism until its introduction into that field by Bünning and Joerrens in 1960. Since then, this model, or variations of it, have dominated the field of photoperiodism—even in the absence of sound experimental evidence in favour of *any* involvement of the circadian system in this form of time measurement. Much of this chapter, therefore, is concerned with evidence for some kind of "Bünning's hypothesis".

A number of different types of "oscillator" clock have been proposed to account for time measurement in insect photoperiodism. These will be described in detail in this chapter, but two are worthy of mention here. The first is a more explicit version of Bünning's general hypothesis, and is due to Pittendrigh and Minis (1964). This model, like its original, depends on the temporal interaction between light and a light-sensitive phase-point of an endogenous oscillator; it is therefore referred to as an "*external* coincidence model" (Pittendrigh, 1972). An alternative proposition, first discussed in a general way by Pittendrigh (1960) and later, in a more explicit fashion, by Tyshchenko (1966), comprises two (or perhaps more) oscillators, one phase-set by dawn and the other phase-set by dusk; induction or non-induction then occurs as a function of the phase-angle between the two. Since in this model light has no direct role in the inductive process, it has been referred to as an *"internal* coincidence model" (Pittendrigh, 1972). In recent years the approach to the problem of the role of circadian oscillations in time

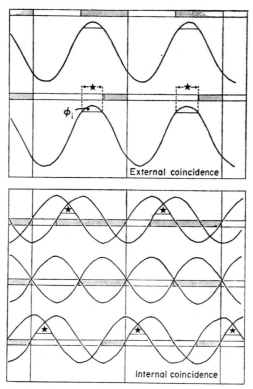

FIG. 7.2. Two current models for the photoperiodic clock in which the circadian system is causally involved. *Upper panel:* the "external coincidence" model of Pittendrigh and Minis (1964) based on that of Bünning (1936) in which an oscillation is phase set by the light-cycle and photoperiodic induction is or is not effected by the coincidence of a particular light-sensitive phase point (ϕ_i) with the environmental light-cycle. Note, therefore, that light plays a dual role. *Lower panel:* the "internal coincidence" model of Pittendrigh (1960) and Tyshchenko (1966) with two independent "dawn" and "dusk" oscillators. Photoperiodic induction is a function of the mutual phase relationships of the oscillators.

measurement has been concerned with both external and internal coincidence types of clock. These two models are illustrated in Fig. 7.2.

Despite the attention paid to oscillator models in the last decade they are not ubiquitous. Indeed, the green vetch aphid *Megoura viciae* measures night length by means of an "hour-glass" type of clock (Lees, 1966, 1968) which shows no evidence of having an oscillatory component. It appears that the mechanisms insects use to measure day- or night-length are products of convergent evolution: having been faced with the same problem, different insect species have developed different types of clock to deal with it. Some have "hour-glasses" as in *M. viciae*, others such as *Nasonia vitripennis* (Saunders,

1970, 1974) and *Sarcophaga argyrostoma* (Saunders, 1973b) have oscillatory (circadian) devices. Moreover, within those species which possess circadian clocks a wide variety of type is possible and, indeed, occurs.

Whilst observation and experiment are limited to a study of responses in a 24-hour cycle the exact nature of time measurement cannot be determined. The best evidence for oscillators, or hour-glasses, therefore, comes from experiments in which the length of the light cycle is abnormal or is experimentally manipulated. The study of photoperiodic responses in cycles with a very protracted "night", for example, reveals one of the fundamental differences between the two types. In the extended dark phase an oscillator "resets" itself, or repeats itself with circadian periodicity, whereas an hour-glass executes only one time-measuring period and then stops; it requires to be "turned over" by another light pulse before it can operate again.

A. Evidence for the Involvement of the Circadian System in Time Measurement

Taken at their face value photoperiodic response curves present no clear evidence of being related to an internal oscillator. There are, of course, some superficial similarities: both photoperiodic induction and circadian rhythms of activity are controlled by the natural light cycle, and both are, to a varying extent, functions of photoperiod.

The purpose of this section is to compare what is known about the mechanisms of time measurement in photoperiodism with the observed properties of overt rhythmic processes as reviewed in Chapters 2 and 3, and to examine the evidence to support some form of Bünning's general hypothesis. The principal difficulty in this type of exercise is that an overt rhythmic expression of the photoperiodic clock is lacking, and one has to obtain results, second-hand as it were, from an analysis of the proportion of a particular insect species entering or escaping diapause in a particular photoperiodic treatment. For this reason Pittendrigh and Minis (1964) advocated an alternative method of investigation based on the simultaneous analysis of photoperiodic induction and of overt rhythmicity, with the hope that the two mechanisms were the "same clock", were coupled to the same clock, or at least would show similar properties. This method has been applied extensively to the pink bollworm *Pectinophora gossypiella* (Pittendrigh and Minis, 1964, 1971; Pittendrigh, 1966) and some other species. The usefulness and validity of this approach will be discussed.

Most of the evidence for the involvement of the circadian system in photoperiodic time measurement comes from experiments in which the light cycle is altered or perturbed, either by changing the period of the light cycle (T) or by interrupting the dark phase of the cycle with short supplementary light pulses. The latter technique often takes on the form of a "skeleton photoperiod" and the results are comparable to those obtained by the use of such a technique in the study of overt rhythmicities (Chapter 3).

1. *Skeleton photoperiods*

(a) *Asymmetrical skeletons.* "Night interruption" experiments, in which the dark period of the cycle is variously perturbed by an additional light pulse, or "light-break", have long been a useful experimental tool in the analysis of photoperiodic phenomena in plants and birds (Hardner and Bode, 1943; Bünning, 1960; Jenner and Engels, 1952;

Kirkpatrick and Leopold, 1952). These experiments demonstrated that photoperiodic reversals could be obtained by light breaks as short as a few seconds (Parker et al., 1946). The same techniques was later applied with varying results to arthropods, in particular to *Grapholitha molesta* (Dickson, 1949), *Panonychus ulmi* (Lees, 1953b) and *Metriocnemus knabi* (Paris and Jenner, 1959). The light breaks were often placed centrally in the night and, unlike similar experiments with plants, found to have limited effect. Lees (1953b), for example, found that the dark component of an *LD 8: 16* cycle had to be interrupted by at least six additional hours of light when placed centrally (e.g. *LD 8: 5: 6: 5*) before an appreciable reduction in the proportion of winter eggs was obtained. Photoperiodic reactivation of the diapausing larvae of the midge *Metriocnemus knabi*, on the other hand, was achieved with a $1\frac{1}{2}$-hour night interruption (e.g. with *LD $10\frac{1}{2}$: 6: $1\frac{1}{2}$: 6*) (Paris and Jenner, 1959). When long-day effects were obtained they were interpreted as evidence that the length of the dark period of the cycle was more significant than the light, and that the "light reaction" developed more slowly in arthropods than in plants (Lees, 1960).

The systematic pulsing of the night as part of a test of Bünning's hypothesis was first carried out with insect material by Bünning and Joerrens (1960), using *Pieris brassicae*. Two-hour supplementary light pulses were placed at different positions in the long nights of diapause-inducing cycles (*LD 6: 18* and *LD 12: 12*), and maximum inhibition of diapause was observed when the light pulse fell between 14 and 16 hours after dawn (*Zeitgeber* time, Zt 14 to 16) in both instances (Fig. 7.3). Thus the period of maximum

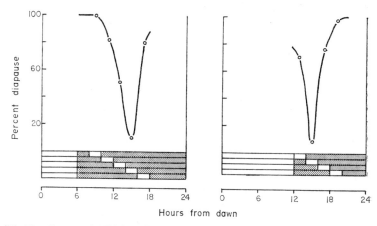

FIG. 7.3. The diapause-inhibiting effects of 2-hour night interruptions in *Pieris brassicae*, showing maximum long-day effect when the supplementary pulse falls 14 to 16 hours after dawn. (Redrawn, from Bünning, 1960.)

sensitivity, considered to constitute the light-sensitive "scotophil", was found to occur during the early part of the night rather than in the middle.

Working with the pink bollworm *Pectinophora gossypiella*, Adkisson (1964, 1966) extended this technique with the systematic interruption of a range of light: dark cycles (*LD 6: 18*, *LD 12: 12* and *LD 13: 11*) with 1-hour pulses. This procedure produced *two* discrete points of diapause inhibiton, the first (peak A) about 14 hours after lights-on (Zt 14) and the second (peak B) about 14 hours before lights-off, in all cycles tested.

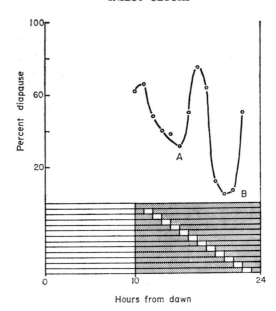

FIG. 7.4. The diapause-inhibiting effects of 1-hour night interruptions in *Pectinophora gossypiella*, showing two maxima of long-day effect. (Redrawn, from Adkisson, 1964.)

(Fig 7.4, 7.5). Light pulses placed in the middle of the night had little effect.* Diapause inhibition was also obtained by Beck (1962) in *Ostrinia nubilalis* using appropriately timed supplementary pulses of 30 minutes. In *Pieris rapae*, similar effects were noted with light pulses as short as 5 minutes (Barker, 1963), or even as short as that provided by an electronic photoflash with a discharge half-life of 0.0008 seconds, provided that the supplementary pulse or flash was correctly timed and of sufficient intensity (Barker *et al.*, 1964). These studies showed that "light reactions" in insects were just as rapid as those previously described in plants. Since these early papers, diapause inhibition in night interruption experiments has been obtained for a number of insect species, including *Megoura viciae* (Lees, 1965, 1966), *Nasonia vitripennis* (Saunders, 1967, 1968, 1969), *Heliothis zea* and *H. virescens* (Benschoter, 1968), *Adoxophyes reticulana* (Ankersmit, 1968), *Drosophila phalerata* (Tyshchenko *et al.*, 1972), *Pieris rapae crucivora* (Kono, 1970), and *Sarcophaga argyrostoma* (Saunders, 1975b). Most of these studies revealed two points of apparent light sensitivity, but considerable specific differences were also apparent. Since these differences affect the interpretation of the phenomenon they will be described in detail later in this section.

The demonstration of two periods of sensitivity to light in the dark component might have been difficult to interpret in terms of Bünning's hypothesis which envisaged a single light-sensitive "scotophil". However, Pittendrigh and Minis (1964) noted the striking parallel between this technique (specifically Adkisson's results with *P. gossypiella*) and the use of asymmetrical skeleton photoperiods to entrain the pupal eclosion rhythm in *Drosophila pseudoobscura* (Chapter 3). They suggested that the results were powerful, if circumstantial, evidence for the involvement of the circadian system in photoperiodic

_____

* This probably explains earlier failures by Lees (1953b) to obtain reversals by centrally placed light breaks unless the light breaks became long enough to encroach upon the two sensitive times in the nigh.

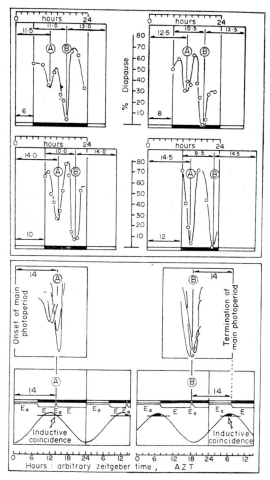

FIG. 7.5. The action of night interruptions in the photoperiodic inhibition of diapause in *Pectinophora gossypiella*, and the "external coincidence" model of Pittendrigh and Minis (1964). *Top panel*: inhibition of diapause by 1-hour night interruptions placed systematically in the nights of LD *6* : 18, *8* : 16, *10* : 14 and *12* : 12 (data from Adkisson, 1964). Night interruptions are most effective at two points, A and B. At both points the pulse creates, in conjunction with the main photoperiod, an asymmetric skeleton of about 12 to 15 hours. At A the pulse functions as dusk; at B it functions as dawn. *Lower panel*: interpretation of the data in terms of the "external coincidence" model. For details see the text. (From Pittendrigh and Minis, 1964.)

time measurement, and proposed an explicit version of Bünning's general hypothesis to account for the data.

In its original form (Fig. 7.5) the external "coincidence model" (Pittendrigh and Minis, 1964) comprised a circadian oscillation of a hypothetical substrate (S) and a phytochrome-like pigment or enzyme (E). In the model the substrate oscillation was entrained by the whole of the photoperiod, or by the asymmetrical skeleton formed by the main photoperiod and the pulse, so that its highest concentration (S-max) fell in the night. The concentration of S above a certain threshold constituted the light-sensitive phase comparable to Bünning's "scotophil". The enzyme (E) was conceived to be in an active state (E_a) when the light was on, but in its inactive state (E_i) in the dark.

In night-interruption experiments the oscillator achieved steady-state entrainment to the asymmetric skeleton photoperiod formed from the main light period and the pulse. When the pulse fell early in the night it was "read" as a terminator, or new dusk; when it fell late in the night it was "read" as an initiator, or new dawn. Between these two points it could be accepted as neither and a phase-jump (ψ-jump) occurred. Very similar behaviour can be seen in the entrainment of the pupal eclosion rhythms of D. pseudoobscura and P. gossypiella to similar skeletons (see Chapter 3). In terms of photoperiodic induction the model predicts diapause-avoidance, or long-day effects, at two points. One is when the pulse acts as a terminator and coincides with S-max early in the night; the other is after the ψ-jump when the pulse, now acting as an initiator, phase-sets the oscillator so that S-max coincides with the end of the main photoperiod.

Since there was no concrete evidence for a substrate oscillation as such, and still less for a phytochrome-like pigment in animals, the final form of the coincidence model (Pittendrigh, 1966) was reduced to a light-sensitive oscillator in which a photoperiodically inducible phase (ϕ_i) assumed a definite phase-relationship to the light cycle when in steady state. The important details of the model, such as the mechanism of entrainment and the ψ-jump, however, were retained.

Although two points of "light-sensitivity" were produced in response to light breaks, Pittendrigh (1966) pointed out that only one could represent the position of ϕ_i: after the ψ-jump ϕ_i was probably a full half-cycle away from its first position. However, since data for the eclosion rhythm in D. pseudoobscura (see Chapter 3) demonstrated quite clearly that the driving oscillation was "damped out" by light periods in excess of about 12 hours in such a way that the onset of darkness always corresponded to Ct 12, he suggested that any extension of the main light component above 12 hours would effectively cause "dawn" to move backwards in the preceding night. On the basis of this argument, Pittendrigh (1966) suggested that ϕ_i lay late in the night (at B) rather than at A. The reason why diapause-inhibition in P. gossypiella was greater at B than at A, however, was not explained, although the photoreceptors might be fully dark-adapted after a prolonged period of darkness.

The external coincidence model draws attention to three important points: (1) the term "scotophil" is restricted to a much shorter light-sensitive phase than an entire 12-hour half-cycle, and is thus in accord with the known facts from night-interruption experiments; (2) light has a *dual* role, serving to entrain the oscillation *and* to effect induction by the temporal coincidence of light and ϕ_i; and (3) it draws a close parallel between the photoperiodic clock and the known properties of circadian rhythms.

Before going on to examine tests for the external coincidence model it is necessary to examine results for light-break experiments in the various species to determine differences or common features. When this is done it is found that large and sometimes important differences occur, some of them affecting the general applicability of the model to insects.

In *Nasonia vitripennis* (Saunders, 1968), *Megoura viciae* (Lees, 1965, 1966), *Drosophila phalerata* (Tyshchenko *et al.*, 1972) and *Sarcophaga argyrostoma* (Saunders, 1975b) night interruption experiments produced two "peaks" of long-day effect with peak B greater than peak A (Fig. 7.6). These species, therefore, are similar in this respect to P. gossypiella. However, the similarities are often superficial. In D. phalerata, for example, peak A occurred about 11 to 12 hours after dawn (Zt 11–12) whilst peak B occurred at Zt 19–20, or about 13 to 14 hours before dusk. There is thus a difference between the

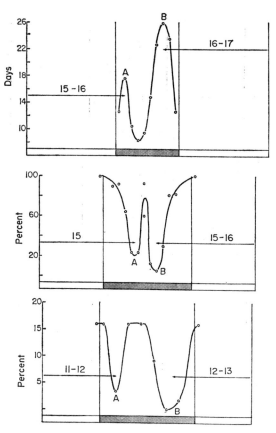

FIG. 7.6. The diapause-inhibiting effects of night interruptions in three species in which peak B is greater than peak A. The arrows and the figures show the times, in hours, of the peaks from dawn and dusk respectively. *Top panel:* larval diapause in *Nasonia vitripennis* at LD *14*: 10 with 1-hour pulses. (From Saunders, 1968.) *Middle panel:* pupal diapause in *Sarcophaga argyrostoma* at LD *10* : 14 with 1-hour pulses. (Saunders, 1975b.) *Lower panel:* ovarian diapause in *Drosophila phalerata* at LD *9* : 15 with 1-hour pulses. (From Tyshchenko *et al.*, 1972.) The ordinate for *N. vitripennis* plots the number of days to the switch to the production of diapause larvae; long-day effects are therefore at the top. The two other panels plot percent diapause on the ordinate.

two effective "skeletons". In *S. argyrostoma* the results agreed with the model quite closely, but in *LD 14*: 10, when the two peaks ought to have shown their maximum separation diapause was eliminated wherever the pulse was applied. Lastly, in *Megoura viciae*, experiments in which the duration of the main light component was varied (Lees, 1966), demonstrated that time measurement began with dusk rather than with dawn; the clock, in fact, was an hour-glass rather than an oscillator.

In *Pieris brassicae* (Goryshin and Tyshchenko, 1968; Bünning, 1969), *P. rapae* (Barker, 1963; Barker *et al.*, 1964; Kono, 1970), and in the codling moth *Carpocapsa pomonella* (Peterson and Hamner, 1968), the response at peak A was much greater than at B, and in some cases peak B was negligible or indistinguishable (Fig. 7.7). Only one peak (A), for example, is readily discernible in *P. rapae* and *C. pomonella*. Lastly, working with *Ostrinia nubilalis*, Beck (1962) examined the effect of a 1-hour pulse applied systematically throughout the night of a *LD 7* : 17 cycle (Fig. 7.8). An uninterrupted cycle of this length gave less than 5 per cent diapause. Pulses placed early in the night induced dia-

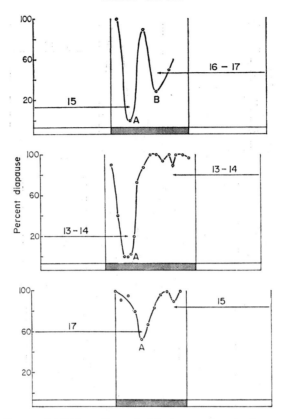

FIG. 7.7. The diapause-inhibiting effects of night interruptions in three species in which peak B is smaller than peak A, or is almost non-existent. The arrows and the figures show the times, in hours, of the peaks from dawn and dusk respectively. *Top panel:* pupal diapause in *Pieris brassicae* at *LD 13* : 12 with 2-hour pulses. (From Goryshin and Tyshchenko, 1968.) *Middle panel:* pupal diapause in *Pieris rapae* at *LD 10* : 14 with interrupting photoflashes with a discharge half-life of 0.0008 second. (From Barker *et al.*, 1964.) *Lower panel:* larval diapause in *Carpocapsa pomonella* at *LD 13* : 11 with 1-hour pulses. (From Peterson and Hamner, 1968.)

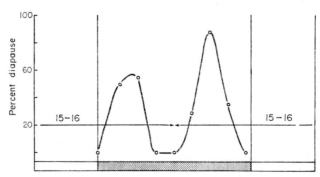

FIG. 7.8. The diapause-inhibiting effect of 1-hour night interruptions in *Ostrinia nubilalis* in LD 7 : 17, showing the coalescence of peaks A and B in a central point of long-day responsiveness. (Redrawn, after Beck, 1968.)

pause (51 to 54 per cent), as did those placed towards the end (28 to 88 per cent). Pulses placed in the middle, however, did not. At first sight these results are not in accord with those for *P. gossypiella*. However, the early and late interruptions form diapause-promoting skeletons of about 10 to 14 hours and the central zone of inhibition probably represents the coalescence of the two peaks expected. The pulse applied at Zt 15 to 16, for example, is 15 to 16 hours after dawn *and* 15 to 16 hours before the dawn of the next light period. Further differences between these various species will be noted in night interruption experiments in which the length of the light cycle (T) is extended to 36, 48 and 72 hours. Some of these differences appear irreconcilable with the external coincidence model—or, indeed, with any oscillator model for the clock.

(b) *Symmetrical skeletons*. In the entrainment of the *Drosophila pseudoobscura* pupal eclosion rhythm, a symmetrical skeleton photoperiod (PP_s), consisting of two brief pulses of light n hours apart, is known to simulate many of the effects of a *complete* photoperiod (PP_c) of n hours duration (Pittendrigh and Minis, 1964) (Chapter 3). Similarly, symmetrical skeletons are known to mimic many *photoperiodic* effects, almost as effectively as the corresponding complete light pulse. These studies demonstrate the importance of the "on" and "off" signals in the light-cycle, and in some instances a more fundamental similarity: that is, a common mechanism between photoperiodism and circadian rhythmicity.

Danilevskii *et al.* (1970), for example, maintained larval cultures of *Acronycta rumicis* in complete photoperiods (PP_c) of *8*: 16, *12*: 12 and *18*: 6 for 7 days and then released them into the corresponding skeleton photoperiods (PP_s) consisting of two 1-hour pulses of light. Control groups were either maintained throughout in the complete photoperiods or released in DD. All of the larvae exposed to PP_c *or* PP_s *8*: 16 and *12*: 12 entered diapause as pupae, but those released into DD never produced more than 50 per cent diapause. This shows that the skeletons of the shorter photoperiods were just as effective as normal short daylengths and suggests, by analogy with the *D. pseudoobscura* data, that the clock in *A. rumicis* consists of an endogenous oscillator. Moreover, whilst a PP_c *18*: 6 produced only 8 per cent diapause, over 80 per cent of the larvae ceased development when released into the corresponding skeleton (PP_s *18*: 6), suggesting that the photoperiodic rhythm became re-entrained to the shorter "interpretation" of PP_s *6*: 18 (or strictly, if the duration of the pulses is also taken into account, PP_s *8*: 16).

Bünning and Joerrens (1960) subjected the larvae of *Pieris brassicae* to skeleton photoperiods also consisting of two 1-hour pulses of light. Pulses 6 hours apart (PP_s *6*: 18) and 8 hours apart (PP_s *8*: 16) induced a greater incidence of pupal diapause (about 42 to 45 per cent) than the DD control (about 39 per cent) (Fig. 7.9), whereas a skeleton of PP_s *12*: 12 gave a lower incidence of diapause, and PP_s *15*: 9 only about 18 per cent. Pulses placed 18 and 21 hours apart, however, were once again diapause-promoting. Following Pittendrigh and Minis (1964), these results may be interpreted as evidence for the operation of an oscillating system. Each skeleton photoperiod is clearly open to two "interpretations", depending on which pulse is regarded as simulating the initiator, or the dawn signal. As in *D. pseudoobscura* and *A. rumicis* the oscillator accepts the shorter of the two intervals. Thus, skeletons of PP_s *6*: 18 and PP_s *8*: 16 are "read" as the complete photoperiods PP_c *6*: 18 and PP_c *8*: 16, and promote diapause. Pulses 18 and 21 hours apart, however, are "read" as the shorter interpretation, namely PP_s *8*: 16

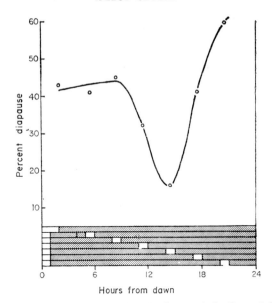

Fig. 7.9. The effect of "symmetrical skeleton" photoperiods (formed from two 1-hour pulses of light) on diapause induction in *Pieris brassicae*. When the two pulses form skeletons of 6 hours (PP_s 6) and 8 hours (PP_s 8) diapause induction is high, likewise when the pulses are 18 and 20 hours apart when the skeletons are "interpreted" as PP_s 6 and PP_e 4. For details see the text. (Redrawn, after Bünning, 1960.)

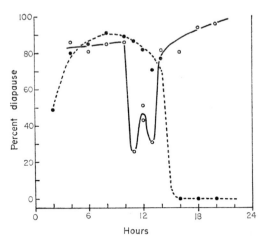

Fig. 7.10. *Sarcophaga argyrostoma:* diapause induction in complete (PP_s) and symmetrical skeleton (PP_e) photoperiods formed from two 1-hour pulses of light. ●– – –● complete photoperiods; o——o skeleton photoperiods. Note that PP_s 4 to PP_s 10 mimic PP_e 4 to PP_e 10 very closely. PP_s 14 to PP_s 20, however, induce diapause, whereas the complete photoperiods of the same length cause diapause aversion. A low level of diapause induction occurs with skeleton photoperiods between PP_s 11 and PP_s 13. For details see text. (Saunders, 1975b.)

and PP_s 5: 19, because the oscillator executes a ψ-jump before attaining its steady state; the results, therefore, are once again diapause-promoting. With PP_s 12: 12 and PP_s 15: 9, however, the skeleton light regime is clearly ambiguous, as it is with the eclosion rhythm in *D. pseudoobscura*, and diapause induction is less effective.

Very similar results have been obtained for *Nasonia vitripennis* and for *Sarcophaga argyrostoma*. Those for *S. argyrostoma* are shown in Fig. 7.10; those for *N. vitripennis* in Table 7.1.

TABLE 7.1. INDUCTION OF DIAPAUSE IN *Nasonia vitripennis* WITH SYMMETRICAL "SKELETON" PHOTOPERIODS FORMED FROM TWO 1-HOUR PULSES OF LIGHT
The type of progeny produced (diapause or developing) was determined on days 15 to 17 of adult life. Temperature = 18°C.

Treatment	Females (number)	Progeny produced on days 15–17			Per cent diapause
		Developing	Diapause	None*	
DD	40	29	2	9	6.4
PP_c 8:16	40	3	34	3	91.9
PP_s 8:16	43	3	38	2	92.7
PP_c 12:12	23	3	15	5	83.3
PP_s 12:12	40	7	26	7	78.8
PP_c 16:8	25	21	0	4	0.0
PP_s 16:8	41	8	33	0	80.5

* The number of experimental females that failed to lay eggs during the test period (days 15 to 17 of adult life); hence the empty host puparia. PP_c — complete photoperiods; PP_s — symmetrical skeleton photoperiods. Note how the PP_s of 8:16 and 12:12 are good or reasonable simulations of the corresponding complete photoperiods (PP_c), but PP_s 16:8 induces almost as much diapause as its shorter "interpretation" (PP_s 8:16).

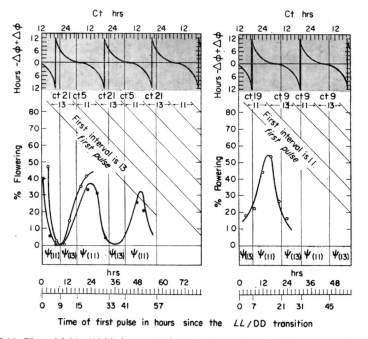

FIG. 7.11. Pittendrigh's (1966) interpretation of Hillman's (1964) data on the induction of flowering in *Lemna* by symmetrical skeletons of PP_s 11 and PP_s 13. Flowering maxima occur when the initial conditions (the initial relationship between oscillation and light-cycle) are such that ψ_{11} is predicted. Whether ψ_{11} or ψ_{13} develops is determined by (1) the duration of the first interval, and (2) the phase-point of the oscillation illuminated by the first pulse. The oscillation begins at Ct 12 (the beginning of the subjective night) when the system is transferred from *LL* to DD. (From Pittendrigh and Minis, 1971.) (See also Fig. 3.18.)

One further type of experiment utilizing skeleton photoperiods must be mentioned here, even though it has only been applied successfully to the short-day plant *Lemna perpusilla* (Hillman, 1964*). Hillman maintained the plants in *LL* prior to their release into DD; after release they were exposed to 15-minute pulses of light defining skeleton photoperiods of either 11 or 13 hours. The results showed that the plants could distinguish between PP_s *11*: 13 and PP_s *13*: 11 depending on which interval they "saw" first after the *LL*/DD transition. Whether the ensuing skeleton simulated an 11- *or* a 13-hour photoperiod, the proportion of the plants initiating flower formation was a periodic function ($T \sim 24$ hours) of the time at which the first pulse was seen. However, if the first *interval* was 11 hours the curve was almost a mirror image of that for an interval of 13 hours (Fig. 7.11). Not only does the periodic nature of the result support some form of Bünning's hypothesis, but it parallels the remarkably similar behaviour of the eclosion rhythm in *D. pseudoobscura* for skeletons of 11 and 13 hours which Pittendrigh (1966) has called the "bistability phenomenon" (Chapter 3).

This technique was extended to *P. gossypiella* using the egg-hatch rhythm as an indicator of phase and 15-minute light pulses to create the skeleton (Pittendrigh and Minis, 1971). The two alternative steady states for PP_s *11*: 13 and PP_s *13*: 11 were shown by the egg-hatch rhythm, but there was almost 100 per cent diapause in both regimes. In hindsight it was considered that 15-minute pulses, although sufficient to entrain the rhythm, were probably too short to bring about the photochemical changes associated with the inhibition of diapause. On the other hand, it may also be that *P. gossypiella* is unsuitable for such an experiment since a 13-hour photoperiod is so close to the critical daylength that intermediate, or unstable, amounts of diapause would be obtained. In this view the bistability phenomenon is only applicable to species, such as *Lemna*, which have a critical value very close to 12 hours.

Hillman (1973) has since applied a similar technique to *Megoura viciae;* his results confirm the conclusion that this species possesses a non-circadian timer, and will be dealt with in the relevant section.

2. *Night interruptions in cycles with an extended "night"*

Some of the strongest evidence for the hypothesis that photoperiodic time measurement is a function of the circadian system comes from experiments in which the "nights" of extended light/dark cycles (i.e. $T = 48$ or 72 hours) are systematically interrupted by light pulses. The results of such experiments frequently show periodic maxima of long- or short-day effect recurring with circadian frequency. In the long-day plant *Hyoscyamus niger*, for example, 2-hour light breaks placed systematically in the long night of an *LD 9*: 39 cycle ($T = 48$) revealed maxima of flowering at *Zt* 16 and *Zt* 40 (Claes and Lang, 1947), and in the Biloxi variety of soybean maintained in a light/dark cycle of *LD 8*: 64 ($T = 72$), flowering was induced when the light pulses fell at *Zt* 16, *Zt* 40 and *Zt* 64 (K. C. Hamner, 1960). Similar results were obtained for *Kalanchoë blossfeldiana* in 48-hour (bidiurnal) and 72-hour (triurnal) cycles (Bünsow, 1953; Melchers, 1956; Bünning, 1960), and for testicular growth in the house finch *Carpodacus mexicanus* (W. M. Hamner, 1964).

Similar experiments have now been performed with a number of insect species and produced a variety of responses. In the aphid *Megoura viciae*, for instance, the results

---

* A similar result has now been obtained with *S. argyrostoma* (Saunders, 1975b).

constituted strong evidence for a dark-period interval timer or hour-glass (Lees, 1966), and in *Pieris brassicae* (Bünning, 1969) and *Carpocapsa pomonella* (Peterson and Hamner, 1968; W. M. Hamner, 1969) the results were either inconclusive or indicative of a considerable degree of complexity in the response. Only in the parasitic wasp *Nasonia vitripennis* has a clear circadian component emerged (Saunders, 1970).

Females of *N. vitripennis* were maintained in 48-hour and 72-hour cycles containing a short "main" photoperiod of between 10 and 14 hours and a supplementary pulse of 2 hours. The wasps were supplied with two fresh puparia of *Sarcophaga argyrostoma*

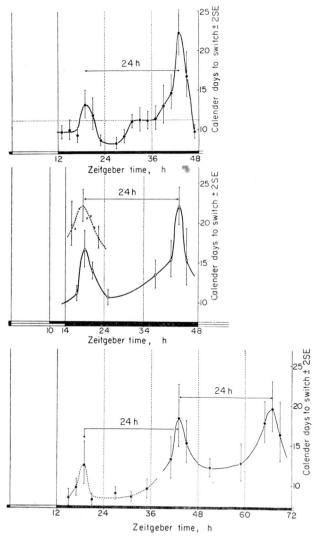

FIG. 7.12. The effect of light breaks (night interruptions) in 48- and 72-hour cycles on diapause production by females of *Nasonia vitripennis*. *Top panel:* 2-hour light breaks in an LD *12* : 36 cycle. *Middle panel:* 1-hour light breaks in LD *14* : 34 (dashed line) and 2-hour light breaks in LD *10* : 38 (solid line). *Lower panel:* 2-hour light breaks in a cycle of LD *12* : 60. Peaks of long-day effect (diapause inhibition) are produced at 24-hour intervals in the extended night, with the first always 19 hours after dawn (*Zeitgeber* time, *Zt* 19). Each point represents the mean age at the "switch" from developing to diapausing offspring for about forty females. (From Saunders, 1970.) (Copyright 1970 by the American Association for the Advancement of Science.)

each day and the type of progeny produced (diapause or non-diapause) scored for each day of the experiment. In Fig. 7.12 short-day effects (a high incidence of diapause) are indicated by a rapid switch to the production of diapausing progeny, whereas long-day effects are indicated by a delayed switch. In a cycle of LD 12:36 ($T = 48$), long-day effects, or points of apparent sensitivity to light, were observed when the pulses fell at Zt 19 and Zt 43—24 hours apart. Similar experiments with LD 10:38 and LD 14:34 produced peaks in the same positions indicating that time measurement begins with dawn rather than dusk. In a 72-hour cycle (LD 12:60) three such peaks were produced, one at Zt 19, one at Zt 43 and a third at Zt 67—once again, 24 hours apart. This result indicates that points of sensitivity to light recur with circadian frequency in the extended night. The central point is particularly informative because it is unlikely to arise from a direct interaction between the pulse and the "main" photoperiod.

Although these results clearly implicate the circadian system in the insect photoperiodic clock, they are difficult to explain in terms of a simple external coincidence model. What, for example, has happened to the two peaks of diapause inhibition observed in night interruption experiments with a 24-hour cycle? It seems likely that the observed peaks correspond to peak A + modulo 24 hours, despite the fact that peak A occurs at Zt 15–16 in 24-hour cycles, but at Zt 19 + $n.24$ hours in cycles where $T = 48$ or 72 hours. This raises the question of the "missing" peak B which would be expected to occur at multiples of 24 hours before dusk. It is also difficult to determine whether the light pulses are coinciding with ϕ_i, or phase-setting the rhythm so that ϕ_i coincides with one of the main light components.

In other insect species the results are less clear-cut. Bünning (1969), for example, examined the effects of 30 minute pulses in the "nights" of 24-hour (LD 12:12), 36-hour (LD 12:24) and 48-hour (LD 12:36) cycles on larvae of the cabbage white butterfly Pieris brassicae (Fig. 7.13). In LD 12:12 two peaks of diapause inhibition were observed, one (A) at Zt 15 and the other (B) at Zt 21 (15 hours before lights-off). The response at peak A was greater than at peak B. In LD 12:24 ($T = 36$) two peaks were again evident, this time at Zt 17 and Zt 33. Two peaks were also observed in the longest cycle (LD 12:36), at Zt 17 and Zt 45. Thus in each cycle the peak in the early part of the night (A) occurred 15 or 17 hours after dawn, whereas the peak late in the night (B) occurred 15 hours before dusk. In the longest cycle (LD 12:36) a peak 24 hours after the first—at Zt 41— was not observed.

These results, therefore, provided no evidence that the circadian system was involved. Bünning considered that only a single cycle of the oscillator was evident, and that the result in no way militated against a circadian clock. On the other hand, the results might easily be interpreted as evidence for an hour-glass which does not repeat itself with circadian frequency in the extended dark period.

In the codling moth Carpocapsa pomonella night interruption experiments in extended cycles produced complex and variable results. In a 48-hour cycle (LD 8:40) peaks of diapause inhibition were observed at Zt 16 and Zt 40 (Peterson and Hamner, 1968). The first was 16 hours after dawn and the second 16 hours before dusk of the next light period; the peaks were thus 24 hours apart. When a similar experiment was conducted using an LD 8:64 cycle ($T = 72$) (Hamner, 1969), maximum inhibition of diapause was observed again at Zt 16, but this peak did not recur 24 hours later at Zt 40. Hamner concluded that at least two types of timing device were involved in this species, one a dusk hour-glass associated with the response at Zt 16 (8 hours after dusk) and the other

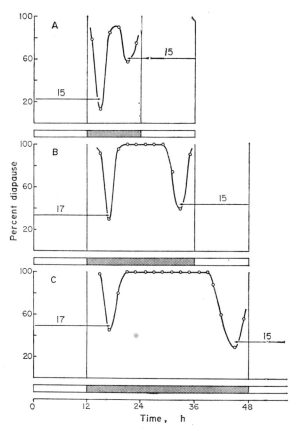

FIG. 7.13. The diapause-inhibiting effects of 30-minute night interruptions in *Pieris bras-sicae*. A — in *LD 12 : 12 (T = 24)*; B — in *LD 12 : 24 (T = 36)*; C — in *LD 12 : 36 (T = 48)*. Note that two peaks of long-day response are observed in each regime, the first 15 to 17 hours after dawn (*Zt* 15–17), the second about 15 hours before dusk. There is no evidence of a peak 24 hours after the first which would be expected (in C) at *Zt* 41 if the circadian system were involved in photoperiodism in this species. (Redrawn, after Bünning, 1969.)

a rhythmic (circadian) response associated with dawn. However, it is clearly not justifiable to regard the first peak as being timed from dusk rather than dawn since the data are restricted to light cycles containing an 8-hour photoperiod, and such a relationship can only be revealed by varying the light period within cycles of the same length. Nevertheless, it is apparent that photoperiodic time measurement in *C. pomonella* is complex and may involve more than one kind of device; the exact mechanism, however, is still open to question.

3. Tests of the external coincidence model

When formulating the external coincidence model Pittendrigh and Minis (1964) suggested that light had a dual role: (1) the entrainment of the rhythm, and (2) bringing about photoperiodic induction by the coincidence of light and an inducible phase (ϕ_i). Consequently they stressed the importance of studying the entrainment of the hypothetical oscillator. However, this cannot be done directly, and it is not known whether ϕ_i is a phase-point of a driving oscillation, of which there are undoubtedly several, or whether it is part of one of the even more numerous driven systems. The method adopted,

therefore, was the parallel study of photoperiodic induction and the entrainment of an
overt rhythm in the same insect species. Most of the tests for the coincidence model have
used this approach, and have been performed with *Pectinophora gossypiella*. Its validity
will be discussed at the end of this section.

Three overt rhythms have been studied in *P. gossypiella;* these are oviposition, egg
hatch and pupal eclosion (Pittendrigh and Minis, 1971). The supposed phase relationship
of ϕ_i to these three rhythms is shown in Fig. 7.14; it is thought to lie late in the subjective
night, about 5 hours before ϕ_r for eclosion, a position which corresponds to "peak B"

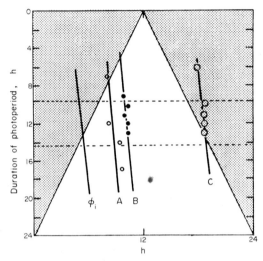

FIG. 7.14. Phase-reference points ($\psi_{R, L}$) for egg hatch (A), pupal eclosion (B) and oviposition (C) rhythms of *Pectinophora gossypiella* as a function of photoperiod. The inducible
phase (ϕ_i) is postulated to lie about 5 hours earlier than $\psi_{R, L}$ for the pupal eclosion
rhythm. Dashed lines mark the range of natural photoperiods at El Paso, Texas, the
source of the strain used. (From Pittendrigh and Minis, 1971.)

in night-interruption experiments. This assumption has enabled a number of testable
predictions to be made about photoperiodic induction in this species. Some of the tests
have produced results which support the external coincidence model, others have not.

(a) *Chilling, and the use of concurrent temperature cycles.* Pittendrigh and Minis (1971)
examined the effects of a light cycle (*LD 8*: 16) and a concurrent sinusoidal temperature
cycles (20° to 29°C) on *P. gossypiella*, the results being assayed simultaneously by diapause induction and by the pupal eclosion rhythm. In a cycle of *LD 8*:16 at constant temperature the peak of the eclosion rhythm (ϕ_r) occurred soon after dawn (Fig. 7.15). With
the addition of the temperature cycle, however, ϕ_r occurred a few hours after the point
of lowest temperature. Thus, as the low point of the temperature cycle was displaced to
the right (later relative to the light cycle) the phase of the eclosion rhythm was shifted
in a similar manner. Since ϕ_i was assumed to occur about 5 hours ahead of ϕ_r for eclosion it, too, would be displaced to the right. Figure 7.15 shows that when ϕ_i was drawn
into the light diapause inhibition occurred. The results therefore were in accord with
prediction from the external coincidence model.

Using pulses of low temperature it has been shown that the diapause response of
Nasonia vitripennis can be reversed (Saunders, 1967) (see Chapter 6). The concurrent use

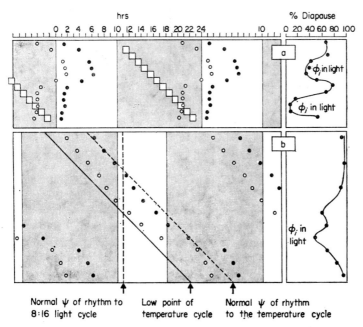

FIG. 7.15. Entrainment and induction by asymmetric skeleton photoperiods in *Pectinophora gossypiella*. *Top panel:* the phase-reference point of the pupal eclosion rhythm is shown by solid circles; the inducible phase (ϕ_i), 5 hours earlier, by open circles. Panel at right plots the inductive effect (prevention of diapause) of the same skeleton regimes. *Lower panel:* entrainment and induction by concurrent cycles of light (*LD 8* : 16) and temperature (20–29°C, sinusoidal). The position of the rhythm's phase-reference point in *LD 8* : 16 at constant 20°C is indicated by the dashed line 1 hour after dawn. Its position in a sinusoidal temperature regime in constant dark is also indicated by a dashed line 7 hours after the lowest point on the temperature curve. Solid circles mark the rhythm's phase-reference point and open circles mark ϕ_i for the twelve environments characterized by twelve different phase relations between the light and the temperature. Panel at right plots the inductive effect (prevention of diapause) of these same environments. (From Pittendrigh and Minis, 1971.)

of night interruptions and periods of chilling has indicated how these reversals may occur. For example, when the wasps were chilled at 2°C for 4 hours daily at the beginning of the 14-hour light period of a short-day cycle (*LD 14*: 10), the first peak of diapause inhibition (A) was observed to "move" further into the night from Zt 15–16 to Zt 17–18 (Saunders, 1967, 1968) (Fig. 7.16). If such a movement had occurred in an *LD 16*: 8 (long-day) cycle, the first peak of sensitivity would have moved from the light into the dark, thereby converting the response from diapause inhibition to diapause promotion. It was assumed that the photoperiodic oscillator, of which ϕ_i was a part, was "stopped" during the period at low temperature so that the 16-hour photoperiod was effectively shortened to 12 hours. A comparable result was obtained by using a cycle of *LD 11.3*: 10 (T = 21.3) in which the first peak also occurred later in the night (Fig. 7.16).

(b) *Entrainment and induction when T is close to* τ. Pittendrigh and Minis (1964, 1971) reported another test of the external coincidence model which again involved the simultaneous assay of circadian phase and diapause induction in *P. gossypiella*. Earlier studies on the entrainment of the *D. pseudoobscura* system to single light pulses per cycle

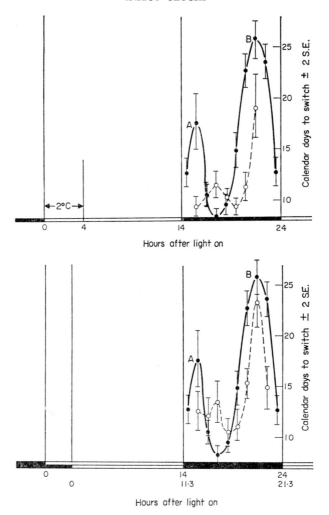

FIG. 7.16. *Nasonia vitripennis. Upper panel:* the effect of chilling (2°C) in the light component of *LD 14* : 10 on the positions and amplitudes of the peaks of diapause inhibition as revealed by night-interruption experiments. ●——● *LD 14* : 10 unchilled; o– – –o *LD 14* : 10 chilled. Vertical lines represent the 95 per cent fiducial limits of the geometric means of the number of days to the production of diapause larvae; long-day effects are the top of the ordinate. Note that when the wasps were chilled at 2°C for 4 hours daily at the beginning of the light component (*Zt* 0–4) the first peak of diapause inhibition (A) moves about 2 hours further into the night (to *Zt* 17–18) and is reduced in amplitude. *Lower panel:* the effect of shortening the light component from 14 to 10.3 hours on the position of peak A as revealed by night-interruption experiments, showing the relation of this peak to dawn. (From Saunders, 1969.)

(Chapter 3) had shown that, within certain limits, the driving oscillation assumed the period (*T*) of the entraining light cycle, the discrepancy between τ and *T* being overcome by a discrete phase-shift. Furthermore, when *T* < τ the pulse fell in the late subjective night and caused a phase advance, whereas when *T* > τ the pulse fell in the early subjective night to cause a phase delay. Therefore simply by changing the period (*T*) of the light-cycle the pulse came to illuminate different phase-points of the oscillator.

An early application of this test to *P. gossypiella* involved the oviposition rhythm as an assay of phase and the use of single recurrent 15-minute light pulses to define cycles ranging from $T = 20$ h 40 min to $T = 25$ h (Pittendrigh and Minis, 1964; Minis, 1965). These values of T encompassed values both greater than and less than τ for this species (22 h 40 min). The prediction was that since the light pulse would illuminate different phase points of the oscillator in different regimes it should coincide with ϕ_i in some. However, little difference in the incidence of diapause was observed, and 15-minute pulses were considered, in hindsight, to be inadequate to effect induction.

The experiment was later repeated using light cycles varying from $T = 20$ to $T = 27$, each containing an 8-hour photoperiod, and the pupal eclosion rhythm as a measure of phase (Pittendrigh and Minis, 1971). Once again, in the shorter cycles ($T = 20$ and $T = 21$) the light pulses, in steady state, should fall in the late subjective night, whereas in the longer cycles ($T = 24, T = 26, T = 27$) they should fall in the *early* subjective night. Therefore, if ϕ_i occurs about 5 hours ahead of the peak of eclosion it should be illuminated when $T < 24$ hours but not when $T > 24$ hours, thus causing diapause inhibition in the shorter cycles and diapause promotion in the longer cycles. The results of this experiment (Fig. 7.17) showed that although the phase of the eclosion rhythm assumed a steady

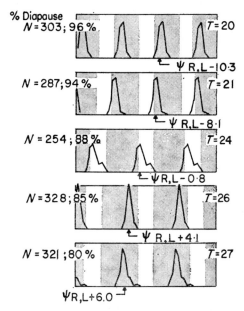

FIG. 7.17. The dependence of induction on T when T is close to τ (non-resonance effects) in *Pectinophora gossypiella*. The entrained steady states of the pupal eclosion rhythm effected by light cycles, all involving an 8-hour photoperiod, whose period (T) ranges from 20 to 27 hours. The photoperiodic induction (measured as percent diapause) effected by each cycle is indicated at the left. (From Pittendrigh and Minis, 1971.)

state as predicted, the proportion of larvae entering diapause was 80 per cent or over in all light-cycles. Indeed, although the overall range was small (80 to 96 per cent) the *sign* of the dependence of diapause incidence on T was the reverse of that expected. Clearly the results provided no clear support for the external coincidence model in *P. gossypiella*. It should be noted, however, that a similar experimental design with *Sarcophaga argyrostoma* (Fig. 7.38), although without concurrent analysis of an overt rhythm as an independent measure of phase, showed practically no diapause at $T = 19.2, T = 21$ and $T = 23$, but practically 100 per cent diapause at $T = 24, T = 26$ and $T = 30$.

(c) *Selection for "early" and "late" eclosion strains in* Pectinophora gossypiella *and its effect on induction.* Artificial selection of those adults of *P. gossypiella* which emerged from their pupae earlier or later than the stock culture resulted in two strains ("early" and "late") which differed in their ψ_{RL} by about 5 hours (Chapter 3). These strains have been used in a test for the external coincidence model (Pittendrigh and Minis, 1971), once again assuming that ϕ_i occurs some 5 hours ahead of ϕ_r. Basing their predictions on the assumption that a correlated shift in ϕ_i had also occurred as a consequence of selection, and the knowledge that the *Pectinophora* system, like that in *D. pseudoobscura*, begins from a fixed phase-point at the onset of darkness, the "late" strain was expected to show less diapause (or more induction of development) at a 12-hour photoperiod than either "early" or "stock". In other words, the critical daylength for "late" ought to be shorter than that for the other two strains. Theoretically this test has an additional importance. Selection for "early" and "late" had demonstrated that the phase relationship of the rhythm to the oscillator (ψ_{RO}) had been altered, but not the phase relationship of the oscillator to the light (ψ_{OL}). Consequently, if a change in the photoperiod response was obtained it might indicate that ϕ_i was part of a *driven* system; if it was not, it might indicate that ϕ_i was part of a driving oscillator.

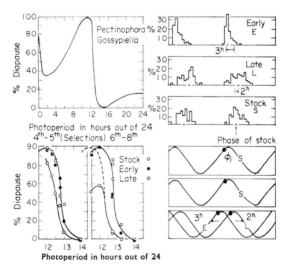

FIG. 7.18. Photoperiodic induction and $\psi_{R, L}$ in *Pectinophora gossypiella*. *Upper left:* the photoperiodic response curve. *Upper right:* the effect of eight generations of selection on $\psi_{R, L}$. *Lower right:* top and middle panels illustrate the coincidence of ϕ_i and light in *LD 14* : 10 and its non-coincidence in *LD 12* : 12; lower panel illustrates the coincidence of ϕ_i and light in *LD 12* : 12 in the "late" strain on the assumption that $\psi_{R, L}$ of the clock rhythm has also been changed by selection. *Lower left:* photoperiodic response curves for early and late as selection progressed. The dashed line portion of the curve for late (right panel) indicates the normal form of the curve — how it would appear had it merely been displaced. The observed value for *LD 11* : 13 indicates, however, that the curve is depressed and not displaced. (From Pittendrigh and Minis, 1971.)

The results of this experiment (Fig. 7.18) showed that the critical daylength was indeed shorter in "late" than in "stock" or "early" after only four to five generations of selection. However, after six to eight generations the response for "late" also showed that the photoperiodic responses at "strong" short daylengths (*LD 11* : 13 and *LD 12* : 12) were lower in *amplitude*. This may mean that the apparent shift in the critical daylength was due to a lowered response at *all* photoperiods (see Chapter 6). Therefore the results of this experiment remain equivocal and provide no clear evidence either for or against the external coincidence model. As Pittendrigh and Minis (1971) point out, however, it is of considerable interest that genetic selection for the parameters of one system (the eclosion rhythm) should have such a profound effect upon another (photoperiodic induction).

(d) *Current status of the external coincidence model.* The several experiments designed by Pittendrigh and Minis (1964, 1971) as tests for external coincidence in *P. gossypiella* were all based on the assumption that ϕ_i was a part of a driving oscillator, or a driven rhythm, closely related to that involved in the rhythms of oviposition and pupal eclosion, in such a way that ϕ_i had a fixed temporal relationship to the overt systems. In hindsight, this assumption does not appear to be valid and the clear initial association has become increasingly unclear. The three overt rhythms studied in *P. gossypiella* (egg hatch, pupal eclosion and oviposition) show somewhat different properties which appear to be fundamental. For example, the two "adult" rhythms (eclosion and oviposition) show an endogenous periodicity (τ) of about 22.5 hours, whereas τ for the egg-hatch rhythm is closer to 24 hours (Minis and Pittendrigh, 1968). In addition, although red light fails to entrain either of the adult rhythms it perceptibly shortens τ for one of them (eclosion). Lastly, and most significantly, although none of the rhythms can be initiated or entrained by red light (>600 nm), the moth can distinguish, in a photoperiodic sense, between 12 and 14 hours of such red light (Pittendrigh *et al.*, 1970) (Chapter 6), even though the constituent oscillators of the circadian system must be in an unsynchronized condition. Therefore, unless the photoperiodic clock consists of a separate driving oscillation coupled to the light-cycle by quite a different pigment (absorbing in the red), time measurement may not be a function of the circadian system in this particular species. Further evidence against the role of the circadian system in *P. gossypiella* is seen in a later section in which "resonance" experiments are discussed.

The initial attraction of the external coincidence model was two-fold. Firstly, it was a simple model comprising a *single* oscillator phase-set to the environmental light cycle. Secondly, it served to point out the similarities between the photoperiodic system and circadian oscillations, particularly in night interruption experiments. Although as an explicit formulation of Bünning's general hypothesis it failed to survive close scrutiny in *P. gossypiella*, it adequately acounts for the data obtained for other species—such as *Sarcophaga argyrostoma*—which will be presented in the next section.

4. *Abnormal light/dark-cycles*

The second main approach to the problem of time measurement has been the study of photoperiodic induction in cycles where $T \neq \tau$ and the hours of light and dark are varied independently of each other. Most of the earlier experiments of this type were confined to cycles in which T was close to τ, and were designed to determine whether the light component was "more important" than the dark, or vice versa. Later experiments—variously called T-experiments, or resonance experiments—consisted of much longer cycles (up to 72 hours or more) containing a short light component and a protracted night. This design, particularly, has produced evidence for the internal coincidence model.

A model of this type was first suggested by Pittendrigh (1960) who proposed, in very general terms, that photoperiodic induction might involve two or more oscillators. He observed (1) the striking dependence on phase of the *Drosophila pseudoobscura* adult eclosion rhythm, and (2) that a change in photoperiod is a change in the phase-angle between dawn and dusk. He also noted that "there is some dynamic equilibrium attained by the multioscillator system responding to opposed perturbations—advance, delay—at the dawn and dusk transitions".

One of the more explicit internal coincidence models is that proposed by Tyshchenko in 1966 (Danilevskii *et al.*, 1970). Tyshchenko proposed that the photoperiodic clock comprised two independent oscillators, each of circadian periodicity, one phase-set by dawn and the other by dusk. As the photoperiod changed the phase-angle between the two oscillations also changed, and induction or non-induction occurred when "active" phase-points of the two oscillators either coincided or failed to coincide (Fig. 7.19). Thus, at short days coincidence did not occur and diapause supervened,

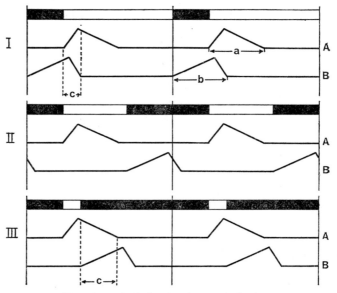

Fig. 7.19. The two-oscillator hypothesis for the photoperiodic time measurement ("internal coincidence" (of Tyshchenko. I — at long-daylength temporal coincidence occurs between the active phases of the two oscillators. II — at short-daylengths temporal coincidence does not occur. III — at ultra-short daylengths the active phases overlap once again. A — oscillator phase-set by dawn; B — oscillator phase-set by dusk; C — areas of temporal coincidence. (Reproduced, with permission, from "Biological rhythms in terrestrial arthropods", *Ann. Rev. Entomol.* **15**, 232. Copyright 1970 by Annual Reviews Inc. All rights reserved.)

whereas at long-days—and at ultra-short daylengths—temporal coincidence occurred and diapause was averted. This model differs from that of external coincidence (Pittendrigh and Minis, 1964) in two important respects. Firstly, light only serves to entrain and phase-set the constituent oscillators, and not to effect any specific photochemical process directly associated with induction. Secondly, it can "explain" the *entire* shape of the photoperiodic response curve, including the fall in diapause incidence at very short daylengths, much more adequately than the latter. With the addition of further assumptions—changes in the duration of the "active" phases, and that one of the oscillators is temperature-sensitive— it can also be manipulated to "explain" many other aspects of the photoperiodic response, such as the long- and short-day forms, and the effect of temperature on the critical daylength. Beck (1964, 1968) independently proposed a similar model, again involving dawn and dusk oscillators, but with an 8-hour periodicity.

Support of a general nature for models of this type comes from observations that the circadian system in insects and mammals comprises a population of driving oscillations, some of which are coupled to the light cycle by the dawn signal and others by dusk (Chapter 4). More specific evidence for internal coincidence comes from the analysis of the photoperiodic clock in *Nasonia vitripennis* (Section 4b).

(a) *Relative importance of the light and dark periods.* Experiments in which the light and the dark components of the cycle are independently varied have frequently shown that the length of the "night" is more critical than the length of the "day". Consequently many earlier investigators considered that nightlength was of central importance in time measurement. In some species, particularly the green vetch aphid *Megoura viciae*, nightlength measurement is performed with little reference to the length of the accompanying light period, and the "clock" has been interpreted as a dark-period interval timer or hour-glass (Less, 1966, 1968) (Chapter 7, Section B). In most species both light and dark periods have been shown to require a definite duration if induction (of diapause) is to occur.

Dickson (1949) investigated the induction of larval diapause in the Oriental fruit moth *Grapholitha molesta* in a wide range of abnormal cycles. In cycles where T varied from 18 to 30 hours, diapause was only induced when the dark component (D) was in excess of 10 to 11 hours, but less than 16 hours. Holding D constant at 11 hours he then found that the light period (L) had to be between 8 and 15 hours. The strongest diapause-inducing effects were noted when D and L between these limits added up to about 24 hours. Since the range for D was smaller than that for L he concluded that dark-period measurement was more important. Abnormal cycles in which the ratio of D to L was held at 1:1 were less informative, but showed that diapause was only induced when T was between 20 and 26 hours.

Similar results which indicate the importance of the dark period have been obtained for other insect species. In *Acronycta rumicis*, for example, the dark period had to be in excess of 9 hours (Danilevskii and Glinyanaya, 1949), and almost all of the pupae of *Antheraea pernyi* entered diapause if D was greater than 11 hours, even if L was as long as 59 hours (Tanaka, 1950, 1951). In *Pieris rapae*, pupal diapause was induced only if D was greater than 12 hours in cycles where $T = 20$ to 36; within these limits the length of the light was much less important (Barker and Cohen, 1965).

Working with *Ostrinia nubilalis*, Beck (1962) studied the effects of 10, 12 and 14 hours of light in conjunction with a wide range of dark periods. Maximum incidence of diapause (90 per cent and over) occurred with dark periods of 10 to 14 hours. Dark periods of this length, however, produced a high incidence of diapause when combined with a wider range of light (5 to 18 hours). Once again maximum induction occurred when D + L was close to 24 hours. This relationship is well illustrated in the "isoinduction surface" calculated by Pittendrigh (1966, 1972) from Beck's data (Fig. 7.20).

Tyshchenko *et al.* (1972) compared diapause induction and entrainment of the ovi-

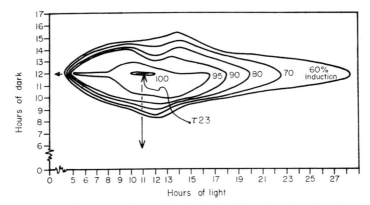

FIG. 7.20. *Ostrinia nubilalis:* a surface of "iso-induction" lines, based on the data given by Beck (1962). The lines are defined by (1) experimentally observed points reported by Beck and (2) interpolations from the curves he draws through these points. The "surface" is biased along the coordinate defined by the 12-hour dark period. But the dominant role played by that dark period is qualified by the duration of the associated light period. The surface peaks to a maximum in inductive effectiveness when T (the sum of the light and dark periods of the cycle) approximates to 24 hours. (From Pittendrigh, 1966).

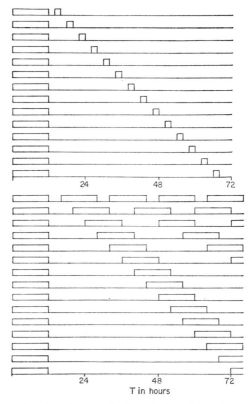

FIG. 7.21. Experimental designs for investigating photoperiodic phenomena. *Upper panel:* a long cycle ($T = 72$) with a short "main" photoperiod (12 hours in this case) and a scanning pulse. *Lower panel:* a series of cycles, each with a short "main" photoperiod (12 hours), but with the period of the cycle (T) varying over wide limits (in this case from $T = 16$ to $T = 72$). With both approaches, the organism is subjected to the experimental light-cycles throughout its sensitive period.

position rhythm of *Drosophila phalerata* in a series of cycles between $T = 20$ and $T = 34$ in which $L = D$. The range of effective cycles was similar in both cases.

These data show that although dark period measurement frequently occupies a "central" role, the duration of dark *and* light are important, and the most inductive cycles are often those in which $L + D$ are close to 24 hours. Dickson (1949) suggested that time measurement in *G. molesta* involved two reactions, one requiring dark and the other requiring light. Beck (1962) was inclined to interpret his data for *O. nubilalis* in terms of an "interval timer". The over-riding importance of a cycle length of $T \sim 24$ hours in this species (and many others), however, led Beck to interpret the results as "lending at least feeble support to the hypothesis that photoperiodic induction of diapause involves a circadian function". Later in this chapter we will see that dark-period measurement may only become important in certain cycles.

(b) *Resonance experiments.* A logical extension of the type of experiment described above is to hold the light-period constant (at, say, 6 to 12 hours) and to vary the dark component over a very wide range to provide environmental light cycles (T) up to 72 hours or more in length. This experimental design (Fig. 7.21) was first used for plants by K. C. Hamner and his associates (Blaney and Hamner, 1957; Hamner, 1960; Takimoto and Hamner, 1964), and later for birds (W. M. Hamner, 1963). The results for plants and birds have shown periodic maxima of induction when $T = 24$, 48 or 72 hours, but minima of induction when $T = 36$ or 60 hours. The interpretation of such "resonance" experiments is that if the photoperiodic clock incorporates a circadian oscillation (i.e. with an endogenous periodicity, τ, close to 24 hours), the product of induction (e.g. flower induction, testis growth, or diapause) is observed to be high when T is close to τ or modulo τ (i.e the two oscillating systems "resonate"), or low when T is *not* close to modulo τ (i.e. the two oscillating systems do not "resonate").

This type of experiment has been applied to at least four species of insect, only two of which showing a clearly rhythmic response. In *Pectinophora gossypiella* 8-hour light pulses (Adkisson, 1966) and 12-hour light pulses (Adkisson, 1964; Pittendrigh and Minis, 1971) induced a high incidence of diapause at $T = 24$, a drop in the incidence of diapause at $T = 36$, but very little in $T = 48$ and over (Fig. 7.22A). In *Carpocapsa pomonella* resonance experiments were equally ineffective: LD $8:4$ ($T = 12$) was non-inductive, LD $8:16$ ($T = 24$) was highly inductive, but longer cycles ($T = 36$, 48, 60 and 72) showed none of the periodicity expected (Peterson and Hamner, 1968) (Fig. 7.22B).

Resonance experiments have produced "positive" results in two other species, however. In both *Nasonia vitripennis* (Saunders, 1974) and its flesh-fly host *Sarcophaga argyrostoma* (Saunders, 1973b) the periodic response in diapause induction with T clearly shows that photoperiodic time measurement is a function of the circadian system in these species; in *N. vitripennis* the data also constitute experimental evidence in favour of a two-oscillator clock of the internal coincidence type.

Females of *N. vitripennis* were exposed for 19 days to different environmental light cycles in which the photoperiods ranged from 4 to 28 hours, and the period of the cycle (T) from 12 to 72 hours, different experimental groups receiving repeated cycles of a different duration. The wasps were supplied throughout with puparia of their host species, *S. argyrostoma*. Between the 19th and 21st days of the experiment the surviving females were enclosed separately with two fresh host puparia, and the type of progeny produced (diapause or non-diapause) during this period assessed by incubating these parasitized pu-

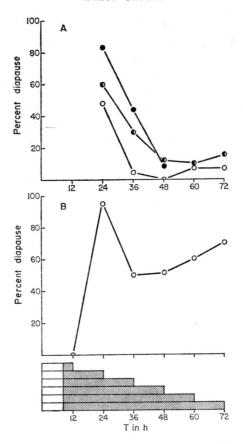

Fig. 7.22. The effect of varying T (the "resonance" technique) on diapause induction in two insect species. A— *Pectinophora gossypiella*: ●——● 12-hour photoperiod (from Pittendrigh and Minis, 1971); ◖——◖ 8-hour photoperiod (from Adkisson, 1966); ○——○ 12-hour photoperiod (from Adkisson, 1964). B — *Carpocapsa pomonella*: 8-hour photoperiod (redrawn from Peterson and Hamner, 1968). Note the absence of a "resonance effect" in both species.

paria for a further 10 days. For each light cycle the results were calculated as the percentage of wasps producing diapausing progeny during the test period.

When a 12-hour photoperiod was employed, maxima of diapause induction were observed when the wasps were "driven" at $T = 24$ to 28, and at $T = 48$ to 52; a third peak was also evident at about $T = 72$. Minima of diapause incidence were observed at $T = 36$ and 60 (Fig. 7.23). When the photoperiod was increased from 12 hours to 14 and 16 hours the "descending slopes" of the peaks remained in the same positions (at $T = 28$ to 32, and at $T = 52$ to 56), but the "ascending slopes" moved to longer T values in such a way that the peaks became narrower. When a 20-hour photoperiod was employed diapause was practically eliminated. A similar but opposite trend was observed when the photoperiod was shortened to 8 and 4 hours: the ascending slopes moved steadily to the left whereas the descending slopes remained at the same point, except for the 4-hour photoperiod where a perceptible movement to the left was also apparent.

Figure 7.24 shows the data replotted as a "circadian topography" in which all points of equal diapause incidence are joined by "contours" to give an "isoinduction surface" (Pittendrigh, 1966, 1972). This figure clearly shows the ascending and descending slopes

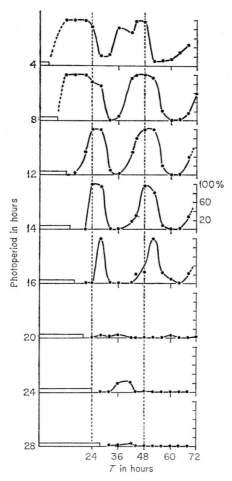

FIG. 7.23. Production of diapause larvae by females of *Nasonia vitripennis* in resonance experiments (17°C), showing the periodic maxima about 24 hours apart. As the photoperiod increases from 4 to 16 hours the "ascending slopes" of the diapause maxima move to the right (i.e. follow dusk), whereas the "descending slopes" of the maxima retain a fixed phase relationship to dawn. With a 20-hour photoperiod diapause is eliminated at all *T*-values, but it reappears in a peak at *Zt* 36 to 42 with a 24-hour photoperiod. The "ascending" and "descending" slopes are interpreted as the manifestations of a dawn and dusk oscillator, respectively. (From Saunders, 1974.)

of the diapause maxima, and their relationship to the photoperiod which is shown at the left of the diagram as a "light wedge". It is clear that the ascending slopes are parallel to dusk and the descending slopes are parallel to dawn; both repeat themselves in the extended dark period with circadian (∼24 hour) frequency. The two slopes are interpreted as reflecting two independent components in the photoperiodic clock, one obtaining its time cue from dawn and the other from dusk. Since they reset themselves with a 24-hour frequency they are both regarded as circadian oscillators, rather than hour glasses which would not recur in such a manner. With a 20-hour photoperiod the two components appear to coalesce and diapause is eliminated at all *T*-values. On the other hand, with a 4-hour photoperiod the two components appear to diverge so that the central maximum of diapause induction divides into two distinct peaks. Finally when a 24-hour photoperiod was used, an additional discrete peak of diapause induction (peak d) appear-

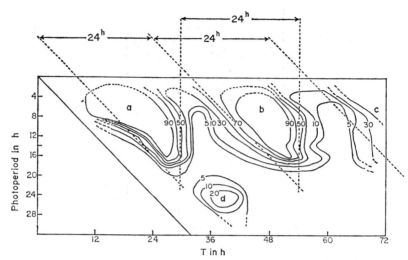

Fig. 7.24. Data from Fig. 7.23 redrawn as a "circadian topography" in which all points of equal diapause induction are joined by contours. This figure shows the relationship of the dawn and dusk oscillators to the photoperiod which is drawn as a "light-wedge". (From Saunders, 1974.)

ed at $T = 36$ to 42. This suggests that the two oscillators enter a "forbidden" mutual phase-relationship and one—probably the dawn oscillator—then executes a 180° ψ-jump, thereafter continuing to function as the dawn component. The fact that an additional peak appears with very long light periods (> 24 hours) forces one to conclude that the dawn oscillator is not damped out by protracted illumination, as is the oscillator controlling adult eclosion in *D. pseudoobscura* (Pittendrigh, 1966) (Chapter 3).

Resonance experiments with *Sarcophaga argyrostoma* have similarly produced evidence that photoperiodic induction is a function of the circadian system, although significant differences exist between this species and *N. vitripennis*. Larvae of *S. argyrostoma* were raised at 17°C and in a range of light cycles from $T = 20$ to $T = 72$, each containing from 4 to 20 hours of light. Newly formed puparia were collected daily from the cultures either during the light component of the cycle, or under red light (> 600 nm) which they apparently do not "see" (Chapter 6). The puparia were opened about 2 weeks later to ascertain the diapause or non-diapause status of the intrapuparial stages. The results showed periodic maxima in diapause induction, but the positions of the peaks varied according to the duration of the photoperiod (Fig. 7.25). The results are also presented in the form of a "circadian topography" (Fig. 7.26) to illustrate the temporal relationship of the diapause maxima to the photoperiod, which is drawn in the figure as a "light wedge". The second peak reached its maximum at $T = 48$ for a 4-, 8- or 12-hour photoperiod, but "moved" to the right by 2, 4 and 8 hours, respectively, for the 14-, 16- and 20-hour photoperiods. This suggests that the circadian system, or that part of the circadian system associated with photoperiodism, attains a phase-relationship to the light cycle in such a way that it receives its principal time cue from *dusk* once the photoperiod exceeds about 11 to 12 hours. This behaviour is reminiscent of the pupal eclosion rhythm in *Drosophila pseudoobscura* in DD free-run after a last photoperiod of more than 12 hours (Pittendrigh, 1966) (Chapter 3). For photoperiods greater than 12 hours duration the two diapause maxima were about 24 hours apart, and for those photoperiods where T was extended to 72 hours there was evidence of a third peak 24 hours after the second.

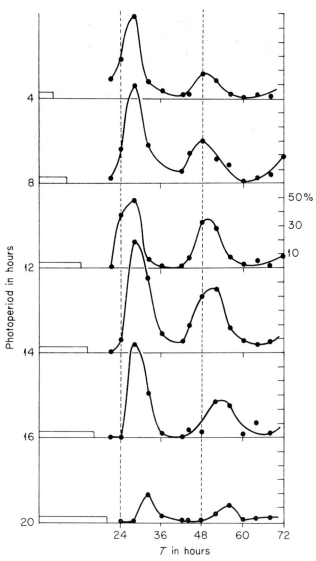

Fig. 7.25. Induction of pupal diapause in *Sarcophaga argyrostoma* when the larvae are reared in resonance experiments (17°C), showing the periodic maxima about 24 hours apart. (From Saunders, 1973b.)

For photoperiods shorter than 12 hours, however, the first two maxima were *less* than 24 hours apart, the first peak always being at $T = 28$ rather than at $T = 24$. This may indicate the presence of an additional component, separate from the clearly circadian one which produces "resonance" in the system, and which, in effect, "depresses" the expected high incidence of diapause at $T = 24$. Since this effect does not recur in the second peak, it might represent an hour-glass component associated with dawn.

These results leave little doubt that the circadian system is somehow involved in photoperiodic time measurement in *S. argyrostoma*. Unlike similar data for *N. vitripennis* (Fig. 7.24), however, one cannot describe the clock in terms of dawn and dusk oscillators. One thing, however, is clear: the circadian system behaves in a manner remarkably similar to that shown by the pupal eclosion rhythm in *D. pseudoobscura* (Pittendrigh,

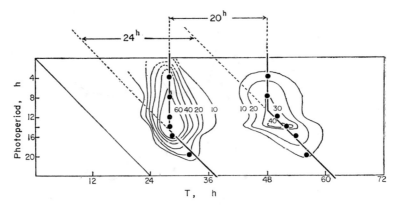

Fig. 7.26. Data from Fig. 7.25 redrawn as a "circadian topography" in which all points of equal diapause induction are joined by contours. The positions of the diapause maxima are marked by closed circles. With photoperiods longer than 12 hours the maxima are about 24 hours apart and appear to take their principal time cue from dusk. With photoperiods shorter than 12 hours, however, the maxima are only 20 hours apart because the first maximum appears at $T = 28$ rather than $T = 24$. (From Saunders, 1973b.)

1966), in that its principal time cue is taken from dusk once the photoperiod exceed about 12 hours. In other words, the oscillator(s) involved in time measurement are "damped out" in photoperiods longer than about 12 hours and then restart at dusk. With longer photoperiods, therefore, the photoperiodic clock in *S. argyrostoma*, although oscillatory, exhibits an hour-glass property and apparently measures nightlength. This illustrates the importance of a "critical nightlength" but suggests that the duration of the dark phase in *S. argyrostoma* only assumes a central role when accompanied by a light component of minimum length. The similarities between *S. argyrostoma* and those species reviewed in Section A.4 (a) now become obvious.

5. *Tests of the internal coincidence model*

Results for resonance experiments in *Nasonia vitripennis* suggest that photoperiodic induction involves the interaction of separate dawn and dusk oscillators and therefore support a model of the internal coincidence type in which light plays no direct role in induction. Two experiments have been devised as tests of this model. The first of these was the demonstration that "photoperiodic" induction can take place in the complete absence of light (Saunders, 1973a); the second involved the transfer of insects from LD into DD.

Females of *Nasonia vitripennis* were raised from the egg stage in total darkness but subjected as adults to daily thermoperiods ($T = 24$) consisting of different ratios of a higher temperature (23°C) to a lower temperature (13°C). It was found that females exposed to fifteen daily thermoperiodic cycles with less than about 13 hours at the higher temperature produced nearly all of their offspring as diapausing larvae whereas those exposed to thermoperiods with a more protracted period at 23°C produced none (Fig. 7.27). This result shows that a thermoperiod can simulate the effects of a photoperiod in diapause control, and demonstrates that light is not required for the operation of the metabolic switch. Since temperature cycles and pulses are known to entrain and phase-set circadian rhythms as effectively as light pulses and cycles (Zimmerman *et al.*, 1968) it

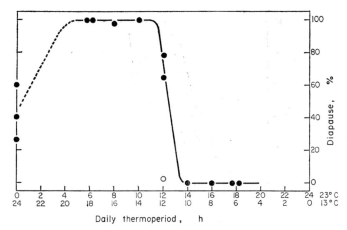

FIG. 7.27. The effect of a daily temperature cycle (23°/13°C) on the induction of larval diapause in *Nasonia vitripennis* kept in continuous darkness. Note the sharp discontinuity
between short and long thermoperiods (the critical thermoperiod). The ordinate plots the
proportion of female wasps producing diapause broods during the test period (days 15 to
17 of adult life). The open circle shows the proportion producing diapause larvae in the
dark at 18°C; this is considered to be the most meaningful control for the group maintained
for 12 hours at 23° and 12 hours at 13°C (mean temperature 18°C). (From Saunders, 1973a.)
(Copyright 1973 by the American Association for the Advancement of Science.)

was concluded that the thermoperiods were entraining constituent rhythms in a manner
comparable to light. Consequently any model for the *Nasonia* clock in which light plays
a direct role in induction (i.e. "external coincidence") must be ruled out. Although a
specific temperature-sensitive phase is possible, a more likely explanation for this
phenomenon in *N. vitripennis* is a model of the internal coincidence type, involving the
mutual interaction of "dawn" and "dusk" oscillators which can be phase-set by light
or temperature cycles.

In the second test groups of *N. vitripennis* were raised in either *LD 8* : 16, *LD 12* : 12
or *LD 16* : 8 for 5 days and then "released" into DD. Fresh puparia of their host species,
Sarcophaga argyrostoma, were supplied daily, then removed and incubated to determine
the type of progeny produced on each day of the experiment. Control groups of wasps
were maintained throughout in the respective light cycles.

Figure 7.28 shows that those wasps maintained throughout at short daylength (*LD
12* : 12) began to "switch-over" to the production of diapausing progeny after 5 to 6
days and had completed this process by about the 12th to 13th day; thereafter they continued to produce all-diapause broods. In those groups transferred from *LD* to DD after
five cycles the "switch-over" continued for about 9 days in continuous darkness before
it became erratic. In the long-day group, the controls produced most of their progeny as
non-diapause larvae, whereas the experimental group continued to produce non-diapause offspring for 12 days after the transfer to DD, before any diapause larvae were
produced. These results show that diapause promotion and inhibition can occur in the
absence of *continued* light-cycles, and may be interpreted in terms of internal coincidence.
In other words, the dawn and the dusk oscillators were entrained and phase-set by the
initial five light-cycles and then "free-ran" after release into DD. Their mutual phase-
relationship was then maintained for a number of endogenous circadian cycles, and diapause induction or inhibition continued during this period. The eventual breakdown

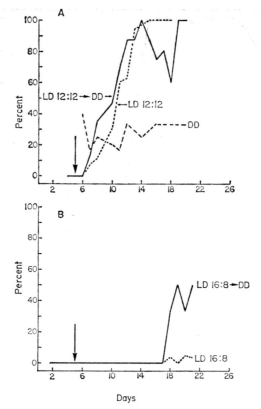

FIG. 7.28. The release of adult females of *Nasonia vitripennis* from *LD* to DD and its effect on diapause induction. A — release from *LD 12* : 12 to DD after five light-cycles; B — release from *LD 16* : 8 to DD after five light-cycles. Note that (in A) the rate of switching to diapause larva production continues to rise after transfer to DD in the same manner as in those females maintained in short days throughout. The DD control, on the other hand, is erratic and lacks a clear upward trend. This behaviour can be attributed to the separate "dawn" and "dusk" oscillators of the photoperiodic clock free-running in DD, but maintaining their mutual phase relationship as entrained in *LD 12* : 12, and hence causing diapause induction. Conversely, exposure to five cycles of *LD 16* : 8 (in B) "programme" the insects for a long period in which they produce no diapause larvae. (Saunders, unpublished.)

probably reflects a breakdown in this phase relationship; whether this indicates that τ for the dawn oscillator differs from τ for dusk is not known. A time-lag in the physiological system associated with diapause control is expected and probably occurred, but the size of this lag is unlikely to be as great as 9 or 12 days.

6. *Summary: external and internal coincidence*

In the foregoing sections we have examined two types of clock model based on the circadian system: (1) the *external* coincidence model which comprises a single oscillator and the temporal coincidence of a light-sensitive phase (ϕ_i) with the environmental light cycle (Pittendrigh and Minis, 1964), and (2) the *internal* coincidence model which depends on the phase angle between separate "dawn" and "dusk" oscillators (Pittendrigh, 1960; Tyshchenko, 1966). Data for the parasitic wasp *Nasonia vitripennis* tend to support the latter. Thus, results for abnormal cycles in which T is close to τ (Saunders, 1968) and chilling (Saunders, 1967; Saunders and Sutton, 1969) show that both light and dark

periods are important. Resonance experiments (Saunders, 1974) demonstrate the existence of "dawn" and "dusk" components, and thermoperiods in the *total absence of light*
(Saunders, 1973a) clearly eliminate the concept of ϕ_i *for this species*. In *Sarcophaga
argyrostoma*, however, the dark period is "more important" than the light and, in resonance experiments, assumes a "central" role once $L > 12$ hours (Saunders, 1973b), when
the oscillatory system appears to restart on entry into the dark. Chilling experiments and
thermoperiodic cycles have not been applied to this species, principally because the larvae,
during their most sensitive stage, are within or under a large chunk of meat, rendering
this approach impracticable. Nevertheless, the available data for *S. argyrostoma* seem
to suggest that a single oscillator is involved and that the clock is of the *external* coincidence type. Thus we may say that both types of circadian clock are probably represented
in the insects.

B. Evidence for an Hour-Glass in Time Measurement

There is now a substantial body of experimental evidence to suggest that photoperiodic
time measurement in some insects—especially the parasitic wasp *Nasonia vitripennis* and
its flesh-fly host, *Sarcophaga argyrostoma*—is a function of the circadian system. However,
one of the most intensively investigated species—the green vetch aphid *Megoura viciae*—
seems to accomplish the measurement of the dark period of the daily cycle by means
of a non-repetitive interval timer or hour-glass. Strong evidence for such a clock in *M.
viciae* has been presented in a series of papers by Lees (1965, 1966, 1968, 1973), using
experimental methods almost identical in plan to those described in the foregoing
sections.

1. *Abnormal light/dark cycles*

The aphid *Megoura viciae* produces successive generations of wingless, parthenogenetic
virginoparae during the summer months when the days are long. In the autumn when
days fall below a critical point ($LD\ 14\frac{1}{4} : 9\frac{3}{4}$) oviparae are produced which subsequently
lay diapausing eggs. Experiments with abnormal light cycles have demonstrated that the
dark period occupies a central role in time measurement (Lees, 1965). Consequently
in most of this section a long-day effect (virginopara-production) will be referred to as a
short-night effect, and vice-versa.

Figure 7.29 shows that when a night of inductive length (12 hours) was combined with
a range of different light periods up to $T = 76$, a high rate of ovipara-production (the
long-night effect) was observed even with "days" up to 32 hours in duration; only with
a light period of 40 hours was the proportion of ovipara-producers reduced to 50 per
cent. In the converse experiment a short light period (8 hours) was combined with
extended nights up to $T = 60$. This experiment is therefore of the resonance type and
should be compared with those in Section A.4 (b). In *Megoura* long-night effects (ovipara-
production) were not observed until the dark period reached $9\frac{1}{2}$ to 10 hours ($T = 18$);
thereafter the response was "saturated" and all aphids became ovipara-producers. This
second experiment is important for two reasons. Firstly, it demonstrates that once a
critical nightlength ($9\frac{1}{2}$–10 hours) has elapsed all aphids are committed to the long-
night response. Secondly, it shows that the response is the same with all cycles over
$T = 18$: there is no hint of a "resonance effect" which would have produced a high level

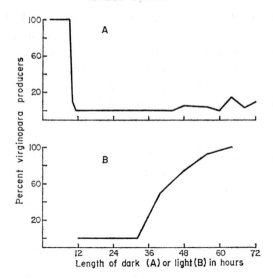

F<small>IG</small>. 7.29. *Megoura viciae:* the production of wingless virginoparae in photoperiodic regimes in which the length of the cycle was varied, but the length of the light component held at 8 hours (A) or the length of the dark component held at 12 hours (B). In A, virginoparae (the long-day response) are produced once the dark period exceeds the critical nightlength (9.5 hours). Note the absence of periodic maxima of short-day effect which would be expected if the *Megoura* clock was a function of the circadian system. In B, a long night is fully inductive (i.e. 100 per cent oviparae, or 0 per cent virginoparae) until the accompanying light period exceeds about 32 hours. (Redrawn from Lees, 1965.)

of the long-night response at $T = 24$ and $T = 48$, but a low level at $T = 36$ and $T = 60$. This constitutes some positive evidence, therefore, that time measurement in *M. viciae* is not a function of the circadian system, but comprises a non-repetitive dark period interval timer or hour-glass.

2. Night-interruption experiments

Further evidence for an hour-glass is provided by light-break experiments. Using a light cycle ($T = 24$) containing a night just longer than the critical (*LD 13.5 : 10.5*), Lees (1965, 1966) showed that 1-hour light breaks introduced systematically into the night produced two "peaks" of short-night effect, or virginopara-production (Fig. 7.30). One of these occurred about 2 hours after dusk; the other was more pronounced and occurred when the light pulse fell during the last 6 hours of the night. Pulses applied in the middle of the night were without effect. These results are superficially similar to those for *Pectinophora gossypiella* (Adkisson, 1964) and a number of other species (Section A.1 (a)). However, Lees (1965, 1966) showed that the response was unaltered if the light period accompanying the 10.5-hour night was extended to 25.5 hours (*LD 25.5 : 10.5*, $T = 36$) or shortened to 8 hours (*LD 8 : 10.5*, $T = 18.5$) (Fig. 7.30). These results were interpreted as evidence that time measurement began at dusk rather than at dawn, and were considered to be inconsistent with a circadian hypothesis.

"Early" and "late" light breaks seem to have quite different modes of action in *M. viciae*. Figure 7.31a shows the effect of a pulse applied $1\frac{1}{2}$ hours after dusk. In the first regime (*LD 13.5 : 1.5 : 1 : 9.0*) the pulse completely reversed the long-night effect and

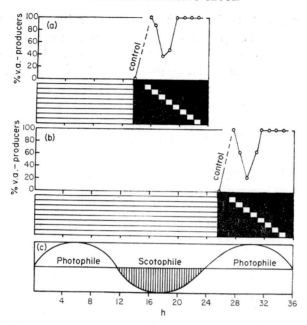

FIG. 7.30. Night-interruption experiments in *Megoura viciae*, using a "main" photoperiod of variable length. The circadian change in light sensitivity, as predicted by Bünning's hypothesis, is shown below. Note that the positions of the peaks of virginopara production (long-day effect) are unaffected by the duration of the preceding light. (From Lees, 1970.)

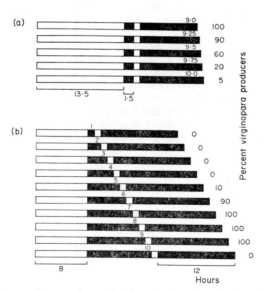

FIG. 7.31. Night-interruption experiments in *Megoura viciae* showing the differences between "early" and "late" light breaks. In (a) a 1-hour light break 1.5 hour safter dusk is followed by a variable dark period. When this dark period exceeds the critical nightlength (9.5 hours) short-day effects (low virginopara-production) ensue, demonstrating the reversibility of the inductive effects of a light pulse early in the night. In (b) a 1-hour light pulse scans the night but is in all cases followed by a dark period greater than the critical night length. When the pulse falls early in the night its long-day effect (virginopara-production) is overridden by the following 12 hours of dark; when it falls later in the night, however, the long dark period which follows it fails to reverse the long-day effect. (From Lees, 1970.)

since it was followed by a short night (9.0 hours) resulted in 100 per cent of virginopara-producers. As the terminal dark period became greater than the critical night (9.5 hours), however, the effect of the early light break was overridden and long-night effects were achieved. This result demonstrates once again the importance of an uninterrupted long night, and shows that the effects of an early pulse are *reversible*. In another series of experiments (Fig. 7.31b) a night interruption was followed by a constant dark period of more than inductive length (12 hours), but the hours of dark preceding the pulse were systematically varied. It can be seen that the terminal 12 hours of darkness function as a long night (100 per cent ovipara-producers) until the light pulse fell between the 7th and 10th hours after dusk, where it functioned as a *late* interruption. The effects of such a late interruption were *irreversible* because the short-night effect so produced could not be undone by the terminal 12 hours of dark. Early and late points of light sensitivity also differed in their action spectra (Chapter 6): the early point showed a pronounced blue sensitivity (470 nm), whereas the late night point was also sensitive in the red (Lees, 1971a).

Night interruption experiments in cycles with a greatly extended night (i.e. *LD 8* : 64, $T = 72$) have also failed to produce evidence for an oscillatory clock (Lees, 1966). Figure 7.32 shows the effect of introducing a 1-hour pulse of light at 4-hour intervals into the 64-hour "night". In all but one position, strong long-night effects (100 per cent ovipara-

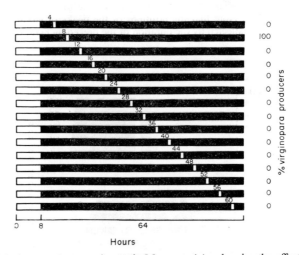

FIG. 7.32. Night-interruption experiments in *Megoura viciae* showing the effect of a 1-hour pulse scanning a very long night ($T = 72$). Long-day effects are produced only when the pulse falls 8 hours after dusk, a position which constitutes a "late" interruption, which is irreversible. Note also the lack of a "resonance" effect with alternating maxima and minima of long and short day effect which would be observed if the *Megoura* clock was a function of the circadian system. (From Lees, 1970.)

production) were produced. The single exception was when the pulse fell 8 hours after dusk; here it completely reversed the response. There was no evidence of further points of sensitivity at circadian intervals (i.e. at Zt 40 or Zt 64), so that, once again, the results are not consistent with an oscillatory hypothesis. At first sight the results are also inconsistent with an hour-glass, since all regimes contained a "night" in excess of 9.75 hours. The pulse falling 8 hours after dusk, however, presumably acted as a late night interruption and rendered the remaining 55 hours of darkness ineffective. The pulse falling at

Zt 64 might have been expected to act either as an "initiator" of an asymmetric skeleton, or as a "main" photoperiod, and thereby also reverse the long-night effect. Lees (1971b) has shown, however, that a pulse falling in this position can only act as a "main" photoperiod when it becomes longer than 4 hours (Fig. 7.33). The function of this "main" light component is not merely to delimit the accompanying dark, but also to prepare the system for the timing process which begins at dusk; in other words, it serves to "turn the hour-glass over". In order to do this it has to be in excess of four hours, otherwise the accompanying "critical nightlength" becomes longer.

The hour-glass in *M. viciae* is thought to comprise a linked sequence of four reactions distinguished on the basis of their responses to light breaks (Lees, 1968). These are:

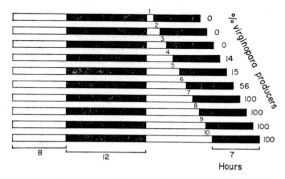

Fig. 7.33. Experiments with *Megoura viciae* showing the effect of increasing the length of an interruption in the late night. When this interruption exceeds 4 or 5 hours it begins to function as a "main" photoperiod. (From Lees, 1970.)

(1) The first 3 hours of the night, during which a timing reaction proceeds which can easily be reversed by low-intensity blue light. If a pulse at this stage is followed by a dark period of more than critical length, the "photoperiodic reversal" can itself be reversed. (2) From the 3rd to the 4th hours of the night the photoreceptor is insensitive, perhaps because a necessary substrate disappears. (3) Between the 5th hour and the end of the critical night (9.75 hours) the system is again photosensitive and responds to both blue and red light. Its effects, however, are irreversible by subsequent exposure to a dark period of more than critical length. (4) The last stage comprises those hours of darkness which occur in excess of the critical value (9.75 hours) in a long night cycle. Light pulses applied during this period are ineffective because they leave an uninterrupted residue longer than 9.75 hours. The hour-glass reaction, as Lees sees it, therefore, is a complex one, presumably operated in nature by the photoperiod extending until dawn falls in stage 3, whereupon virginoparae are produced.

3. Symmetrical skeleton photoperiods

Hillman (1973) has recently investigated the *Megoura* system using ambiguous symmetrical skeleton photoperiods in experiments of the "bistability" type, previously applied to the short-day plant *Lemna perpusilla* (Hillman, 1964; Pittendrigh, 1966) (Chapter 3). Stock cultures of virginoparae were raised in *LL* and then transferred to a number of skeleton regimes formed from two 1-hour pulses of light per 24-hour cycle. The two main skeletons used were *LD 1 : 9 : 1 : 13* and *LD 1 : 13 : 1 : 9* (PP$_s$ *11 : 13* and

PP$_s$ *15* : 9). In *Lemna perpusilla* and *Drosophila pseudoobscura* the steady state phase adopted by the circadian oscillation when presented with such "ambiguous" regimes was found to depend in part upon which *interval* was seen first. Therefore, in the present experiments—and if *M. viciae* possesses an oscillatory timer—those aphids transferred from *LL* into *LD 1* : 9 : *1* : 13 should "read" the 9-hour interval of dark as "night" and, because 9 hours is below the critical, produce a high proportion of virginoparae. Conversely, those transferred from *LL* into *LD 1* : 13 : *1* : 9 should respond with a low incidence of virginoparae. In additional experiments the aphids were transferred from *LL* into an initial 12-hour period of darkness before experiencing the first pulse, the pre-

TABLE 7.2. PHOTOPERIODIC RESPONSE OF *Megoura viciae* TO "AMBIGUOUS" SYMMETRICAL SKELETON PHOTOPERIODS AFTER TRANSFER FROM CONTINUOUS LIGHT

Light : dark regime	Predicted response if clock is oscillatory (per cent virginopara-producers)	Observed response (per cent virginopara-producers)
LD 1: 9: *1*: 13 (PP$_s$ 11: 13)	High	100.0
LD 12: 12 one cycle; then PP$_s$ 11: 13	Low	100.0
LD 1: 13: *1*: 9 (PP$_s$ 15: 9)	Low	100.0
LD 12: 12 one cycle; then PP$_s$ 15: 9	High	100.0

(From Hillman, 1973.)

diction being that the phase of the oscillator would be displaced by a full half cycle, thereby reversing the response to the following skeleton. Table 7.2 shows that each of the four regimes produced 100 per cent of virginopara-producers; consequently observation did not agree with prediction. Other experiments, in which the duration of the pulses forming the skeletons was increased from 1 hour to 2, 3 or 4 hours, also failed to confirm predictions based on an oscillatory hypothesis. For this reason the results were interpreted as evidence against even the most rudimentary form of circadian timing in *Megoura*.

C. Oscillators and Hour-glasses

In this chapter we have examined the evidence for two apparently different kinds of timer, one based on one or more endogenous oscillators, the other on a non-resetting dark period hour-glass. Since the two types of clock may show common properties, and since some species appear to possess a clock with both hour-glass *and* oscillatory components, it seems legitimate to ask how distinct they really are.

1. *Clocks with hour-glass and oscillatory components*

In at least two insect species the data suggest that both types of timer are involved in the photoperiodic response. In *Carpocapsa pomonella*, for example, Hamner (1969) proposed an interaction between a rhythmic dawn timer and a dusk hour-glass. In *Sarcophaga argyrostoma*, there was evidence for a separate, possibly hour-glass, component associated with dawn and modifying the first resonance peak, and the observation that the photoperiodic oscillator measures night length when the light period exceeds 12 hours (Saunders, 1973b). In addition, Truman (1971b) has proposed an almost entirely theoretical model to account for diapause termination in *Antheraea pernyi* which shows how a circadian rhythm and an hour-glass might interact.

This model, like the external coincidence model, is based on a "parallel" rhythm which is amenable to study. *A. pernyi* has a pupal diapause which is terminated by the long days of spring during which the moths eclode in a rhythmic fashion. Since the clock governing the termination of diapause (Williams and Adkisson, 1964; Williams, 1969a) and the clock governing the eclosion rhythm (Truman and Riddiford, 1970) both reside in the brain (Chapter 10) and have similar properties, it is possible that a single clock controls both processes.

The rhythm of eclosion in *A. pernyi* shows many similarities to the better-known rhythm in *Drosophila pseudoobscura* (Pittendrigh, 1966) (Chapter 3). When transferred from a synchronizing *LD* cycle to DD the rhythm free-runs with a period (τ) of about 22 hours, and since the first eclosion peak always assumes the same phase-relationship to lights-off after light periods of 12 to 17 hours, the timing process is assumed to start at dusk. In 24-hour light cycles the peak of eclosion assumes a definite phase relationship to the photoperiod. With *LD 8* : 16 and *LD 12* : 12, for example, eclosion occurs just before dusk, and in *LD 17* : 7 it occurs late in the "afternoon", in all cases with a narrow

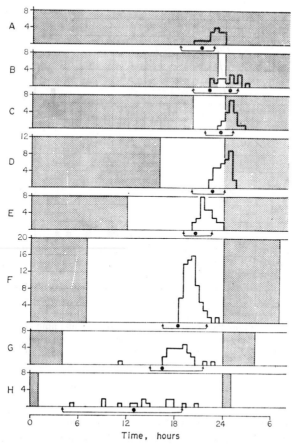

Fig. 7.34. The ecdysis of *Antheraea pernyi* under various photoperiodic regimes. A — first day in DD from an *LD 12* : 12 regime; B — *LD 1* : 23; C — *LD 4* : 20; D — *LD 8* : 16; E — *LD 12* : 12; F — *LD 17* : 7; G — *LD 20* : 4; H — *LD 23* : 1. (Redrawn, from Truman, 1971b.)

gate width of 3 to $3\frac{1}{2}$ hours. In very short days the gate width becomes somewhat wider and occurs partly before lights-on, and in very long photoperiods synchronization breaks down so that eclosion takes place over a very wide period of time. In *LL* the rhythm is inapparent (Fig. 7.34).

A study of the time of eclosion when the dark period was interrupted at different times by the onset of light suggested that the process consisted of two alternate pathways. The first is a dark-dependent process with a duration of about 22 hours (the scotonon) which is seen as an oscillation of a substance (S) which starts at dusk (Fig. 7.35). The

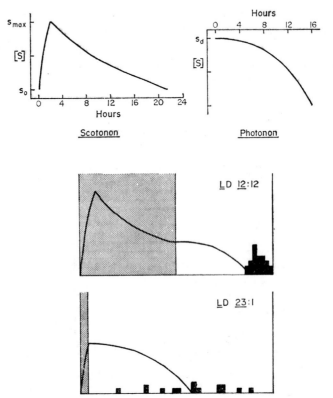

FIG. 7.35. *Upper panels:* a schematic representation of the components of the eclosion clock in *Antheraea pernyi*. The components are represented as the fluctuation of a hypothetical substance (S) versus time. The scotonon occurs only during darkness. The photonon is triggered by a light interruption of the scotonon; S_d is the concentration of S reached by the scotonon at the time of the interruption. The timer initiates emergence behaviour as S approaches S_0. *Lower panels:* application of the model of the ecdysis timer in *Antheraea pernyi*. A light interruption occurring during the gradual falling limb of the scotonon produces a relatively narrow gate (as in *LD 12* : 12). An interruption of the steep rising phase produces a wide gate (as in *LD 23* : 1). (From Truman, 1971b.)

first 2 hours of the scotonon consist of a "synchronization period" in which S rises from S_0 to S_{max}. This is followed by a 20-hour period of "dark decay" during which S returns to S_0. The free-running oscillation, therefore, is represented by a "saw-tooth" rhythm. When the light goes on the second pathway (the photonon) commences and the oscillation continues according to photonon kinetics. The duration of the photonon depends on how much of the scotonon has been completed by lights-on. When the photonon reaches S_0 the neurotropic hormone controlling eclosion behaviour is triggered. If the photonon terminates in the dark the next scotonon restarts immediately so that the process behaves as an oscillator. If the process is completed in the light, however, it stops until dusk when the scotonon can continue. The behaviour of this oscillator, therefore, may be compared to that for *D. pseudoobscura*, particularly in the damping action of (prolonged) light.

In order to use this model to account for diapause control an additional, and entirely theoretical, assumption had to be made. To provide "yes-or-no" control deciding whether brain hormone is to be released, it had to be assumed that the first part of the 24-hour period after dawn is *inhibitory* to hormone release whereas the second part allows it. The signal for the release of the brain hormone (and therefore for diapause termination) is thought to occur when S reaches S_0. Therefore, in short photoperiods the signal falls entirely within the inhibitory zone so that the insects remain in diapause. At long daylengths it falls outside the inhibitory zone, consequently hormone is released and diapause terminated (Fig. 7.36). At very short photoperiods part of the signal occurs before

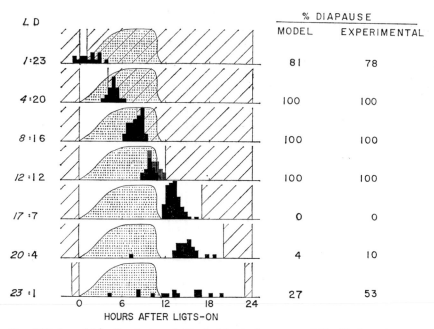

FIG. 7.36. A model for the photoperiodic clock in *Antheraea pernyi*. The black peaks represent the times dictated by a photosensitive clock for brain hormone release. (To compensate for sexual differences in ecdysis times, only the data for males are considered in the LD *12* : 12 regime.) The stippled areas represent the period during which brain hormone release is inhibited. The experimentally determined values are interpolated from the data of Williams and Adkisson (1964). (From Truman, 1971b.)

dawn so that a proportion of the pupae remain inactive, and at very long photoperiods the gating mechanism "breaks down" so that some of the pupae escape from diapause but others do not.

Although the "inhibitory zone" is entirely hypothetical, the model includes data from the known pattern of entrainment of the eclosion rhythm, and invites comparison with Tyshchenko's model (Section 4) because it involves the "internal" temporal relationship between two components, rather than the coincidence of a phase-point with light. Furthermore, since the inhibitory zone seems to possess the properties of an hour-glass rather than an oscillator, this model also illustrates the possible interaction between two different kinds of timer. Truman (1971b) has also shown that prediction from this model closely matches the experimental data of Williams and Adkisson (1964).

2. Oscillators with hour-glass properties

Pittendrigh (1966) has pointed out that the two types of timer—hour-glass and oscillator—although different, are not necessarily mutually exclusive. The oscillator controlling the pupal eclosion rhythm in *Drosophila pseudoobscura*, for example, is suppressed or "damped out" by photoperiods in excess of 12 hours so that, on its release into darkness, it resumes its motion at a fixed phase point (Ct 12). Consequently in 24-hour cycles where $L > 12$ hours the oscillator "measures" the duration of the dark period as an hour-glass (Fig. 7.37). Only when the oscillator is allowed to free-run in DD does it reveal its endogenous periodicity.

A very similar behaviour has been seen with the oscillator governing pupal diapause in *Sarcophaga argyrostoma* (Saunders, 1973b) (Fig. 7.26). In resonance experiments the first peak of diapause induction occurs a fixed number of hours after dark, once the photoperiod exceeds about 12 hours. Clearly in this species the clock is measuring the dark

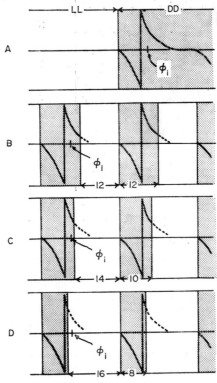

FIG. 7.37. *Drosophila pseudoobscura:* oscillation as an hour-glass. A—constant light (*LL*) suppresses the oscillation; on transfer to DD it resumes its motion starting from Ct 12. B, C and D—photoperiods of 12 hours or more similarly damp the oscillation which starts afresh from Ct 12 with the onset of darkness each night. The light at dawn falls further back into the late subjective night as the photoperiod increases. For illustrative purposes ϕ_i has been represented as occurring at Ct 21: it is not illuminated by photoperiods of 12 and 14 hours duration (short daylengths); a 16-hour photoperiod (long daylength), however, illuminates it. (From Pittendrigh, 1966.)

period as an hour-glass, although, once again, its oscillatory nature is revealed in longer cycles. The damping action of extended photoperiods provides an explanation for many of the observations that the duration of the dark period is of "central" importance in photoperiodic induction (Section A.4 (a)). In many instances this only occurs if the dark period is combined with a photoperiod of "critical length". Even in *Megoura viciae* the critical nightlength becomes longer with photoperiods shorter than about 4 hours (Lees, 1970). Only in *Nasonia vitripennis* is this hour-glass property not apparent, and one is forced to conclude that the dawn oscillator is not damped out by protracted light periods (Saunders, 1974).

Truman (1971 a, b) showed that the oscillator controlling adult eclosion in *Antheraea pernyi* started from dusk after light periods of almost any length. Light pulses as short as 30 minutes applied early in the night (during the synchronization period) served to "reset" the start of the oscillator which, in 24-hour cycles, also measured nightlength as an hour-glass. In *M. viciae* early night interruptions similarly serve to "reset" the clock; a long-night effect (ovipara-production) is then only observed when the hours of darkness following such a break exceed the critical nightlength.

The main difference between the hour-glass in *M. viciae* and the oscillator clock in *S. argyrostoma* is that the latter resets itself and repeats itself with circadian frequency in protracted dark periods whereas the former does not. Even this difference, however, is open to question. The photoperiodic response is clearly a very complicated one and includes a number of intermediate "steps" between the reception of the photoperiodic signals and the programming of the endocrine system governing diapause induction or morph determination. Temperature, in particular, has a marked effect on the summation of inductive cycles (see Chapter 8) and therefore on the *level* of the photoperiodic response. In *S. argyrostoma* "positive" resonance is obtained at 20°C, but at 16°C all *T*-values over about 23 hours gave a saturated (i.e. 100 per cent) diapause response (Fig. 7.38). At this lower temperature a 12-hour photoperiod resulted in diapause promotion in all cycles in which a "critical nightlength" exceeded about 9.5 hours. This result is very reminiscent of the response of *M. viciae* to resonance experiments at 15°C (Fig. 7.29). In two respects, therefore, the oscillator in *S. argyrostoma* can behave as an hour-glass, and the distinction between the two kinds of timer becomes blurred.

The mechanism of time measurement in insects is clearly very diverse. Some species have clocks which behave as hour-glasses and show no clear oscillatory properties. Some, such as *Nasonia vitripennis*, are at the opposite extreme: they show no evidence of hour-glass behaviour. Others occupy an intermediate position. The clock in *S. argyrostoma* is oscillatory but reveals hour-glass properties with extended photoperiods or at low temperature; that in *Pectinophora gossypiella* entrains to asymmetrical skeleton photoperiods in a manner which is consistent with an oscillator, but resonance experiments with this species have failed to produce more direct evidence for an endogenous rhythmicity. Two overall views of the nature of the photoperiodic clock are possible, therefore. In one, as suggested at the outset of this chapter, the diversity of clocks is a consequence of convergent evolution. In the other, clocks are of monophyletic origin but have diverged to give rise to a continuous "spectrum" with "pure" hour-glasses *(M. viciae)* at one end, "pure" oscillatory clocks *(N. vitripennis)* at the other, and a range of intermediate types between these two extremes. Which of these two views is correct is highly speculative and may remain so. Only a comparative study of the clocks of a large number of species— especially in resonance experiments—will help resolve this particular problem.

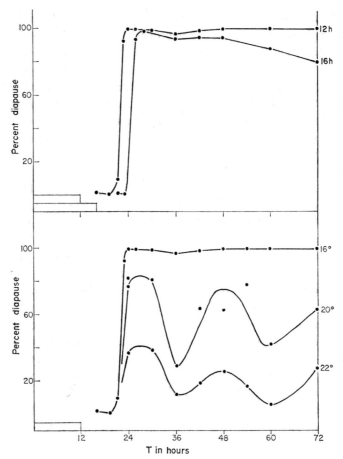

FIG. 7.38. *Top panel:* the effect of varying the period of the driving cycle (T) whilst holding the photoperiod constant (at 12 or 16 hours) on the proportion of larvae of *Sarcophaga argyrostoma* entering pupal diapause at 16°C. Note the lack of a clear "resonance" effect at this low temperature, and the appearance of a "critical nightlength" (9 to 10 hours) before diapause induction is saturated. *Lower panel:* the results of similar experiments using a 12-hour photoperiod, but raising the temperature from 16° to 20° and 22°C. As the temperature is raised a clear "resonance" effect becomes apparent, and the amplitude of the diapause maxima becomes lower. (From Saunders, 1973b.)

Annotated Summary

1. The photoperiodic response curve shows that the insect is able to "measure" either daylength or nightlength (or both) with a "clock" mechanism.

2. Two classes of model have been proposed to account for time measurement. In one, the nightlength is measured by a non-oscillatory interval timer or "hour-glass"; in the other there is an oscillatory circadian component similar to that involved in overt physiological and behavioural rhythmicities. There is also a number of different types of oscillatory model, the most important being (1) "internal coincidence" in which photoperiodic induction is a function of the internal temporal phase-relationship of two or more oscillators, and (2) "external coincidence" in which induction of development is controlled by the coincidence of the environmental light cycle with a photoperiodically inducible phase (φ_i).

3. Methods available for investigating time measurement are presently limited to an indirect "black-box" approach. Most useful are experiments in which the light and dark components are independently and systematically varied, or the night portion of the cycle scanned with a supplementary light pulse.

4. Evidence available from the rather small number of species so far investigated suggests that hourglass *and* oscillatory components are present in the photoperiodic clock. When oscillators are involved

they may act like "hour-glasses" in a natural 24-hour cycle because they are unable to oscillate freely before being reset by the next light period.

5. Evidence for a circadian component in the clock has been obtained from two sources: (1) the interruption of long (48- and 72-hour) cycles with a scanning supplementary pulse, and (2) from "resonance" experiments in which a short day (~ 12 hours) is variously coupled with a dark period to give cycle lengths up to 72 hours in duration. Both techniques may reveal a 24-hour periodicity in the diapause response.

6. More circumstantial evidence for a circadian component is available from responses to asymmetrical and symmetrical "skeleton" photoperiods, which may be similar to those for an overt rhythmicity.

7. Evidence for "internal coincidence" has been obtained for *Nasonia vitripennis* in which the clock consists of separate "dawn" and "dusk" oscillators, and environmental light plays no direct role in induction. Consequently, diapause and development-promotion can occur in the complete absence of light, or after transfer from *LD* to *DD*.

8. Evidence for some sort of "external coincidence" has been obtained for *Sarcophaga argyrostoma*. In this species resonance experiments demonstrate a "single" oscillatory component which behaves in almost exactly the same way as the oscillator governing the rhythm of pupal eclosion in *Drosophila pseudoobscura*. Thus, it is "damped out" by photoperiods in excess of about 11 to 12 hours, but restarts at dusk to measure nightlength as an "hour-glass". Only in greatly extended nights does the photoperiodic oscillation reveal its endogenous circadian periodicity. In natural 24-hour cycles containing a day longer than 12 hours, diapause is promoted when the accompanying dark period is greater than the critical nightlength ($9\frac{1}{2}$ hours), but diapause is eliminated when the dawn transition extends "backwards" to coincide with a particular light-sensitive phase (ϕ_i) occurring $9\frac{1}{2}$ hours after dusk. In many other insect species nightlength also occupies a "central role" in photoperiodism.

9. In *Pectinophora gossypiella*, the species for which external coincidence was first proposed, many of the experimental results are in disagreement with the model, and a circadian component has not been demonstrated.

10. In the aphid *Megoura viciae* results are consistent with a non-oscillatory clock (an hour-glass) which measures nightlength. When nights are long, egg-laying aphids (oviparae) are produced, but when dawn extends "backwards" to curtail the night it coincides with a particularly sensitive region and viviparous, parthenogenetic aphids (virginoparae) are produced. The similarities and differences between this hour-glass and the oscillation that behaves like an hour-glass in *S. argyrostoma* are noted.

11. Although time measurement in insect photoperiodism is clearly very diverse, it is possible that a continuous "spectrum" of clocks has evolved. All show some evidence of being two-component systems, and clocks with at least two components are theoretically necessary to account for all aspects of the phenomenon. At one end of this proposed "spectrum" lies *N. vitripennis* with separate "dawn" and "dusk" oscillators. In the middle are species like *S. argyrostoma* with an oscillator commencing at dusk and a possibly hour-glass component associated with the light. At the other end of the spectrum are insects like *M. viciae* in which nightlength is measured by an hour-glass which shows no oscillatory properties. How similar or how different these mechanisms really are, and whether the diversity represents convergence or a common ancestry, are questions reserved for the future.

CHAPTER 8

THE PHOTOPERIODIC COUNTER

THE photoperiodic response enables insects and other organisms to distinguish between a long day and a short day (or between a short night and a long night) and produce
a seasonally appropriate switch in metabolism. Which of the alternate pathways constitutes the "active" response—in the physiological sense—however, is open to question.
Most authors agree that it is the long-day response (i.e. the induction of *development*, or
the production of aphid virginoparae) which is the *actively* induced state, despite the fact
that homodynamic development must be the ancestral type of life-cycle, and the diapause
mechanism an acquired character facilitating dormancy during unfavourable periods
of the year. Much of the evidence suggests that long days result in the release of brain
hormone which, in turn, results in the avoidance of diapause, or the escape from diapause
in the spring; consequently most of the models for the photoperiodic clock are based on
this principle. Similarly, the production of virginoparae in the aphid *Megoura viciae* is
regarded as a mechanism "actively" operated by the extension of dawn into stage 3
(5 to 9.75 hours after dusk) of an hour-glass mechanism (Lees, 1968). The induction of
development in the external coincidence model is similarly operated by the "movement"
of dawn backwards into the preceding night to coincide with a photoinducible phase
(Pittendrigh, 1966) or, in the internal coincidence model, by the temporal interaction
of dawn and dusk oscillators, again in long days (Tyshchenko, 1966; Saunders, 1973a,
1974). In each of these models the short-day response is thought to be "passive",
caused by the failure of light to coincide with a photosensitive part of the night, or by
the failure of the active phase points of constituent oscillators to coincide in time.
Nevertheless, the concrete nature of the inductive mechanism remains unknown. In view
of this uncertainty, much of the data discussed in this and other chapters is presented
in terms of the *short-day* induction of diapause (and the long-day induction of reactivation) if only to preserve a degree of uniformity with earlier publications.

In some species a single "inductive" cycle is sufficient to produce a measurable
effect on the photoperiodic response. In *Chaoborus americanus*, for example, a single
long-day cycle (*LD 17*: 7) is sufficient to reactivate about 30 to 35 per cent of the diapausing larvae (Bradshaw, 1969). In most, if not all, species, however, a number of
such cycles is required. It is clear that the programming of the central nervous system
for subsequent development or diapause involves not only the measurement of nightlength or daylength by the photoperiodic clock, but the summation of successive cycles
to a point at which induction can occur. The summation of light cycles is therefore an
integral part of the photoperiodic response, and the observation that the number of
cycles required is temperature-compensated (Saunders, 1966a, 1971; Goryshin and
Tyshchenko, 1970) has led to the concept of a "photoperiodic counter". The interaction

between this "counter" and the rate of development during the sensitive stage—and hence with environmental factors such as temperature and nutrition—has facilitated the formal analysis of many features of the photoperiodic response.

A. The Cumulative Effects of Photoperiod

The reciprocal transfer of insects from long to short-day cycles during development may be used to map out the sensitive period and to determine the number of inductive cycles required. It may also provide information about the relative importance of long-day and short-day cycles in the inductive process. It should be noted, however, that the reversal of photoperiod from long to short, or vice versa, may produce different results from the exposure to a different number of light cycles against a non-periodic background such as continuous darkness. Examples of both types of experiment are to be found in the literature.

Early data for the oriental fruit moth *Grapholitha molesta* (Dickson, 1949) clearly indicate the cumulative effects of photoperiod. In experiments in which only part of the larval population entered diapause those larvae which emerged from the fruit first were found to be less likely to enter diapause than those which emerged later (Table 8.1.).

TABLE 8.1. THE LENGTH OF THE LARVAL FEEDING PERIOD IN *Grapholitha molesta* AND THE PROPORTION ENTERING DIAPAUSE, SHOWING THAT THOSE LARVAE EMERGING FROM THE FRUIT FIRST ARE LESS LIKELY TO ENTER DIAPAUSE THAN THOSE WHICH EMERGE LATER

Length of larval feeding period (days)	Number of larvae	Diapause (per cent)
12	10	10.0
13	63	15.9
14	76	28.9
15	46	32.6
16	40	45.0
17	22	45.5
18	17	64.7

(From Dickson, 1949.)

Working with the induction of pupal diapause in *Antheraea pernyi*, Tanaka (1950a) transferred the larvae from long daylength (*LD 18*: 6) to short daylength (*LD 9*: 15), and vice versa, at different stages of development. The results (Fig. 8.1a) indicated that the sensitive stage extended at least back to the second instar, and that diapause inhibition by long days was "stronger" than diapause promotion by short days. A cumulative effect of both long- and short-day cycles was also apparent. Williams and Adkisson (1964) demonstrated a similar cumulative effect of long days in the termination of diapause in this species: previously chilled diapausing pupae maintained for 16 weeks at *LD 12*: 12 did not break diapause, whereas those maintained *LD 16*: 8 for 1 week showed almost 50 per cent reactivation and those maintained at *LD 16*: 8 for 4 weeks showed about 90 per cent reactivation. Similar results for the induction and termination of diapause in *Ostrinia nubilalis* have been recorded by Beck *et al.* (1962) (Fig. 8.1b),

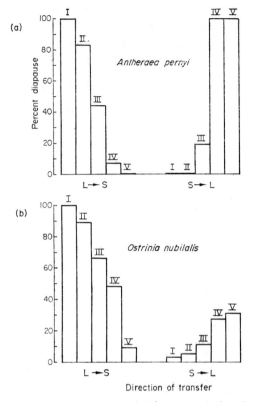

FIG. 8.1. The effect of reversing the photoperiod from long daylength to short daylength, and vice versa, on the induction of diapause in two insect species. The histograms show the cumulative effects of photoperiod and, in *Ostrinia nubilalis*, that short days are "less effective" than long days. ((a) from Tanaka, 1950; (b) from Beck, 1968.)

and by Beck and Alexander (1964); the rate of "diapause development" in long days was found to be five times as rapid as in short days. The rate of reactivation of long-day larvae was also the same when kept throughout at *LD 16*: 8 as when maintained at *LD 16*: 8 for the first 10 days and then transferred to DD (Table 8.2). This result, like

TABLE 8.2. REACTIVATION FROM DIAPAUSE IN *Ostrinia nubilalis* IN LONG DAYS (*LD 16*: 8) AND AFTER TRANSFER FROM LONG DAYS TO SHORT DAYS (*LD 12*: 12) OR TO DARKNESS (DD). All larvae were reared in short days (*LD 12*: 12) and were 22 days old at the beginning of the experiment

Photoperiodic treatment		Post-treatment photoperiod	Average time to pupation (days)
1st 10 days	2nd 10 days		
LD 16: 8	*LD 16*: 8	*LD 16*: 8	31
LD 16: 8	*LD 16*: 8	DD	27
LD 16: 8	DD	*LD 16*: 8	29
LD 16: 8	DD	DD	30
LD 16: 8	*LD 12*: 12	*LD 16*: 8	42
LD 16: 8	*LD 12*: 12	DD	>50
LD 16: 8	*LD 12*: 12	*LD 12*: 12	>50

(From Beck and Alexander, 1964.)

that described for *Nasonia vitripennis* (Chapter 7), is cogent evidence for an endogenous oscillator controlling the inductive process.

Evidence for the summation of both long- and short-day cycles has been obtained for *Acronycta rumicis* (Tyshchenko *et al.*, 1972). The reciprocal transfer of larvae from short to long photoperiods at different stages of development has shown that the "critical number" of short- or long-day cycles (i.e. the number of cycles required to produce 50 per cent diapause) was about 10 to 11 for a population from Belgorod (50°N.) (Fig. 8.2).

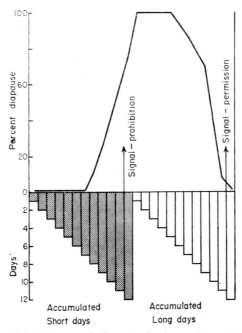

FIG. 8.2. The summation of photoperiodic cycles in *Acronycta rumicis* (Belgorod strain), showing the photoperiodic counter. The graph shows changes in the number of diapausing pupae against the accumulation of short days (dark columns) or long days (light columns). The abcissa shows the duration of larval development in days (24 days at 21°C). After summing the effects of 11 short-day cycles the photoperiodic counter gives the signal for diapause (signal—prohibition); after summing 11 long days the counter gives the signal for development without diapause (signal—permission). Only the signal which is sent last is effective. (From Tyshchenko *et al.*, 1972.)

For a population from Sukhumi on the Black Sea coast (43°N.), the critical number of long days was 6 to 7, whereas the critical number of short days was 16 to 18. Once again, long days are clearly more "effective" than short days.

The parasitic wasp *N. vitripennis* is particularly useful for the analysis of cumulative effects because the adult females constitute the sensitive stage, the effects of photoperiod are transmitted through the eggs to the larvae, and eggs are deposited on practically every day of adult life. Since the development of the progeny—either diapause or non-diapause—is fully determined by the time the eggs are deposited within the host puparium, the daily batches of offspring can be used to monitor the physiological state of the parent female, particularly with regard to the programming of the wasps by the photoperiodic regime.

Figure 8.3 shows the effect of photoperiod on the production of diapause larvae by females of *N. vitripennis* maintained in a variety of photoperiods at 18°C. Females ex-

TABLE 8.3. THE EFFECT OF PHOTOPERIOD ON THE PRODUCTION OF DIAPAUSE LARVAE BY FEMALES
OF *Nasonia vitripennis* (C STRAIN) AT A CONSTANT TEMPERATURE OF 18°C

Light per day (h)	Number of females	Mean adult life-span (days±S.E.)	Mean age at "switch"† (days±S.E.)	Mean number of offspring per female (±S.E.)	Percentage of offspring in diapause
6	20	25 1±1.44	9.4±0.40	488.4±23.26	58.9
8	24	22.4±0.93	8.4±0.34	484.7±13.66	65.2
10	19	22.3±1.64	9.6±0.33	436.3±34.78	54.7
12	24	20.5±0.92	8.5±0.34	481.7±19.80	64.2
14	19	22.9±1.34	7.9±0.36	467.6±16.88	66.4
14.5	39	24.2±1.17	9.3±0.35	604.2±29.15	50.6
14.75	39	24.3±1.23	11.7±0.50	684.3±24.49	35.6
15	39	25.8±1.25	13.4±0.82	652.2±24.17	31.8
15.25	59	27.0±0.93	14.8±0.66	691.2±19.87	24.4
15.5	20	24.2±1.23	21.5 (10)*	591.3±44.03	4.8
15.75	19	23.6±1.22	22.3 (6)*	585.8±36.64	2.6
16	44	22.5±0.94	23.5 (11)*	546.2±21.77	3.5
18	12	25.5±1.71	23.6 (5)*	530.3±40.67	2.8
20	23	22.5±0.78	20.2 (4)*	543.0±14.08	1.8

* At long daylength only a small number of females survive long enough to show a "switch" to the production of diapause larvae.

† Required day number (RDN).

(From Saunders, 1966a.)

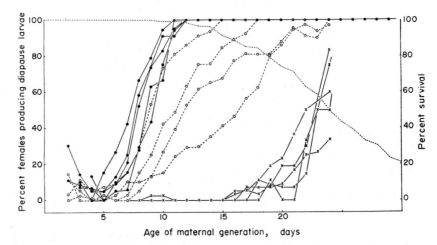

FIG. 8.3. The summation of photoperiodic cycles in *Nasonia vitripennis* at 18°C. ●——● at "strong" short daylengths (6, 8, 10, 12 and 14 hour/24); ○– – –○ at intermediate daylengths (from left to right: $14\frac{1}{2}$, $14\frac{3}{4}$, 15, $15\frac{1}{4}$ hour/24); ×——× at "strong" long daylengths ($15\frac{1}{2}$, $15\frac{3}{4}$, 16, 18 and 20 hour/24). The dashed line shows the survival rate for the 400 females used in the experiment. (From Saunders, 1966a.)

posed to "strong" short daylengths (LD 6: 18, LD 8: 16, LD 10: 14, LD 12: 12 and LD 14: 10) produced developing progeny for the first few days of adult life and then switched one by one to the production of diapause larvae. In a group of females the switch started on about the fifth day and was generally completed by the eleventh; the females then continued to produce diapausing offspring until they died (Saunders, 1966a). There was no significant difference between the various short-day groups with respect to the number of short-day cycles required to effect the switch.

At daylengths of $LD\ 15\frac{1}{2}:8\frac{1}{2}$, $LD\ 16:8$, $LD\ 18:6$ and $LD\ 20:4$ the wasps produced very few diapause larvae, the "switch", if occurring, being delayed until the end of imaginal life. These photoperiods, therefore, were strong long daylengths. Nevertheless, the fact that the wasps at long daylength did produce diapausing offspring after a suffi-cient number of such cycles (>20) is of considerable interest. At photoperiods close to the critical daylength ($LD\ 14\frac{1}{2}:9\frac{1}{2}$, $LD\ 14\frac{3}{4}:9\frac{1}{4}$, $LD\ 15:9$ and $LD\ 15\frac{1}{4}:8\frac{3}{4}$) a rapid change occurred in the mean age of the females at the switch and therefore in the overall propor-tion of the offspring produced as diapause larvae (Table 8.3). Figure 8.4a shows the "required day number" (RDN) plotted as a function of photoperiod. The required day

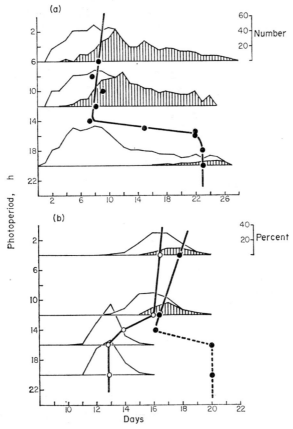

FIG. 8.4. The interaction between the required day number (RDN) and the sensitive period (SP) in two insect species. (a) In *Nasonia vitripennis* the whole of adult life con-stitutes the sensitive period and is represented by the oviposition curve; it is essentially the same at all daylengths. The number of photoperiodic cycles required to raise the proportion of diapause to 50 per cent (the RDN), however, is low at short daylength, and becomes abruptly greater as the critical daylength is passed. Consequently, at short daylengths ($<15/24$) the proportion of diapausing offspring is high, whereas at long daylength ($>15/24$) it is low. (Data from Saunders, 1966a.) (b) In *Sarcophaga argyrostoma* the period of larval development constitutes the sensitive period (SP). At short daylengths lar-val development is protracted but at long daylengths it is significantly shorter. The required day number (RDN) at short daylengths, however, is small so that a high proportion of the larvae enter diapause in the pupal instar. At long daylength the RDN is presumed to be high, and none of the larvae become dormant. O——O sensitive period, SP; ●——● required day number, RDN. The polygons show the proportion of the eggs produced (a) or puparia formed (b) per day; the shaded portions represent those insects in diapause. (Data from Saunders, 1971, 1972.)

number is defined as the number of calendar days or photoperiodic cycles (where $T = 24$)
required to raise the proportion of diapause larvae in that day's batch to 50 per cent;
it is equivalent, therefore, to the "critical day number" of Tyshchenko *et al.* (1972).
The RDN for short photoperiods varied between about 7 and 9 and the overall proportion of diapause larvae was consequently high. At long daylength the RDN was over
20 days and the overall proportion of diapause larvae was low. At the critical daylength
the RDN showed an abrupt transition.

The flesh-fly *Sarcophaga argyrostoma* is also convenient for the analysis of cumulative
effects. Since the sensitive period comes to an end at puparium formation, and a batch
of larvae form puparia over a period of several days, it follows that those larvae which
form puparia first experience fewer light-cycles than those which form puparia later
(Saunders, 1971). Figure 8.5 shows that the incidence of pupal diapause rises with an

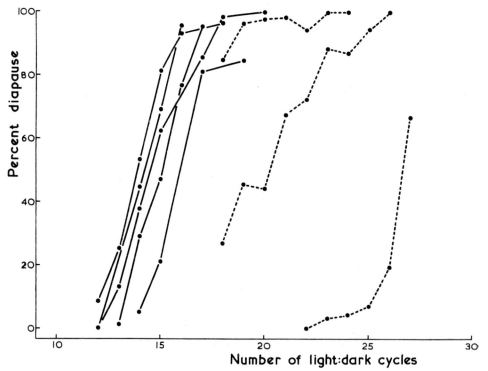

FIG. 8.5. The summation of photoperiodic cycles in *Sarcophaga argyrostoma*. ●——● at 20°C,
from left to right: *LD 12* : 12, *10* : 14, *13* : 11, *8* : 16 and *14* : 10; ●----● at 15°C, from
left to right: *LD 10* : 14, *14* : 10 and *15* : 9. (From Saunders, 1971.)

increasing number of light cycles experienced by the larvae. With short-day cycles
(*LD 8*: 16, *LD 10*: 14, *LD 12*: 12, *LD 13*: 11 and *LD 14*: 10) at 20°C those larvae
forming puparia after 12 to 13 cycles became non-diapausing pupae, whereas those
experiencing 17 to 19 cycles all became dormant; the required day number was about
14 to 15. As the photoperiod passed the critical value the RDN became higher until at
long photoperiods (>14 hr/24) it could no longer be measured because all the larvae
had formed puparia before the critical number of long-day cycles had been experienced.
As in *N. vitripennis*, however, it is likely that long-day cycles are also inductive (in the

diapause-induction sense) if sufficient of them are seen before the end of the sensitive period.

Subsequent experiments (Saunders, 1972) have shown that larvae of *S. argyrostoma* raised in long daylength (>14 hr/24) develop more rapidly than those raised at short daylength (Chapter 4). Consequently, the sensitive period is shorter in the long-day larvae. Figure 8.4b shows the "antagonistic" interaction between the sensitive period and the required day number. At short daylength (*LD 4*: 20 and *LD 12*: 12) the duration of the sensitive period (SP) was relatively long (>16 days) and the RDN short. The proportion of pupae entering diapause was therefore high because a high proportion of them were able to experience a sufficient number of photoperiodic cycles before puparium formation. Conversely, at long daylength (*LD 16*: 8 and *LD 20*: 4) the SP was relatively short (~13 days) and the RDN (presumably) too great to measure; consequently diapause pupae did not appear. Further evidence for this interaction between the sensitive period and the required day number will be described in the next section.

B. Interactions between the Photoperiodic Counter and Environmental Variables

1. *Temperature*

With long-day insects, low temperature and short daylength tend to complement each other and increase the incidence of diapause, whereas high temperature and long daylength work together to avert diapause (Chapter 6). In at least three insect species, *Nasonia vitripennis*, *Acronycta rumicis* and *Sarcophaga argyrostoma*, this relationship between temperature and photoperiod can be interpreted in terms of an interaction between the photoperiodic counter and the rate of development. This relationship is probably widespread if not universal.

Females of *N. vitripennis* incubated at 15°, 20°, 25° and 30° in *LD 12*: 12 all reacted to the strong short daylength and switched to the production of diapause larvae within 11 or 12 days (Saunders, 1966a). The required day numbers (RDN) for the four groups were 8.4, 7.6, 8.4 and 6.9 days, respectively, and therefore showed a high degree of temperature compensation ($Q_{10} = 1.04$). Life span and the rate of oviposition, however, showed a more normal relationship to temperature. Thus, at 30°C, the wasps showed a short life (11.5 days) and a rapid rate of oviposition with a peak 3 to 5 days after emergence, whereas at 15°C the wasps showed a protracted life (32.7 days) and a slower rate of oviposition with a peak about 14 days after emergence (Fig. 8.6). The interaction between the temperature-dependent rate of oviposition and the temperature-compensated mechanism accumulating light-cycles resulted in a high incidence of diapause amongst the progeny produced at 15°C (90.9 per cent), somewhat lower at 20°C (71.2 per cent) and 25°C (61.0 per cent), and a low incidence at 30°C (27.4 per cent).

A very similar relationship between developmental rate and RDN was observed with *S. argyrostoma* (Saunders, 1971). Batches of larvae were set up at short daylength (*LD 10*: 14) and at 16°, 18°, 20°, 22°, 24° and 26°C. As the larvae formed puparia they were collected, separated from the rest of the group and incubated in the dark at 20°C. Figure 8.7 shows the pattern of puparium formation at the various temperatures with the shaded portion of the polygons representing the diapausing pupae in each batch. The length of

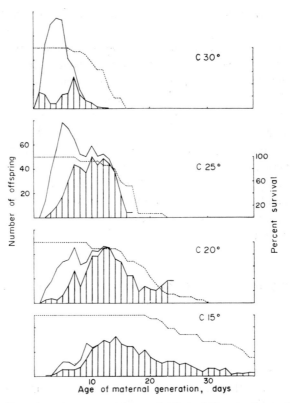

FIG. 8.6. The effect of temperature on the production of diapause larvae by females of *Nasonia vitripennis* maintained at short daylength (*LD 12* : 12), showing its relation to egg production. The solid line shows the rate of egg production, with the shaded portion of each polygon the proportion of the larvae subsequently entering diapause. The dotted line shows the survival rate for the females in each group. Note that temperature has a marked effect on survival and on oviposition rate (equivalent to the sensitive period in this species), whereas it has a negligible effect on the number of short-day cycles needed to effect the switch to diapause (the RDN). Consequently, the proportion of offspring produced as diapause larvae falls as the temperature rises. (From Saunders, 1966a.)

larval development (the sensitive period) was clearly temperature-dependent, being about 9 days at 26°C and about 22 to 23 days at 16°C. At 26°C none of the pupae entered diapause whereas at 18° and 16°C practically all did so. At the intermediate temperatures (24°, 22° and 20°C) both developing and diapausing pupae were produced. The curves for puparium formation at these intermediate temperatures were in two cases clearly bimodal, with the developing individuals in the first peak and the diapausing individuals in the second. This suggests that those larvae which are destined for continuous development have a shorter developmental time than those which are destined to enter diapause; this might, in part, account for the increase in diapause incidence with age. Nevertheless, when the incidence of pupal diapause in the individual groups was plotted as a function of the number of light-cycles experienced as larvae (Fig. 8.8) it became clear that there was a "family" of curves all with the same general and upward trend. This is compelling evidence that larvae of *S. argyrostoma* are able to "add up" successive light-cycles and react accordingly. At 24°, 22° and 20°C the first larvae to form puparia (after 9 to 11 cycles) showed continuous development as pupae, whereas those which experienced a greater number of cycles became dormant. The larvae reared at

18° and 16°C experienced 17 or more short-day cycles and all of them became diapausing pupae.

Figure 8.7 shows that the rate of larval development—or the duration of the larval sensitive period—was temperature-dependent, with a Q_{10} of about 2.7; this is normal for

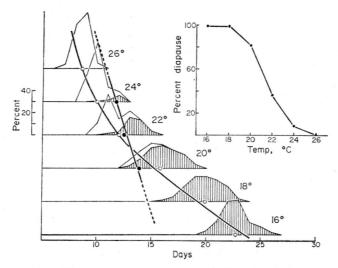

FIG. 8.7. The effect of temperature on the induction of pupal diapause in *Sarcophaga argyrostoma* at LD *10* : 14, showing the interaction between the sensitive period (SP) and the required day number (RDN). The polygons show the proportion of each batch of larvae forming puparia each day; the shaded portion of the polygons those larvae which became diapausing pupae. Note that the SP and RDN have different temperature coefficients. At high temperatures (26 and 24°C) the SP is shorter than the RDN and few, if any, of the pupae enter diapause, whereas the opposite is true at lower temperature (18 and 16°C). *Inset:* the effect of temperature on the proportion of diapause pupae at LD *10* : 14. (Redrawn from Saunders, 1971.)

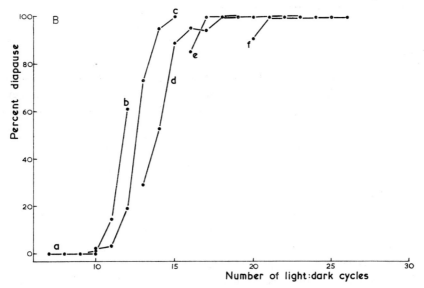

FIG. 8.8. The effect of temperature on the summation of photoperiodic cycles (LD *10* : 14) in *Sarcophaga argyrostoma*: a—26, b—24, c—22, d—20, e—18, f—16°C. (From Saunders, 1971.)

a physiological process of this kind. On the other hand, the temperature coefficient for the number of light-cycles needed to raise the proportion of diapause pupae to 50 per cent (the RDN) is much nearer unity ($Q_{10} = 1.4$), and therefore shows a high degree of temperature-compensation. The mechanism controlling diapause induction in *S. argyrostoma*, therefore, is similar to that in *N. vitripennis*. It depends on an interaction between a temperature-dependent process (the length of the sensitive period, SP) and a temperature-compensated process (the summation of light-cycles, or the RDN). At 26°C larval development is so rapid that the larvae experience too few short-day cycles before the sensitive period is terminated by puparium formation; consequently none of the pupae enter diapause. Conversely, at 18° and 16°C, the sensitive period is so protracted that more than a sufficient number of inductive cycles are experienced before the end of the sensitive period, and practically all of the insects become dormant. Figure 8.8 suggests that the required day number (RDN) at *LD 10*: 14 is about 13 to 14, and that about 17 to 19 are required to complete the switch to diapause.

Later observations on *S. argyrostoma* (D. Gibbs, unpublished) have shown that a transfer of newly formed puparia to a higher temperature results in a reduction of diapause incidence, whereas a transfer to a lower temperature causes an increase. Consequently, in the experiments described above—in which all puparia were incubated at 20°C—the observed incidence of diapause at 18° and 16° was probably *lower* than if the insects had been kept at these temperatures throughout, and the observed incidence of diapause at 24° and 22° was probably higher. It is likely, therefore, that the temperature coefficient for the summation process is much closer to unity.

Goryshin and Tyshchenko (1970) have demonstrated a similar phenomenon with the photoperiodic counter in a Belgorod strain of *Acronycta rumicis*. At temperatures between 18° and 26° C the length of larval development was dependent on temperature, but the number of either long-day (*LD 22*: 2) or short-day (*LD 12*: 12) cycles applied at the end of larval development, and required to produce a "critical" (50 per cent) level of diapause, was roughly the same at all temperatures. The RDN in this species, therefore, is also temperature-compensated.

2. Nutrition

In Chapter 6 it was seen that the photoperiodic response can be modified by both qualitative and quantitative aspects of nutrition. In a few insect species there is evidence to suggest that these nutritional factors affect the balance between the required day number (RDN) and the length of the sensitive period (SP). In *Nasonia vitripennis*, for example, host deprivation causes starvation and delays oviposition whilst the photoperiodic counter continues to function normally, so that the proportion of diapause offspring is increased. The type of host puparium offered to the wasps, however, constitutes a qualitative difference which may alter the number of light-cycles needed to effect the switch. Similarly, starving the larvae of *Sarcophaga argyrostoma* may shorten the larval sensitive period with marked effects on the incidence of pupal diapause.

One of the most important factors affecting the biology of *N. vitripennis* is the availability of hosts, because host puparia provide a protein supply for the adult wasps and a place in which to deposit the eggs (Roubaud, 1917). If blowfly puparia are readily available feeding and oviposition occur without delay. Host shortage, on the other hand, results in starvation and egg retention. In a newly emerged wasp so deprived, the few eggs which develop from larval reserves undergo a slow cycle of resorption in the ovary (Edwards, 1954; King, 1963), and only when hosts are again available can full egg production proceed.

Newly emerged females of *N. vitripennis* maintained at 18°C and at short daylength (*LD 12*: 12) were deprived of host puparia *(S. argyrostoma)* for 3, 5 and 7 days, respectively, before being supplied with two hosts per day for the rest of their lives (Saunders, 1966b). A control group was supplied with hosts daily throughout the experiment. Analysis of the progeny produced by these wasps gave the results shown in Table 6.3 and Fig. 8.9. The females of the control group produced about 630 offspring during a mean life-span of 32.0 days, with a peak in the oviposition rate on about the tenth day. The required day number for this group was 9.1 days, and about 73 per cent of the progeny entered diapause. Being unable to feed, the host-deprived groups starved and were unable to develop and deposit eggs. Under the most severe conditions of starvation

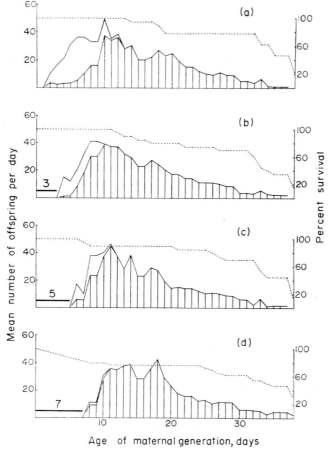

Fig. 8.9. The effect of host deprivation on the production of diapause larvae by females of *Nasonia vitripennis* at 18°C and *LD 12* : 12. (a) Nineteen females provided with two host puparia daily; (b) twenty females deprived of hosts for 3 days; (c) eighteen females deprived of hosts for 5 days; (d) thirty-one females deprived of hosts for 7 days. The polygons show the mean rate of egg production; the shaded portion the proportion of them becoming diapause larvae. The dotted line shows the survival rate for the females of the group. Note that host deprivation delays the onset of oviposition without altering the required day number; consequently the proportion of diapause larvae rises with the increase in the period of host deprivation. (From Saunders, 1966b.)

(7 days without hosts) eight (20 per cent) of the group died before a protein meal could be obtained. The survivors, however, showed no significant reduction in longevity or fecundity once provided access to host puparia.

Figure 8.9 shows that host deprivation had a marked effect on the overall pattern of diapause production. For instance, a 3-day period without hosts raised the proportion of diapause larvae to 86 per cent, and 5 and 7 days without hosts to 91 and 99 per cent, respectively. In other words, oviposition was delayed during the period of starvation but the photoperiodic counter continued to operate normally. Once again, therefore, the degree of the photoperiodic response can be attributed to an interaction between rate of development and the RDN.

The use of puparia other than *S. argyrostoma* as host for *N. vitripennis* had quite a different effect on diapause induction (Saunders *et al.*, 1970). With *Calliphora erythrocephala* (= *vicina*) the wasps showed a much greater survival and in some experiments a higher fecundity. The number of short-day cycles (*LD 13½ : 10½* or *LD 14½ : 9½*) required to effect the switch to diapause production (the RDN), however, was also considerably *increased*, from about 8.0 with *S. argyrostoma* to between 12.4 and 17.7 in various experiments with *C. erythrocephala*. A similar increase in the RDN was observed with wasps supplied with puparia of *Phormia terrae novae*. The net result (Fig. 8.10) was an overall *reduction* in the proportion of the progeny produced as diapausing larvae.

Wasps supplied with the smaller *C. erythrocephala* and *P. terrae novae* were able to drill through the relatively thin puparia of these species in about 9½ and 12½ minutes respectively. Those provided with the large and thick puparia of *S. argyrostoma*, however, needed about 74 minutes to complete the drilling process. The ease in drilling and the consequent ready access to haemolymph almost certainly accounts for the increased

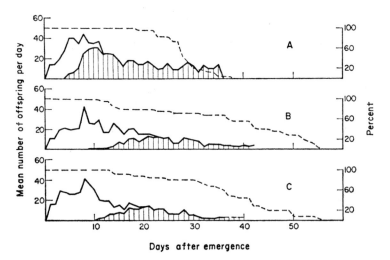

FIG. 8.10. The effect of host blowfly species on the production of diapause larvae by females of *Nasonia vitripennis* at 18°C and short daylength (*LD 14½ : 9½*). A—twenty-three females supplied with two puparia of *Sarcophaga argyrostoma*/day; B—twenty-five females supplied with two puparia of *Calliphora erythrocephala*/day; C—twenty-five females supplied with four puparia of *C. erythrocephala*/day. The polygons show the rate of egg production, the shaded portion the proportion of them becoming diapause larvae. The dashed lines show the survival rates for the females in each group. Note that the use of *Calliphora erythrocephala* as the host species delays the onset of diapause larva production (increases the required day number) and thereby reduces the overall proportion of the progeny which enter diapause. (From Saunders *et al.*, 1970.)

TABLE 8.4a. PUPARIUM THICKNESS AND THE TIME TAKEN BY FEMALES OF
Nasonia vitripennis TO DRILL THROUGH THE PUPARIUM

Host species	Puparium thickness (μ)	Time taken to drill (min$\pm$S.E.)
Sarcophaga argyrostoma (large puparia)	112–138	74.5$\pm$3.31 (45)
S. argyrostoma (small puparia)	34–48	13.9$\pm$0.86 (86)
Calliphora erythrocephala	32–42	9.5$\pm$0.62 (90)
Phormia terrae novae	58–70	12.4$\pm$0.68 (74)

Number of observations in parentheses.

TABLE 8.4b. THE PRODUCTION OF DIAPAUSE LARVAE BY FEMALES OF *Nasonia vitripennis* SUPPLIED
WITH LARGE AND SMALL PUPARIA OF *S. argyrostoma*, PUPARIA OF *C. erythrocephala*, AND
SMALL AND LARGE *S. argyrostoma* ON ALTERNATE DAYS

Host	No. of females	Mean adult life-span (days$\pm$S.E.)	RDN (days$\pm$S.E.)	Mean no. of offspring/ female ($\pm$S.E.)	Offspring in diapause (%)
Large *Sarcophaga argyrostoma*	18	21.3 $\pm$1.66	10.2 $\pm$0.72	313.3 $\pm$36.3	50.9
Small *Sarcophaga argyrostoma*	19	52.1† $\pm$2.10	12.4 0.82	555.3† $\pm$28.4	54.8
Calliphora erythrocephala	18	52.9† $\pm$2.39	15.2† $\pm$1.2	481.8† $\pm$20.1	36.8
*Small *Sarcophaga*/ large *Sarcophaga* on alternate days	18	31.8 $\pm$1.85	13.1 $\pm$0.54	412.3 $\pm$28.4	47.1

† P<0.01. Difference between mean marked with daggers and corresponding mean for the large
Sarcophaga group.

* This group was added at a later date. Since it is not from the same population as the other groups it
is not strictly comparable.

(From Saunders *et al.*, 1970.)

longevity and fecundity shown by the wasps provided with puparia of the two Calli-
phorines. It does not account for the increased RDN, however: wasps supplied with *small*
puparia of *S. argyrostoma* with a puparium thickness equivalent to that in the Calli-
phorines completed the drilling process in less than 14 minutes, showed a correspondingly
high longevity and fecundity, but a relatively low RDN (Table 8.4b). The effect of *C.
erythrocephala* and *P. terrae novae*, therefore, is considered to be a *qualitative* aspect of
nutrition affecting the adult wasps which feed on the haemolymph exuding from the
puncture in the host puparium (see also Chapter 6).

If larvae of *S. argyrostoma* are manually extracted from their larval medium before
the feeding process is completed, or if the larvae are grossly overcrowded, starvation
results and the small-sized larvae undergo premature puparium formation. The shortening of the larval sensitive period causes a reduced incidence of pupal diapause (Fig. 8.11a),
presumably because the number of short-day cycles they are able to experience is reduced

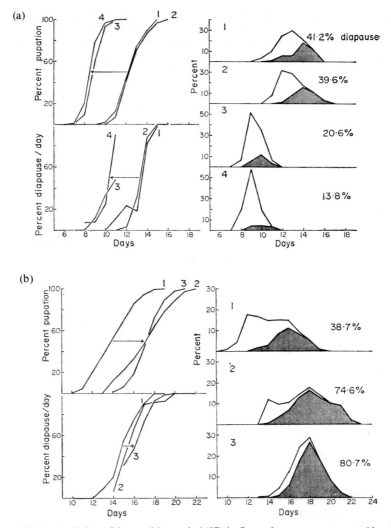

FIG. 8.11. Manipulation of the sensitive period (SP) in *Sarcophaga argyrostoma* and its effect on diapause induction. (a) Overcrowding the larvae within a limited quantity of meat: cultures maintained at LD 12 : 12 and 20 °C. *Upper left:* cumulative per cent pupation of larval cultures, *Lower left:* percentage of each day's pupae entering diapause (arrow at the 50 per cent point indicates the required day number, RDN). *Right-hand panels:* the polygons show the proportion of each batch of larvae pupating per day; the shaded portion represents those larvae which entered diapause in the pupal instar. 1 and 2 were "normal" uncrowded batches of larvae; 3 and 4 were grossly overcrowded. (b) Allowing the mature larvae to disperse into wet rather than dry sawdust; cultures maintained at LD 12 : 12 and 20°C. 1—larvae dispersing into dry sawdust, 2—larvae dispersing into wet sawdust, 3—larvae dispersing into wet sawdust and then transferred to dry sawdust after 4 days. Note that overcrowding shortens larval development (the sensitive period) more than it shortens the RDN, thereby reducing the proportion of the pupae entering diapause. Dispersal into wet sawdust, on the other hand, delays puparium-formation (lengthens the SP) and therefore raises the proportion of pupae entering diapause. (Saunders, 1975 c.)

without a correspondingly great reduction in the RDN (Saunders, 1975 c). Conversely, allowing fully fed larvae to disperse into *wet* rather than dry sawdust, *delays* puparium formation (Ohtaki, 1966, Ohtaki *et al.*, 1968) and causes a corresponding increase in pupal diapause (Fig. 8.11b).

Clay and Venard (1972) have recently described a similar effect on the induction of larval diapause in the mosquito *Aëdes triseriatus*. Larvae maintained at *LD 14* : 10 and provided with an "adequate" diet (40 mg of pulverized Purina chow/week for 20 larvae) entered diapause; the effects of photoperiod were clearly cumulative. The incidence of larval diapause, however, was raised if the larvae were provided with an "inadequate" diet (10 mg chow/week for 20 larvae), or kept at a lower temperature; both of these treatments slowed development and presumably allowed the larvae to "see" a greater number of inductive cycles before the end of the sensitive period.

C. The "Programming" of the Central Nervous System by Photoperiod

The photoperiodic receptors and the clock are located in the brain (Chapter 10) and the brain neurosecretory cells appear to be the endocrine effectors regulating the onset and termination of diapause. We may therefore assume that the photoperiodic "information" is integrated and stored in the brain, but we know very little about the nature or the exact location of this storage. Despite these gaps in our understanding of the concrete physiological processes involved in the photoperiodic regulation of development, we probably know enough of the formal properties of the system to indulge in a little speculation.

Photoperiodic control is a two-stage process. One stage comprises the clock which enables the insect to distinguish between a short and a long day, and the second comprises the photoperiodic counter which "adds up" successive cycles to a point at which induction occurs. The required day number (RDN) for a particular species is constant at a particular photoperiod, and is temperature compensated. The final expression of the photoperiodic response is frequently the result of an interaction between the length of the sensitive period and the required day number. In insects with a circadian clock, time measurement and the discrimation between a long and a short-day cycle, is accomplished by a limited range of steady states of the constituent oscillators. The required day number probably represents the accumulation of the product of the coincidence between an oscillator and light, or the coincidence between constituent oscillators, over successive cycles to a threshold level. There is ample evidence for such product accumulation since all insects require a number of photoperiodic cycles to effect induction. As stated earlier, however, we have no concrete information as to whether long days "actively" promote a substance which accumulates to a threshold enabling the neurosecretory cells to release their product (the most favoured hypothesis), or whether, for instance, short days result in the accumulation of a substance which inhibits the release of neurohormones.

In most insects the sensitive stage and the resulting diapause occur in the same individual, although often in different instars. Therefore, although there may be considerable internal reorganization of tissues and organs during metamorphosis, there is a continuity of the CNS between the two. This continuity, however, is not essential: in *Nasonia vitripennis*, for example, the photoperiodic "information" is transmitted via the undifferentiated egg. There are at least three ways in which this might be achieved: (1) product accumulation occurs at either long or short daylength, and the product is incorporated into the ovarian egg as a cytoplasmic factor; (2) product accumulation results in a substance which operates a genetic "switch" mechanism in the oocyte nucleus; or (3) there is simply a transfer of circadian phase from the mother to the larva via the egg: this, for example, is known to occur in the Queensland fruit-fly, *Dacus tryoni* (Bateman, 1955).

The first two seem most likely in view of the strong evidence for some sort of product accumulation. There is, however, little to enable a choice to be made betwen (1) and (2), unless the probable "dilution" of the cytoplasmic factor during morphogenesis rules out the first.

Williams (1969b) has likened the development of a multicellular organism to the reading" of a genetic "construction manual" inherited from the preceding generation. In the higher insects this construction manual contains information for larval, pupal and adult characters in that order, and Williams sees the advanced forms of metamorphosis as involving the "derepression and acting out of what is little short of successive batches of genetic information". Since developmental characters such as the number of moults and the form of the photoperiodic response (Chapter 6) are "coded for" in the genome as well as morphological characters, it seems possible that the photoperiodic input, if containing the "correct" information, operates a genetic "switch" mechanism directing development down alternate pathways. This hypothesis is obviously attractive for species such as the aphid *Megoura viciae* in which daylength controls morph determination (Lees, 1959), or for the *levana* and *prorsa* forms of the butterfly *Araschnia levana* in which the seasonal morphs are closely linked to the diapause or non-diapause state (Müller, 1955). In the case of diapause induction, the process might involve the derepression of genes opening up different developmental pathways by the ultimate (but presumably not direct) action of photoperiod on the information store in the genome.

Annotated Summary

1. Superimposed on the photoperiodic clock is a second mechanism called the "photoperiodic counter" which accumulates successive long- and/or short-day cycles to a point at which diapause or development are determined.

2. Both long and short days are "added up", although the effects of the former are frequently more pronounced.

3. In at least three species (*Nasonia vitripennis*, *Acronycta rumicis*, and *Sarcophaga argyrostoma*) the summation of photoperiodic cycles has been shown to be a temperature-compensated process.

4. In these species the induction of diapause is controlled by an interaction between the length of the sensitive period (SP) and the number of short-day cycles needed to raise the proportion of diapause in a days' batch to 50 per cent (the required day number, RDN). These two components have different temperature coefficients, with the latter showing a high degree of temperature-compensation. Insects raised at high temperature therefore reach the end of their sensitive period before "seeing" a sufficient number of inductive cycles; consequently they develop without arrest. At lower temperature, however, they see a sufficient number of inductive cycles before the end of the sensitive period and diapause supervenes. It is considered that this phenomenon is of wide occurrence, if not universal.

5. In some insects diet is known to affect diapause incidence by a similar mechanism. "Poor" diet, either in a quantitative or qualitative sense, may protract or curtail the sensitive period, with predictable effects on diapause incidence.

CHAPTER 9

OTHER TYPES OF INSECT CLOCK

FOUR types of clock are described in this chapter: each has a different and usually clear functional significance, and the four types show a variety of controlling mechanisms. The so-called "time-sense" *(Zeitsinn)* or "time-memory" *(Zeitgedächtnis)* of honey-bees (Beling, 1929), and the time-compensated sun orientation of bees and other organisms (von Frisch, 1950; Kramer, 1950) are controlled by, or contain, an endogenous circadian component. The rhythm of emergence in the marine midges, *Clunio* spp. (Neumann, 1963, 1966 a, b) is controlled by a "combination" of endogenous circadian and semi-lunar rhythms entrained by the dominant environmental periodicities in their intertidal environment. Lastly there are several "long-period" timers controlling the onset of diapause or seasonal morphs (Blake, 1958; Lees, 1960) which may be either rhythmic or non-rhythmic in nature, but not clearly related to the circadian or hour-glass timers involved in "classical" photoperiodism.

A. The Time-memory *(Zeitgedächtnis)* of Bees

The ability of honey-bees *(Apis mellifera)* to return to a food source at the same time each day has been known since the turn of the century when the Swiss naturalist August Forel observed bees arriving at his breakfast table for food. Since they always came at the same time—even when food was *not* present—Forel (1910) proposed that the bees had a "memory" for time *(Zeitgedächtnis)*. Similarly, Von Buttel-Reepen (1900) observed that bees only visited a buckwheat field in the morning when the blossoms were secreting nectar; he concluded, therefore, that the insects possessed a "time-sense" *(Zeitsinn)*.

Modern work on the bees' *Zeitgedächtnis* began with Beling in 1929. She trained bees at an artificial feeding place by offering sugar solution at the same time each day. Individual bees were marked with paint when they were feeding on the sugar. During subsequent days (the test period) the feeding place was without sugar, but each visiting bee and its time of arrival was recorded. Beling demonstrated that bees do indeed come back at the same time each day (Fig. 9.1); she also showed that bees could be "trained" to come at *any* time of the day and, moreover, that they could be trained to come at two or more separate periods during the day provided that the interval between two successive training periods was greater than 2 hours.

These experiments confirmed the early observations of Forel and von Buttel-Reepen but did not answer the question whether the bees possessed an innate "time memory" or "clock", or whether they were merely responding to external signals such as the position of the sun in the sky. Beling (1929), however, also conducted an experiment in which obvious time cues, such as the daily cycles of light, temperature and humidity, were

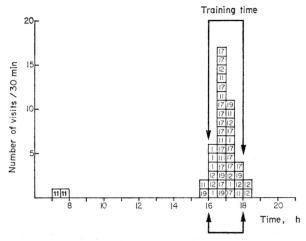

FIG. 9.1. The time-memory *(Zeitgedächtnis)* of bees. The bees were "trained" to come to a sugar source at a fixed feeding position at the same time (16 to 18 hours) during several consecutive training days. As they visited the sugar they were marked individually. On the "test day" the sugar was omitted, but the bees continued to arrive at the dish at the same time of the day. The numbers refer to the individually marked bees. (After Beling, 1929.)

removed. Subsequently, Wahl (1932) took elaborate procedures to exclude even cosmic radiation by conducting the entire experiment in a salt mine 180 m below the surface of the earth. In both experiments the bees returned punctually to the feeding place during the test period, indicating the endogenous nature of the clock involved. Beling (1929) and Wahl (1932) also showed that it was impossible, even after weeks of training, to train bees to a feeding rhythm too far removed from that of the solar day. Attempts to train bees to a 48-hour rhythm, for example, resulted in foraging activity every 24 hours (Beling, 1929). It seemed, therefore, that the bees' time sense was—in modern terminology—controlled by an endogenous circadian oscillation.

The unequivocal test for endogeneity, however, was not performed until the 1950s. Renner (1955, 1957) using essentially the same technique as Beling, trained bees to a food source in Paris (2°E.) between 8.15 and 10.15 local time in a closed chamber in constant light (*LL*) and constant temperature. The bees were then transported overnight to New York (74°W.) and tested the following day under identical conditions. In this now classical "translocation" experiment, the bees had been transported over 76° of longitude, or a difference of about 5 hours in real local time. If an endogenous circadian rhythm was involved the bees should have come to the test dish 24 hours after the training period; if, on the other hand, the bees were responding to subtle local influences, they should forage at the same *local* or sun time. The results showed that the former alternative was the case: the bees came to the feeding dish at 3.00 Eastern daylight time, exactly 24 hours after their last feeding period in Paris. The reciprocal experiment involving a translocation from New York to Paris had an analogous result.

This experiment was subsequently repeated in the open. Renner (1959) trained bees to a sugar source in a field on Long Island, N.Y., between 12.54 and 2.54 p.m. Eastern standard time. They were then flown overnight to Davis, California (a change in longitude of 49°, and a difference in real local time of 3 hours 15 minutes) and tested in the open on successive days. The results showed that foraging activity occurred initially at a time 24 hours after the last feeding period on Long Island, but then showed signs of re-

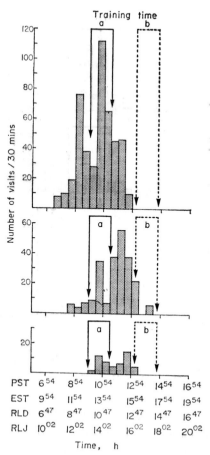

FIG. 9.2. Visiting frequency of bees which were trained in the open air to a feeding time of 12. 54–2. 24 p.m. EST at St. James (Long Island, N.Y.) and which were tested after an overnight translocation over 49° longitude at Davis, California. The three panels show the visiting frequency of the bees on three consecutive test days at Davis. a—24-hour term of the training period, b—day time at Davis which corresponds to training day time at St. James. Abscissa: PST Pacific Standard Time, EST Eastern Standard Time, RLD Real local time at Davis, RLJ Real local time at St. James. Ordinate: number of photoelectrically induced recording impulses caused by the bees which looked for food at the recording boxes. (Redrawn, after Renner, 1960.)

entrainment to the local light cycle by the third day (Fig. 9.2). This behaviour, therefore, was very similar to that of endogenous activity rhythms subjected to similar phase-shift experiments (Chapters 2 and 3). Further evidence for the endogenous and circadian nature of the bees' *Zeitgedächtnis* was obtained by Bennett and Renner (1963), Beier (1968) and Beier and Lindauer (1970) who demonstrated that the rhythm free-ran in *LL* with a natural period (τ) of 23.4 to 23.8 hours.* In addition, despite the failure of earlier workers to entrain the bees to cycles other than 24 hours (Beling, 1929; Wahl, 1932), they showed that entrainment was possible to *LD* cycles between 20 and 26 hours. The primary range of entrainment was therefore similar to other circadian oscillations.

* τ for locomotor activity in *Apis mellifera* is 21.8 hours in DD for workers; 23.7 in DD for drones (Spangler, 1972, 1973).

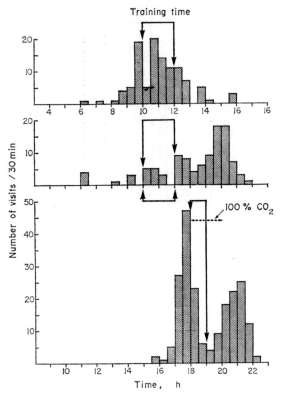

FIG. 9.3. The effect of cold anaesthesia and CO_2 narcosis on the bees' *Zeitgedächtnis*. *Top panel:* control experiment, bees trained to feed between 10 and 12 hours. *Middle panel:* bees kept for $3\frac{3}{4}$ hours at 4–5°C after the last training period. (Redrawn, after Renner, 1960.) *Lower panel:* bees trained to visit food source between 18 and 19 hours, and then narcotized in 100 per cent CO_2 for 2 hours beginning at their usual training time, and kept in constant illumination from narcosis until the next day, when the feeding place was constantly observed. Individually marked bees appeared in both "peaks" of activity; this result suggests that two or more circadian oscillations are involved in the bees' *Zeitgedächtnis*. (Redrawn, after Medugorac and Lindauer, 1967.)

A number of authors have attempted to influence the *Zeitgedächtnis* clock by changing the bees' metabolism (Wahl, 1932; Grabensberger, 1934; Kalmus, 1934; Werner, 1954; Renner, 1957). These experiments showed that increasing the rate of metabolism with thyroxine, or decreasing the rate of metabolism with quinine, had no effect on the time-sense, but chilling to 4 or 5°C for about 5 hours caused a 3- to 5-hour delay in their arrival at the test feeding dishes (Fig. 9.3). After CO_2-narcosis (Medugorac, 1967; Medugorac and Lindauer, 1967) the bees turned up at the original feeding time *and* some time later, the delay of the second peak depending on the duration of narcosis and on the concentration of the CO_2. Medugorac and Lindauer (1967) concluded that at least two "clocks" were involved in *Zeitgedächtnis*, one which could be delayed by CO_2 narcosis, the other which could not. Hoffmann (1971) points out that this result, and its interpretation, is consistent with the multioscillator concept developed for the clocks controlling activity rhythms and for photoperiodism (see Chapter 4).

The "time-memory" of bees is therefore an endogenous circadian rhythm which possesses properties common to all circadian systems. Since bees can be trained to come to a food source at any time of the day, or to more than one such time, *Zeitgedächtnis* comes under Pittendrigh's (1958) designation of a "continuously consulted" clock. It is also probably of Truman's (1971) type-II clock, because of its obvious affinities to an activity rhythm. The adaptive significance of this type of clock is obvious: it enables bees to return to a known food source when nectar and pollen are most readily available (Kleber, 1935), and therefore to maximize productivity. One advantage of the clock being oscillatory is that bees can stay in the hive for a day or two because of bad weather and still "remember" the time when a particular plant species secretes its nectar. The fact that the rhythm is fairly easily extinguished without positive reinforcement, however, is also of biological importance because there is an ever-changing array of nectar sources, and there is little selective advantage in continuing to arrive at flowers long past their best.

B. Time-compensated Sun Orientation

A number of insect species are known to orientate themselves by maintaining a fixed angle to the Sun, to an artificial light source, or to the pattern of polarized light from a blue sky. Santschi (1911, 1913), for example, showed that ants used the position of the Sun (the "light compass reaction") to maintain a straight course in territory poor in landmarks. He also showed that if the ants were shielded from the direct rays of the Sun, but exposed to its reflection in a mirror, they altered their course in a predictable fashion. The Sun, of course, "moves" during the day, and if the light-compass reaction is to be at all useful for longer-term orientation, the animals must be able to compensate for such changes in the Sun's azimuth. Early experiments by Brun (1914) with *Lasius niger* seemed to discount this possibility: if ants were detained for a few hours in a darkened box and then released, they continued at the *same* angle to the Sun and therefore in a different compass direction (but see the work of Jander, 1957, below). Wolf (1927) demonstrated light-compass orientation in the honey-bee but also found no evidence to suggest that the insects could compensate for the Sun's motion.

The fact that some organisms possess an innate biological clock for such time-compensation, however, was demonstrated almost simultaneously and independently by von Frisch (1950) for the honey-bee and by Kramer (1950) for the starling. Von Frisch trained bees to visit a feeding place west of the hive in the evening, then moved the hive during the night to a new site in unfamiliar surroundings. When the bees' behaviour was tested the next morning they were found to forage to the west, even though they now had to fly away from the Sun instead of towards it as in training. A similar experiment was later conducted by von Frisch and Lindauer (1954). Bees were trained to visit a sugar source 180 m north-west of the hive and the whole colony was then moved overnight to a new and unknown territory. Next morning the bees were offered four feeding choices each placed at the same distance from the hive (180 m) but at different compass directions (NE., SE., SW. and NW.). Despite the fact that there were no familiar landmarks and that the sun was now in the east instead of the west, the majority of the bees came to the feeding table in the training direction, i.e. to the north-west (Fig. 9.4). Subsequently Meder (1958) trained bees to a certain direction from the hive, then captured them at the feeding table and kept them for one or more hours in the dark. When they were released they flew unerringly in the direction they had been trained despite the fact that the Sun had "moved" during their captivity. These experiments all indicate that bees are able to compensate for the movement of the Sun allowing them to maintain a constant compass direction.

13*

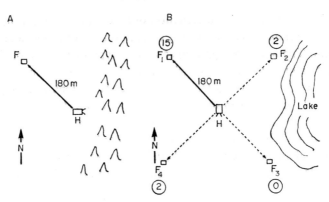

FIG. 9.4. Time-compensated sun orientation in the honeybee. A—a beehive was placed in an unknown region, and a group of bees was fed *in the afternoon* on a feeding table 180 m NW. B—during the night the hive was translocated to another area and *in the morning* the bees had to choose one of four feeding tables 180 m NE., NW., SE or SW. of the hive. The new landscape did not offer any familiar landmarks; the Sun stood at another angle relative to the training line as in the previous afternoon. Nevertheless, most bees (encircled numbers) came to the NW., i.e. the bees had calculated the Sun's movement. (After Lindauer, 1960.)

Using starlings, Hoffmann (1953, 1954, 1960) was able to demonstrate that the clock used in time-compensated sun orientation possesses the same properties as that controlling locomotor activity, and is therefore oscillatory and circadian in nature. In one experiment Hoffmann (1953, 1954) trained starlings to look for food in a given compass direction and then shifted the LD cycle so that it was out-of-phase with that experienced by the birds during training. After 12 to 18 days in the artificial day 6 hours behind local time, the direction in which the birds looked for food was predictably shifted, indicating that the clock involved in orientation had become entrained to the shifted LD cycle. After return to an LD cycle *in phase* with the training programme the birds returned to their original training direction. The clock was also shown to be gradually drawn into phase with an experimental light cycle in a way which was consistent with a circadian oscillator. Later experiments (Hoffmann, 1960) clearly demonstrated that the direction chosen by starlings, and the rhythm of locomotory activity, both free-ran with an endogenous periodicity of less than 24 hours when the birds were transferred to conditions of constant light and constant temperature.

To an observer (or to a bee) in the northern hemisphere the Sun appears to "move" in a clockwise direction from east to west, whereas in the southern hemisphere its movement appears to be anticlockwise. Bees transported from the northern to the southern hemisphere, or vice versa, therefore, are confronted with the task of orientating to sun movements in an apparently opposite direction to that in their original home. Transportation experiments of this nature, designed to test whether the time-compensated sun orientation behaviour in bees is inherited or learned, have produced equivocal results, however. Kalmus (1956) moved bees from North America to Brazil and studied the orientation behaviour of their descendants: he found that the bees continued to orientate themselves as though the Sun was still moving in a clockwise direction. Lindauer (1957) in a repeat of Kalmus' experiment, however, found that the F_1 progeny of bees moved from North America to Brazil orientated correctly to southern hemisphere conditions. Transportation of a colony of *Apis indica* from Ceylon to Germany caused initial disorientation but, after 43 days, bees trained to the south in the afternoon were able to find correct direction the following morning after an overnight removal to unfamiliar territory (Lindauer, 1959). Lindauer (1960) concluded that the time-compensated orientation mechanism was innate, but the bees had to learn the (apparent) direction in which the Sun moves, and its "speed".

The use of a clock to compensate for the changing azimuth of the Sun has also been described for ants (Jander, 1957), the beetle *Geotrupes sylvaticus* (Birukow, 1953; Geisler, 1961), and for the pond skater *Velia currens* (Birukow, 1956; Birukow and Busch, 1957). Jander (1957) found that ants *(Lasius niger)* continued with the same compass direction after being confined in a dark box, showing that they had taken in account the Sun's movement. This result, therefore, was contrary to that of Brun (1914) mentioned above. A similar result was obtained for *Formica rufa*, but only during the summer: in March and April this species was apparently unable to compensate for solar movement and showed an incorrect compass direction after being imprisoned in the dark (von Frisch, 1967, p. 448). Evidently compensation for the changing azimuth has to be leaned anew after the winter.

The pond skater *Velia currens* is apparently able to orientate to an artificial light source, or to the sun or the plane of vibration of the polarized light from a blue sky. According to Birukow (1960), the insects run exactly southwards when placed on dry ground under a blue sky, and compensate for the shifting position of the Sun during the day. Figure 9.5 shows that the angle of orientation to the sun or to an artificial light source decreases

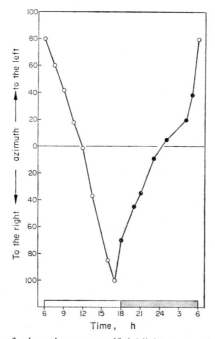

FIG. 9.5. Mean angles of orientation to an artificial light source of ten specimens of *Velia currens* in ten successive trials during "day" and "night" of an *LD 12* : 12 regime. *Ordinate:* angles of orientation. *Abscissa:* clock readings in central European time. (Redrawn, after Birukow, 1960.)

from sunrise to noon on the insect's left side and then increases on the right to sunset. From sunset to midnight the angle decreases on the animal's right side and then from midnight to sunrise it increases again on the left. It appears, therefore, that the underlying process runs through the night but in the opposite direction. Although the *Velia* "clock" appears to show some characteristics of an oscillatory system such as dependence on photoperiod and re-entrainment to a reversed light-cycle, the animals tend to move *directly* to the light source after about 30 hours in *LL* or *DD*, casting doubt on the existence of a persistent endogenous periodicity. Birukow (1960) considered that the orientational clock in *Velia* was mainly regulated by exogenous light signals. In more recent experiments with this species, Heran (1962) was unable to find any compass-true southerly course, and in the open the insects were found to orientate preferentially in relation to the wind. Further examination of orientation in *Velia* would therefore seem to be necessary.

C. Lunar, Semilunar and Tidal Rhythms

Midges of the genus *Clunio* (Chironomidae) are found in the intertidal zones of Atlantic and Pacific shores from temperate areas to the Arctic. They have—for insects— very curious life-cycles. *Clunio marinus* on the Atlantic coast of western Europe, for example, lives in the *lowest* parts of the intertidal zone which are only exposed during the times of spring low water. It is only during these times that the insects are able to emerge (Neumann, 1963). The females of this species are wingless. The males emerge before the females, assist the females in their own emergence, copulate with them and then carry them to the larval habitat. Both sexes are extremely short lived ($\sim$2 hours) and oviposition must occur before the tide rises to cover the larval site.

On the Normandy coast low water occurs twice a day at intervals of 12.4 hours; consequently low and high tides are about 50 minutes later each day and it takes a period of about 15 days before the times of low and high water complete a full cycle. Superimposed on this semi-diurnal tidal cycle is a semilunar tidal range. Tides reach their lowest point (spring low water) twice during each lunar month, once just after the full moon and once just after the new moon. Neap tides also occur twice per lunar month and occur just after half-moon. The period between successive spring low waters is 14.77 days (Fig. 9.6). On the days of the spring tides, low water occurs in the early morning and again in the evening. Emergence of *C. marinus* in this locality is restricted to the evening low water, just following full and new moon. The insect therefore shows a well-marked diurnal (=circadian) and semilunar (=circasyzygic) rhythm of eclosion.

In a series of elegant experiments Neumann (1963, 1966 a, b) has analysed the rhythm of adult emergence in *C. marinus* and shown it to be governed by the superposition of a circadian rhythm controlling pupal eclosion and a semilunar rhythm determining the beginning of pupation. In populations of *C. marinus* reared in the laboratory eclosion occurred towards the end of the photoperiod (i.e. about 12 hours after light-on in *LD 16*: 8), thereby corresponding to the observed time in the natural habitat. Cultures raised in *LL* throughout development showed an arrhythmic pattern of eclosion, but transfer from *LD* to *LL*, or exposure of an *LL*-raised culture to a single dark period, initiated a rhythm of eclosion which free-ran with an endogenous periodicity (τ) of less than 24 hours (Fig. 9.7). These experiments clearly show that pupal eclosion is controlled by a circadian clock similar to that described for *Drosophila pseudoobscura* (Chapter 3).

The semilunar rhythm of pupation which is superimposed on the circadian cycle was shown to be entrained by natural or artificial moonlight (Neumann, 1966b). Cultures of *C. marinus* from Normandy were raised in *LD 12*: 12 or *LD 16*: 8 and then exposed

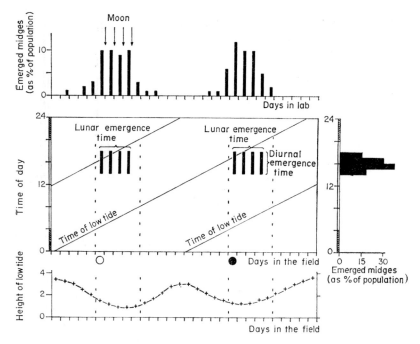

FIG. 9.6. *Middle panel:* the emergence of the marine midge *Clunio marinus* in natural tidal cycles. Population in Normandy, France. *Top panel:* emergence in *LD 16* : 8 with artificial moonlight (every 30 days 4 nights with 0.4 lux). *Bottom panel:* changes in the height of low tide during the month depicted in the middle panel. *Right:* diurnal emergence time in *LD 16* : 8 (light time from 4 to 20 hours). (From Neumann, 1967.)

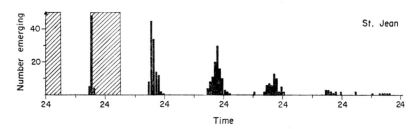

FIG. 9.7. A population of *Clunio marinus* from St. Jean-de-Luz released from *LD 12* : 12 into constant light (LL) showing the free-running rhythm of pupal eclosion. (From Pflüger and Neumann, 1971.)

to pulses of weak light (0.4 lux) during the dark period for 4 to 6 days at intervals of 30 days. This treatment initiated and entrained a semilunar rhythmicity (Fig. 9.8) which was absent from control populations without artifical moonlight. The endogeneity of this rhythm was demonstrated by exposing a population of larvae to a single period of artificial moonlight, whereupon the system "free-ran" for more than three cycles (each of 14.77 days) before all the individuals in the population had completed their development.

Considerable differences are known to occur between populations of *C. marinus* in different localities, and between different species of the genus. One of these differences arises because, although the *dates* of spring tides are the same at all parts of the same coastline, the phase of the 12.4-hour tidal cycle relative to local time shows differences,

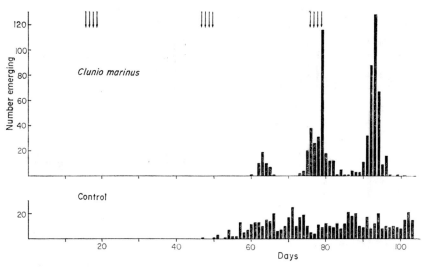

FIG. 9.8. Semilunar rhythm of emergence in *Clunio marinus* induced by artificial moonlight (4 nights with weak—0.4 lux—light every 30 days) in a population from Normandy, France. Above: experimentals; below: controls without additional illumination at night. (From Neumann, 1966b.)

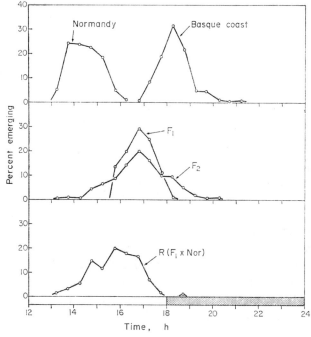

FIG. 9.9. Cross-breeding between two stocks of *Clunio marinus* which differed in their diurnal emergence times. Above: diurnal emergence times of the parental stocks Normandy (Port-en-Bessin) and Basque Coast (St. Jean-de-Luz). Middle: F_1 and F_2 generations. Below: backcrosses between F_1 and Normandy stock (in all curves only the emergence times of the males are presented). Breeding conditions: *LD 12* : 12, light from 6 to 18 hours, 20°C. (From Neumann, 1967.)

even at the same longitude. For this reason different local populations of *C. marinus* show different emergence times. These differences, moreover, have been shown to be genotypic (Neumann, 1966 a, b). Figure 9.9 shows the results of cross-breeding experiments between populations from Normandy (Port-en-Bessin) and the Basque coast (St. Jean-de-Luz). In the natural populations emergence occurs between 14:00 and 16:00 hours in Normandy and between 18:00 and 20:00 hours in the south. Because of the extremely short-lived nature of *Clunio* adults, experimental cross-breeding could only be achieved by manipulating the light-cycles so that the emergence times coincided. The results of such crosses showed that the emergence times of the F_1 and the F_2 progeny were strictly intermediate between the parental times, although the "spread" of emergence was greater for the F_2. Neumann (1966b, 1967) concluded that eclosion time was controlled by a polygenic mechanism, but that a small number of genes was involved. Results for back-crosses between F_1 and Normandy parents were consistent with this view. Between some local populations of *C. marinus* a non-reciprocal cross sterility was found (Neumann, 1971). The cross between St. Jean-de-Luz females and Santander males, for example, was fertile, but the reciprocal cross (Santander ♀ × St. Jean-de-Luz ♂) was not. This unilateral crossing ability was thought to represent a specialized mechanism which in natural populations would be effective in the formation of physiological races.

In a more northerly population of *C. marinus* from Heligoland (54°N.), Neumann (1966b) found only weak entrainment with artificial moonlight. At this latitude the Moon may not be bright enough: its maximum "altitude" is, on average, only 12.5°, and the shorter summer nights are not so dark as those further south. In this population, however, the semilunar cycle was found to be associated with tidal stimulation.

In arctic populations of *C. marinus* from Tromsö, Norway, there was a strictly tidal (~ 12.4 hour) cycle of eclosion in the summer (Neumann and Honegger, 1969; Pflüger and Neumann, 1971), emergence time coinciding with the initial exposure of the larval habitat during each ebb tide. Pflüger and Neumann (1971) showed that populations raised in *LD 16*:8 emerged as adults about 10 to 11 hours after light-on, but when they were "released" into *LL* no persistent rhythm could be detected (Fig. 9.10). In conditions

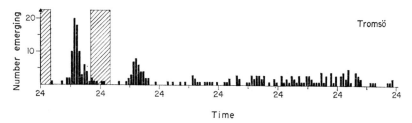

FIG. 9.10. A population of *Clunio marinus* from Tromsö, Norway, released from *LD 16*:8 into constant light (LL) showing one further emergence peak and then arrhythmicity. (From Pflüger and Neumann, 1971.)

of continuous light interrupted by a single 6-hour dark period, or in DD interrupted by a single 6-hour light pulse, a single peak of eclosion occurred about 10 to 11 hours after light-on. It was concluded that an hour-glass mechanism starting at least 10 to 11 hours earlier during the preceding ebb, rather than an endogenous oscillator, was involved in this Tromsö population.

Alternative patterns of eclosion are to be seen in other *Clunio* species. The Japanese species *C. takahashii*, for example, is restricted to the mean intertidal level and is therefore exposed twice daily by the tidal cycle throughout the season; emergence occurs on every ebb tide (Hashimoto, 1966) as with the arctic population of *C. marinus*. Populations of *C. mediterraneus* from Yugoslavia inhabit a coast with a small tidal range, but a diurnal cycle of land- and sea-breezes. During the summer months a relatively high fall of the tide occurs between midnight and sunrise during the strongest influence of the land-breeze. Every 15 days, following both full and new moon, the lowest tides coincide with the times of the strongest land-breezes; larval habitats are then exposed and eclosion occurs (Neumann, 1967).

Lunar rhythms of activity or adult emergence (with a period of about 28 days) have been described under field conditions for a number of insects (e.g. Hartland-Rowe, 1955; Corbet, 1958; Fryer, 1959; Kerfoot, 1967). In at least two examples (Hartland-Rowe, 1958; Youthed and Moran, 1969b) these rhythms have been shown to be endogenous.

Hartland-Rowe (1955, 1958) showed that the mayfly *Povilla adusta* emerged from the waters of Lake Victoria in its greatest numbers just after the full moon. This rhythm was maintained after the nymphs had been kept in the dark for 10 days, and in two individuals, for 6 weeks. The second case concerns the rhythm of pit-building activity by larvae of the ant-lion *Myrmeleon obscurus*. Using mean pit volume as a measure of activity, Youthed and Moran (1969b) showed that maximum activity occurred at the time of the full moon (Fig. 9.11). There was also a clear lunar-day (24.8 hour) rhythm with a peak in activity about 4 hours after moonrise. The authors demonstrated that the ob-

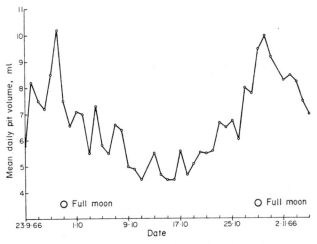

FIG. 9.11. Mean daily pit volume of a group of fifty *Myrmeleon obscurus* larvae (two of these larvae were subsequently shown to be *M. medialis*) subjected to normal daylight conditions, and each larva fed one ant a day. Open circles indicate the times of the full moon. (From Youthed and Moran, 1969b.)

served lunar rhythm (period about 28 days) was a combination of this lunar-day and the solar-day (circadian) rhythm described earlier (Youthed and Moran, 1969a), and which produces a peak in activity shortly after dark. This was achieved by subjecting twelve larvae of *M. obscurus* to a reversed light-cycle in which the 14-hour dark period began at 9.45 a.m. instead of at its normal time. This treatment rapidly reversed the solar-day activity pattern, but the lunar-day activity peak still occurred about 4 hours after moon-rise. Results showed that the lunar rhythm reached its maximum at the new moon rather than at the full moon. The lunar rhythm of pit-building activity was also shown to "free-run" for at least two or three cycles (92 days) in DD, but damp out in *LL*; it was therefore considered to be endogenous. A lunar rhythm was absent, however, in larvae reared since hatching in the absence of moonlight. The functional significance of the lunar rhythm was unclear.

D. Circannual Rhythms and "Long-range" Timers

The use of daylength as an indicator of season is widespread in the insects and other organisms, the "information" so gained being used to synchronize a variety of developmental and physiological phenomena to the appropriate season (Chapter 5). Apart from the winter and summer solstice, however, each daylength occurs twice in a year, once when the photoperiods are increasing in the spring and once when they are decreasing in the autumn. Although responses to the *direction* of such changes are known (Chapter 6), many insects are not "required" to make such a differentiation because (1) the developmental stages sensitive to daylength may only be present at the appropriate season (i.e. autumn), or (2) cold spring weather, especially at higher latitudes, may delay the resumption of activity well beyond the critical daylength. Nevertheless many long-lived animals such as birds, mammals and reptiles have to distinguish between spring and autumn and the direction of seasonal change: one way this is achieved is by utilizing an endogenous "calendar" based on a biological oscillation with a period close to a year (i.e. a circannual rhythm).

Circannual rhythms governing seasonal cycles of migration, moulting and breeding in vertebrates have been postulated for several decades, but only recently has convincing experimental proof of their existence been available. Such rhythms are now known for birds (Gwinner, 1967, 1971), mammals (Pengelley and Fisher, 1963; Goss, 1969 a, b; Heller and Poulson, 1970) and lizards (Stebbins, 1963). Circannual rhythms are also known for the cave crayfish *Orconectes pellucidus* (Jegla and Poulson, 1970). Gwinner (1971) found annual cycles of *Zugunruhe* (migratory restlessness), moulting and body weight in warblers, which persisted in an unchanging photoperiodic regime (*LD 12*: *12*) for several cycles. In the absence of a natural *Zeitgeber* (which is thought by Goss, 1969a, b, to be the annual *changes* in daylength) these endogenous cycles deviate considerably from their entrained period of 12 months. In starlings, for example, the free-running period for the circannual rhythm of testicular size was about $9\frac{1}{2}$ months (Schwab, 1971).

Circannual rhythms governing seasonal cycles are also to be found in some long-lived insects. Indeed, the best-documented example apparently antedates any of those mentioned above but seems to have escaped the attention it deserves. In a series of experiments, Blake (1958, 1959) described an endogenous circannual rhythm controlling diapause and the rate of development of the "carpet" beetle *Anthrenus verbasci*. The range of this work equals any of the later work with birds and mammals and provides some remarkable parallels to the circadian rhythm controlling eclosion in fruit-flies (Chapter 3).

In their natural habitat the larvae of *A. verbasci* feed on material of animal origin and are commonly found in house-sparrows' nests. The life-cycle generally occupies 2 years (semi-voltine) with the first winter spent in diapause as a young larva and the second as a full-grown larva, again in diapause. Some individuals may take three or more years to complete their development. After the second or last diapause the larvae pupate and the adults emerge the following spring. Blake (1958) showed that when the larvae were reared in the laboratory in constant conditions of temperature and humidity, and in continuous darkness (DD), this rhythm of diapause and development persisted, thereby demonstrating its endogeneity. In the absence of its natural *Zeitgeber*, which was subsequently shown to be changing daylength (see below), the period of the rhythm was found to be between 41 and 44 weeks, rather than the 52 weeks found in the entrained condition. Blake (1958) recognized this endogenous 41-week interval as the rhythm's "basic periodicity".

The period of the rhythm in the laboratory remained comparatively unaffected by either stationary photoperiods or by temperature (i.e. it was temperature compensated). Populations of larvae maintained in DD and in different constant temperatures differed, however, in the proportion of the population utilizing each pupation peak (Blake, 1958, 1959). Thus at high temperature (25° and 22.5°C) development was rapid and all larvae pupated in the first peak, whereas at low temperature (15°C) all the larvae underwent two cycles of development and diapause, and pupated in the second peak about 41 weeks after the first. Depending on the temperature, therefore, the insects were either univoltine or semivoltine. Larvae maintained at intermediate temperatures (20° and 17.5°C), however, were "split", some utilizing the first and some the second pupation peak (Fig. 9.12). This remarkable experiment clearly demonstrated that the circannual rhythm of pupation in *Anthrenus verbasci* is a "gated" phenomenon exactly comparable to the similar gated control of pupal eclosion in *Drosophila pseudoobscura* (Pittendrigh,

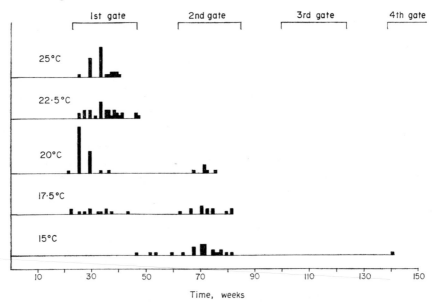

FIG 9.12. The circannual rhythm of pupation in *Anthrenus verbasci*. Frequency of pupation times when larval development has occurred in constant conditions of temperature, humidity and darkness. A black square represents the time of pupation, to the nearest week, of an individual. Note that the larvae at higher temperature are able to utilize the first gate; at lower temperature an increasing proportion of them are required to wait until the next. (From Blake, 1959. Published by the courtesy of the Pest Infestation Control Laboratory, Ministry of Agriculture, Fisheries, and Food, Slough, England. Crown Copyright is reserved by the Controller, Her Britannic Majesty's Stationery Office.)

1966; Skopik and Pittendrigh, 1967). In other words, at the intermediate temperatures, any larva "missing" the first gate had to wait a full cycle of 41 weeks before it could pupate.

In subsequent papers (Blake, 1960, 1963) it was shown that the *Zeitgeber* entraining the circannual rhythm in *A. verbasci* was the natural seasonal *change* in daylength. Both increasing and decreasing daylengths were shown to influence the rate of development. Blake (1960) maintained four groups of larvae at different combinations of outdoor temperature, constant temperature (20°C), natural changing daylengths, and constant dark. Under conditions of continuous darkness and constant temperature (Fig. 9.13D)

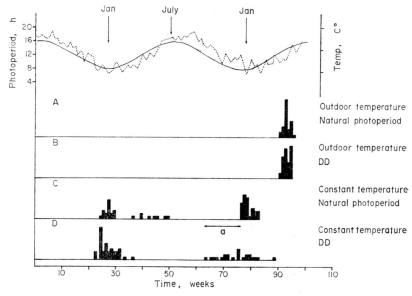

FIG. 9.13. Frequency of pupation times when larval development of *Anthrenus verbasci* has taken place under various combinations of light and temperature. C and D, constant temperature (20°C). One black square marks the time of pupation, to the nearest week, of one individual. Observations were made weekly. a — the delay in pupation (13 weeks) caused by the exposure of the larvae to decreasing photoperiod. (From Blake, 1963.)

the endogenous rhythm free-ran and the second peak occurred, as expected, about 40 to 44 weeks after the first. Under conditions of constant temperature and natural daylength (C), however, the second peak of pupation was delayed by about 13 weeks. It appeared that the naturally decreasing daylengths experienced by the larvae had delayed pupation from October to January, thereby lengthening the second cycle to about 52 weeks. In biological terms it ensured that the subsequent pupal development and eclosion were restricted to the following spring, and represented the entrainment of the free-running circannual rhythm with $\tau \sim 41$ weeks to the environmental year, $T = 52$ weeks. In outdoor conditions of temperature, populations of larvae maintained in natural daylength, or in DD (A and B), pupated at the same time, suggesting that light played no part in synchronization. These results, however, were also a strong indication that the natural seasonal change in temperature acted as an important *Zeitgeber*.

It was later demonstrated that the first cycle, as illustrated by the development of univoltine individuals, was controlled in a different way from the second (Blake, 1963) (Fig. 9.14). During the first cycle the length of the larval period was *decreased* whenever

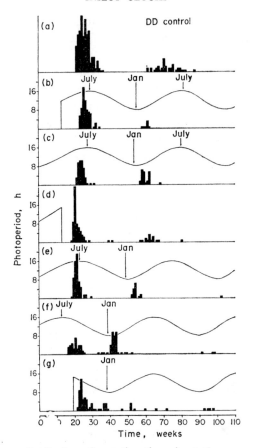

FIG. 9.14. Frequency distribution of pupation times in *Anthrenus verbasci* when larval
development has taken place in various combinations of natural daylength and DD at
20°C, 70 per cent r.h. Experiments (c), (d), (e) and (f) commenced during the months of
increasing daylength; note that the developmental period of the univoltine insects (1st peak)
is shortened. The sine-wave indicates the natural changes in daylength, the area below the
sine-wave the length of the photoperiod. One black square marks the time of pupation
(to the nearest week) of one individual. (From Blake, 1963. Published by the courtesy of the
Pest Infestation Control Laboratory, Ministry of Agriculture, Fisheries and Food, Slough,
England. Crown Copyright is reserved by the Controller, Her Britannic Majesty's Stationery
Office.)

the larvae were reared in increasing photoperiods during early larval life, whereas the second cycle was delayed by decreasing photoperiods so that most individuals pupated in
January and February. Advance and delay phase-shifts are clearly recognizable in these
phenomena.

Short-lived insects such as aphids can also distinguish spring from autumn but by
means of an apparently non-rhythmic "interval timer" which may extend its effects over
several generations, thereby preventing the aphids from responding prematurely to short
days in the spring (Bonnemaison, 1951; Lees, 1960). In the green vetch aphid *Megoura
viciae*, for example, the overwintering diapause egg gives rise to a specialized virginopara
(the fundatrix) which gives birth to successive generations of viviparous and parthenogenetic offspring. Lees (1960) found that if such a clone was exposed continuously to a
short photoperiod (at 15°C) no sexuales (males and oviparae) were produced for at least
90 days, a time which covered several generations of virginoparae. After this period, the

photoperiodic response was suddenly restored and oviparae were produced under the short-day treatment.

Working with *M. viciae*, Lees (1960) bred from either the first-born progeny in each generation, or from the last-born, the generation time of the latter being almost doubled by this procedure. Results showed that the oviparae were produced on almost the same day in both lineages even though the first-born line had been through seven generations and the last-born line only four (Fig. 9.15). Thus the long-range "interval timer" was dependent on the passage of time rather than the number of generations. The timer,

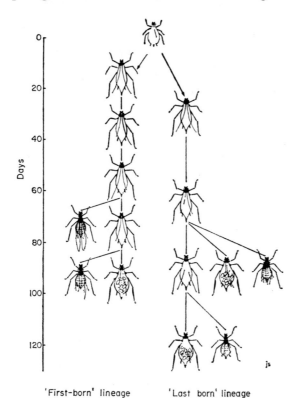

'First-born' lineage   'Last born' lineage

FIG. 9.15. A trans-generation interval timer in *Megoura viciae*. Two lineages, started from a single fertilized egg and a single fundatrix (above) were exposed permanently to a short-day regime. The appearance of males (winged) and oviparae is delayed for 75 to 90 days, irrespective of the generation. (From Lees, 1970.)

however, showed a marked temperature-dependence. At 20°C oviparae appeared in the first-born lineage after 32 to 44 days, at 15°C they appeared after 55 to 76 days, and at 11°C after 114 days. The timer also appeared to be independent of the "hour-glass" clock responsible for nightlength measurement in this insect: in young clones maintained in long daylength (*LD 16* : 8) the capacity to produce oviparae remained latent, but became apparent after transfer to short days provided that the timer had run its course.

Annotated Summary

1. The bees' *Zeitgedächtnis* (time memory) is an endogenous circadian clock which enables them to "remember" the time of the day at which nectar sources are available. The oscillation free-runs in *LL* with a natural period (τ) slightly less than 24 hours, and shows many similarities to the overt behavioural

rhythms described in Chapter 2. It differs, however, in that any phase of the oscillation can be used as a time reference (i.e. it is "continuously consulted"), and two or more such "training periods" can be established per day provided they are separated by intervals greater than 2 hours.

2. Bees and some other insects can orientate themselves to the Sun, to the pattern of polarized light from the sky, or to an artificial light source. A constant compass direction is maintained because the orientational mechanism also contains a circadian clock component which compensates for the Sun's changing azimuth.

3. Rhythms with a tidal periodicity are obviously rare in insects, but the marine Chironomid *Clunio marinus* shows a semi-lunar rhythm of emergence from the pupa which restricts eclosion to a 2 hour period at the lowest spring low waters, occurring roughly twice a month (every 14.7 days). This periodicity is a "combination" of a circadian rhythm controlling eclosion itself, and a semi-lunar rhythm determining the beginning of pupation. Both components may "free-run" in constant laboratory conditions. Arctic populations of *Clunio* may show an "hour-glass" mechanism controlling eclosion on every ebb tide ($\sim$12.4 hours) during the summer season.

4. Circannual rhythms with a natural period (τ) close to a year have been described with certainty only for the "carpet" beetle *Anthrenus verbasci*. This rhythm controls pupation in this essentially semi-voltine species, although all the larvae will pupate in the first year's "gate" at higher temperature. In constant conditions the oscillation free-runs with a period (τ) between 41 and 44 weeks, but is entrained to a strictly *annual* period by the seasonal *changes* in photoperiod.

5. The aphid *Megoura viciae* possesses a non-oscillatory and "trans-generation" timer which prevents them producing premature oviparae during the long nights of spring and early summer. After an interval of several weeks, which is not temperature compensated, the ability to respond to photoperiod is suddenly restored.

CHAPTER 10

THE ANATOMICAL LOCATION OF PHOTORECEPTORS AND CLOCKS

ALTHOUGH individual cells—in protists and in metazoa—probably contain or constitute "clocks" in their own right (Chapter 4), it is probable that particular groups of cells or organs in the metazoan organism have a particular clock function, especially with regard to overt behavioural activities such as locomotion or eclosion. The search for the anatomical location of such driving oscillations, and for the site of the photoreceptors which mediate entrainment by the environmental light cycle, therefore occupies a central and important position in the study of insect clocks, and one which is relevant to further studies on the mechanisms of clock function. In insects most of the research activity in this area has been focused on the clock controlling locomotor activity rhythms in cockroaches, party because their large size and ease of handling makes them suitable experimental subjects, and partly because of the interest generated by the work of Harker (1954, 1955, 1956, 1960a) indicating the presence of an autonomous endocrine "clock" in the suboesophageal ganglion of *Periplaneta americana*. Since Harker's original observations the whole field has become complicated and highly controversial, mainly because subsequent investigators (Roberts, 1966; Brady, 1967 a, b; Nishiitsutsuji-Uwo *et al.*, 1967; Nishiitsusuji-Uwo and Pittendrigh, 1968 a, b) have failed to corroborate her findings. In view of the complicated nature of the field it will be reviewed in a chronological order.

A. Clocks Controlling Rhythms of Locomotor and Stridulatory Activity

1. *The location of the photoreceptors*

The rhythm of locomotor activity in cockroaches free-runs in the absence of environmental *Zeitgebers* (Chapter 2) but becomes entrained to a 24-hour cycle of light and dark so that most of the activity occurs during the dark phase. Candidate photoreceptors for such entrainment are the compound eyes, the ocelli, or a generalized photosensitivity in the central nervous system, particularly in the brain (i.e. a so-called "dermal light sense"). In the search for the photoreceptors in cockroaches attention has been directed at all three.

Working with *Periplaneta americana*, Cloudsley-Thompson (1953) claimed that painting over both eyes and ocelli caused the insects to become arrhythmic. They also became arrhythmic after decapitation. Harker (1955, 1956) extended these observations and focused attention on the ocelli. She found that destroying or covering the ocelli, or cutting the ocellar nerves, resulted in a "gradual loss of the normal rhythm" and also the production of a new activity phase during the light period. After a period in DD cockroaches with cut ocellar nerves would not "take up" a new rhythm under new light conditions. According to Harker, covering the compound eyes but leaving the ocelli intact had no effect. The increase in activity in the light phase noted in insects with their ocelli eliminated was considered to be exogenous because it did not persist in DD and was eliminated by covering the compound eyes. These results seemed to implicate the ocelli as the principal if not the only photoreceptors, and Harker (1956) claimed that the ocelli were "directly connected with the establishment of the rhythm by the external factors of light and darkness". The result one would expect from eliminating the photoreceptor, however, is the *free-running* of the rhythm in *LD*, not the appearance of arrhythmia. The results were therefore equivocal.

Roberts (1965a) painted over the compound eyes and ocelli of the cockroaches *Leucophaea maderae* and *Periplaneta americana* with a mixture of lacquer, carbon black, and beeswax, and came to quite the opposite conclusion. The insects were also maintained in running wheels which gave longer and more clear-cut activity records than the rocking actographs used by Harker. Figure 10.1 shows the record of a specimen of *L. maderae*

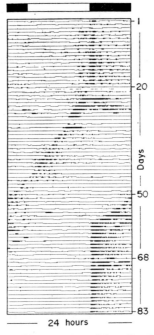

Fig. 10.1. Record of a rhythm of locomotor activity of a single cockroach, *Leucophaea maderae*, maintained in a light/dark cycle (*LD 12* : 12) for 83 days. On day 20 the compound eyes were painted with black lacquer; on day 50 the paint was peeled off; and on day 68 the ocelli were surgically removed. The position of the light and dark fractions of the *LD* regime is indicated at the top of the figure. (From Roberts, 1965a.) (Copyright 1965 by the American Association for the Advancement of Science.)

initially entrained to *LD 12* : 12. On day 20 the compound eyes were painted over but the ocelli left intact; the insect free-ran, even though still in a light cycle, clearly demonstrating that the photoreceptors had been eliminated. On day 50 the paint was removed and the rhythm re-entrained to *LD 12* : 12. Subsequent removal of the ocelli on day 68 had no effect on entrainment. These results clearly indicated that the photoreceptors for entrainment were the compound eyes, not the ocelli, but did not rule out the possibility of *direct* photostimulation of the brain.

These observations were later confirmed and extended by Nishiitsutsuji-Uwo and Pittendrigh (1968a). They found that surgical removal of the ocelli had no effect on entrainment, but that painting over the entire head, or bilateral section of the optic nerves, caused the insects to free-run in *LD 12* : 12 (Fig. 10.2). In order to investigate

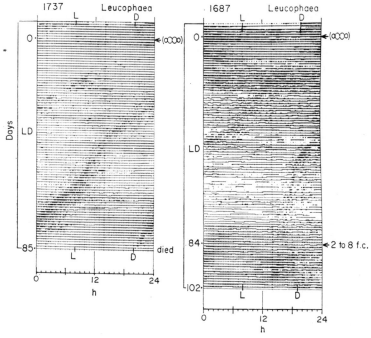

FIG. 10.2. The free-running rhythm of locomotor activity of *Leucophaea maderae* in a light/dark cycle (*LD 12* : 12) following bilateral section of the optic nerves between the optic lobes and the ommatidia of the compound eyes (day 0). The insect died on day 85. (From Nishiitsutsuji-Uwo and Pittendrigh, 1968a.)

the question of direct photostimulation of the brain, insects with their entire head capsule painted had the paint removed from the translucent antennal sockets, or had a glass "window" inserted in the vertex of the head and above the protocerebrum. This procedure caused entrainment to be re-established. It was concluded, however, that the light was probably not absorbed directly by the brain. For example, in animals with a "window" and their optic nerves cut, entrainment failed—suggesting that light entering the glass window scattered within the head capsule and activated the ommatidial elements.

Ball (1971) confirmed that the compound eyes and not the ocelli were the principal photoreceptors in *P. americana*. Since earlier studies had shown that the terminal abdominal ganglion was also light sensitive (Ball, 1965), the role of this organ in the control of rhythmicity was also investigated. When this ganglion was occluded more activity was found to occur in the light phase, but transection of the ventral nerve cord did not alter the locomotor rhythm.

In subsequent experiments with *Blaberus craniifer* maintained in running wheels at *LD 12* : 12, Ball (1972) inserted a small glass window over the protocerebrum and then covered the rest of the head capsule with opaque black wax. In twelve out of fourteen insects so treated entrainment to the light cycle was maintained for 3 to 6 weeks. In the remaining two cases arrhythmia occurred. In view of Nishiitsutsuji-Uwo and Pittendrigh's (1968a) observations with *L. maderae*, however, the result is equivocal: light entering the window could have been activiting the ommatidia. In a further investigation of possible direct stimulation of the brain, Ball and Chaudhury (1973) removed the optic lobes, or the brain and optic lobes, from "donor" cockroaches and inserted them into the abdomen of recipients. The transplants were then covered with a glass window and the entire head covered with opaque wax to eliminate the normal photoreceptors. The majority of insects so treated showed "normal" entrainment to the light cycle. The authors concluded that since the eyes were blinded the "implanted tissue, either brain or optic lobes, when exposed to environmental light stimuli in its new abdominal site, assumes the function of the host's brain in controlling activity rhythms". However, apart from an unsubstantiated statement that some of the insects receiving abdominal implants of optic lobes free-ran in *LD 12* : 12, elimination of the normal photoreceptors by occluding the head was not demonstrated. It could be, for instance, that the insects were not blinded and that the activity records merely represent entrainment to *LD 12* : 12 by the normal photoreceptors (compound eyes). In addition, it ought to be possible to demonstrate that subsequent occlusion of the abdominal window caused the insects to free-run once more in a light cycle. The existence of an extraoptic photoreceptor in cockroaches, therefore, remains to be demonstrated. It seems almost certain that the compound eyes are the only such organs involved.

Working with the house cricket *Acheta domestica*, Nowosielski and Patton (1963) found that entrainment to a "new" light cycle could only be prevented by blacking out both ocelli *and* compound eyes. Out of twenty-seven insects with their compound eyes occluded thirteen showed no distinct re-entrainment, nine of these becoming arrhythmic. The authors concluded that both photoreceptors were necessary but the eyes play the dominant role.

In some insects neither the eyes *nor* the ocelli seem to be involved. Loher and Chandrashekaran (1970), for example, found that the oviposition rhythm of the grasshopper *Chorthippus curtipennis* continued to be entrained by an environmental light cycle when the ocelli and the compound eyes were destroyed or covered with an opaque wax, or even when the entire head capsule was covered with a thick coating of the wax material. They concluded that an extracephalic photoreceptor was in operation. Similarly, Godden (1973) found that ablation of the compound eyes in the stick insect *Carausius morosus* failed to interfere with entrainment to *LD 12* : 12 and suspected an extra-optic receptor. In either case light might impinge directly on the central nervous system, or a true "dermal light sense" as in *Schistocerca gregaria* (Neville, 1965, 1967) (Chapter 4) may be found.

Evidence for an extraoptic photoreceptor has also been found for the entrainment of the stridulatory rhythm in species of *Ephippiger* (Dumortier, 1972). In these insects, covering the entire head with paint or with an aluminium "cowl", or surgical destruction of the compound eyes and ocelli by electrocoagulation, failed to prevent entrainment to an *LD* cycle (*LD $13\frac{1}{4}$: $10\frac{3}{4}$*). Blinded individuals could also re-entrain to a reversed *LD* cycle. The exact location of the photoreceptor in *Ephippiger* is not clear, but it may be in the brain: it is certainly head-located because illumination of the head with an *LD* cycle will entrain, whereas similar illumination of the body will not.

In the cricket *Teleogryllus commodus*, however, the compound eyes are the *only* photoreceptors involved in entrainment (Loher, 1972). The males of this species possess a well-defined stridulatory rhythm with singing commencing about 2 hours after dark. In DD the rhythm free-runs with a natural period (τ) of 23 hours 36 minutes; in *LL* τ is 25 hours 40 minutes. Surgical removal of the compound eyes or severance of the ommatidial nerves caused the insects to free-run in *LD* or in *LL* with a period similar to that occurring in unoperated animals maintained in DD; removal of the ocelli, on the other hand, was inconsequential. These results are directly comparable to those for *Leucophaea maderae* obtained by Roberts (1965) and Nishiitsutsuji-Uwo and Pittendrigh (1968a).

2. *The location of the clock*

In 1954 Harker reported results of some parabiosis experiments with *P. americana* in which an upper cockroach previously maintained in *LD 12* : 12 and therefore rhythmic but with its legs removed was joined by its pronotum to a lower insect which was intact but rendered "arrhythmic" by spending its entire life in *LL*. She found that the motile insect then showed a circadian rhythm of activity corresponding to that of the top insect and concluded that "a secretion, carried either in the blood or tissues, is involved in the production of a diurnal rhythm of activity in the cockroach". In subsequent papers (Harker, 1955, 1956) she identified the source of the proposed hormone as the suboesophageal ganglion. Using a headless and therefore arrhythmic insect as a "test-bed" she implanted a suboesophageal ganglion from a rhythmic donor into the abdomen just lateral to the heart. When the recipient was then kept in any light condition (*LL*, DD or reversed *LD*) it showed a "normal rhythm" with its active phases corresponding to those of the donor. The source of the secretion was later traced to two pairs of neurosecretory cells, one on either side of the ganglion. It was concluded that the suboesophageal ganglion could carry on secreting rhythmically after all nervous connections had been broken; it was therefore an autonomous endocrine "clock". It was also found that allatectomy caused a gradual loss in rhythmicity because the operation broke two small nerves which conducted neurosecretory material from the corpus cardiacum to the suboesophageal ganglion (Harker, 1960 a, b). This supply of material was considered essential for the continued running of the clock. At the time these observations constituted one of the most important findings in the study of insect clocks because they were the first to claim anatomical localization of a driving oscillation. They have, however, generally failed to withstand corroboration by subsequent investigators.

Roberts (1965b, 1966) criticized Harker's experiments on a number of grounds. (1) Her parabiosis experiments were conducted without controls in which cockroaches were united without haemocoel connection, so that apparent "transfer" of the rhythm could have been due to mechanical stimulation. (2) The fact that he (Roberts, 1960) had demonstrated the persistence of rhythmic locomotor activity in constant light of quite high intensity, implying that Harker's lower animals were not, in fact, arrhythmic. (3) That Harker had not paid sufficient attention to the respective phases of the donor and recipient. He pointed out that experiments involving the transfer of suboesophageal ganglia, or any other organ, should be conducted between insects purposely entrained to light cycles out-of-phase with each other. In the absence of this precaution the evidence merely suggests that the suboesophageal ganglion is necessary for the "expression" of

the rhythm. Similar criticisms were later voiced by Nishiitsutsuji-Uwo *et al.* (1967) and
Cymborowski and Brady (1972).

    In his own experiments, Roberts (1966) failed to reinstate the rhythm in decapitated
P. americana by implanting a suboesophageal ganglion from a rhythmic donor (twenty
attempts). He also found that allatectomy or even removal of the entire retrocerebral
complex failed to interfere with the rhythm, and that the rhythm could not be reinstated
with suboesophageal ganglion transplants to decapitated insects which were then subjected to a 24-hour sinusoidal temperature cycle (19° to 27°C). He therefore failed to
confirm Harker's observations on the role of the suboesophageal ganglion. He did,
however, evoke arrhythmicity in both *Leucophaea maderae* and *Periplaneta americana*
by surgical bisection of the pars intercerebralis (Fig. 10.3) thus focusing attention on
the brain itself as the locus of the driving oscillation.

FIG. 10.3. Loss of overt activity rhythm in *Periplaneta americana* following surgical bisection
of the brain. The arrow shows the time that the operation occurred on day 18. In this
operation, the brain is simply cut in half, no tissue is removed. (From Roberts, 1965b.)

    Brady (1967 a, b) carried out a careful study designed to reconcile some of the conflict
between the results of Harker and Roberts. This investigation involved larger numbers of
insects, microcautery of neurosecretory cells, subsequent autopsy by histological examination, and careful attention to phase differences between donors and recipients. He
found that although complete removal of the corpora cardiaca (as revealed by autopsy)
was almost impossible, this operation broke the nervous connection from the corpora

cardiaca to the suboesophageal ganglion. None the less, the rhythm persisted, thereby confirming Roberts (1966). Destruction of the median neurosecretory cells of the pars intercerebralis of thirty-seven insects by microcautery left seven arrhythmic, twelve possibly rhythmic and sixteen clearly rhythmic; the remaining two died. Autopsies revealed that cockroaches were able to maintain a rhythm after massive if not total reduction of their neurosecretory cells. In his experiments involving suboesophageal ganglion transplants, Brady (1967b) maintained donors and recipients both in *LD 12* : 12 before the operation, but with the light to dark transitions for recipients 5 to 10 hours later than that for donors. Suboesophageal ganglia were implanted into forty-eight headless recipients which were then maintained in DD. Of the twenty-nine "acceptable" records, the majority were arrhythmic. Only in two cases was the greatest part of their recorded activity at nearly the same time each day as the donor's previous activity peak. Cautery of the neurosecretory cells in *in situ* suboesophageal ganglia failed to make the cockroaches arrhythmic. Brady's data therefore also failed to confirm Harker's claim.

Nishiitsutsuji-Uwo *et al.* (1967) also used larger numbers of insects and checked the results of their various surgical procedures by post-mortem histological examination. Surgical removal of the pars intercerebralis in forty-seven insects caused arrhythmicity in nineteen cases and some rhythmicity in the remaining twenty-eight. Subsequent autopsy of three of the nineteen arrhythmic cockroaches revealed that no neurosecretory cells remained, whereas the eleven of the twenty-eight partially rhythmic animals so examined were found to contain at least some of these cells. It was concluded, therefore, that ablation of the pars intercerebralis caused arrhythmicity provided that all of the cells were removed. This result —which is in conflict with that of Brady (1967b)—was interpreted as evidence for a relationship between the neurosecretory cells and the circadian rhythm, and therefore for a hormonal link.

In a subsequent paper Nishiitsutsuji-Uwo and Pittendrigh (1968b) produced the best evidence that the driving oscillation is located in the brain. More specifically, they found that cutting the optic nerves (of *L. maderae*) caused the rhythm to free-run in *LD 12* : 12, but cutting the optic tracts (between the optic lobes and the rest of the brain) resulted in arrhythmicity. They concluded, therefore, that the "clock" was located in the optic lobes, and that these structures require connection to the compound eyes for entrainment and connection to the rest of the protocerebrum for the mediation of the locomotory rhythm. Since they also found that destruction of the neurosecretory cells of the pars intercerebralis resulted in arrhythmicity, they favoured an output to the locomotory centres which included a humoral component.

In a review attempting to collate all published observations, Brady (1969) concluded, however, that a neural channel between the optic lobes and the thoracic centres was involved. He pointed out that there were only three routes by which such control could reach the thoracic ganglia and hence the legs. These are (1) by hormones in the blood, (2) by hormones carried in nerve axons, and (3) by normal electrical impulses in the nerves. The fact that virtually all of the cephalic endocrine tissue can be removed without stopping the rhythm makes the first seem unlikely. The crucial experiments to demonstrate nervous control would therefore seem to be cutting the ventral nerve cord at various points along the ganglionic chain. These operations are shown in Fig. 10.4. Cutting the ventral nerve cord behind the thorax (cut C) was found to have no effect on the activity rhythm, as might be expected (Nishiitsutsuji-Uwo and Pittendrigh, 1968b). Cutting the connectives between the thoracic ganglia (cut D) had a greater effect when performed between the pro- and the mesothoracic ganglia than between the meso- and metathoracic ganglia (Nishiitstutsuji-Uwo and Pittendrigh, 1968b). Cutting the nerve cord between the suboesophageal ganglion and the thorax (cut E) caused complete loss of the rhythm (Brady, 1967b; Nishiitsutsuji-Uwo and Pittendrigh, 1968b). The results of cuts A and B, between compound eyes and optic lobes and between optic lobes and brain, respectively, have already been noted. Unfortunately the really crucial test of cutting the circum-oesophageal connectives (cut F) proved equivocal. If the operation is performed with *P. americana* activity becomes so intense that it becomes difficult to interpret the records (Brady, 1967b); with *L. maderae*, however, the same operation clearly causes arrhythmicity (Roberts *et al.*, 1971) (Fig. 10.5). These results all constitute strong evidence that the driving oscillation located in the optic lobes has a nervous (i.e. electrical) rather than a hormonal output.

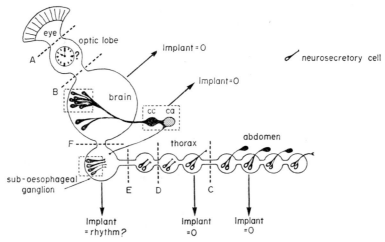

FIG. 10.4. Synopsis of experiments on the control of the circadian rhythm of locomotor activity in cockroaches. The ganglia of the central nervous system are represented by the linked spheres, with neuroendocrine tissue, including known neurohaemal organs, indicated in black. Dotted boxes represent endocrine tissue which can be removed without altering the rhythm. Arrows show organs transplanted from rhythmic donors to headless arrhythmic recipients: O signifies that the host shows no detectable rhythm. Heavy broken lines are cuts made in the nerve trunks: cuts B, E, F, or splitting the protocerebral lobes bilaterally, apparently stop the rhythm; cuts A, D, C, or splitting the pars intercerebralis mid-sagittally, do not. cc, corpora cardiaca; ca, corpora allata. (Redrawn from Brady, 1969.)

Section of the circum-oesophageal connectives, however, was also performed by Nishi-itsutsuji-Uwo and Pittendrigh (1968b) and they found that five of the six insects so treated subsequently resumed their normal rhythm of activity. On the basis of this experiment, together with the claim that destruction of the neurosecretory cells in the pars intercerebralis caused arrhythmia, the authors favoured a hypothesis in which a hormonal link was involved. Both Brady (1969) and Roberts *et al.* (1971), however, considered that the explicit directions provided by Nishiitsutsuji-Uwo and Pittendrigh lead to cutting the *maxillary* nerves rather than the commissures. If this is true, the observations are still consistent with an electrical output.

Recent work by Roberts (1974) has attempted to locate the driving oscillation in *L. maderae* and *P. americana* to a more precise site within the optic lobes. Surgical removal or destruction of different synaptic areas suggested that the two innermost elements (the lobula and the medulla) were crucial in the control of rhythmicity. Removal of the outer synaptic area (the lamina), on the other hand, caused the animals to remain rhythmic but to free-run in *LD*; for this reason the lamina was considered essential in the coupling between the photoreceptors (the compound eyes) and the clock.

The possibility of *some* humoral influence, however, seems to be far from dead. Subsequent to Brady's (1969) review in which he could state that the optic lobes contained no glandular tissue, monopolar neurosecretory cells have been found in the optic lobes of *P. americana* (Beattie, 1971), although their role in the clock mechanism (if any) has not been investigated. More significantly, Cymborowski and Brady (1972) have re-examined Harker's original parabiosis experiments paying particular attention to the criticisms mentioned above. Thus, the recipient animals (*P. americana* and the house cricket *Acheta domestica*) were beheaded to ensure true arrhythmicity, the donors and recipients were both maintained in *LD 12* : 12 but 12 hours out-of-phase with each other

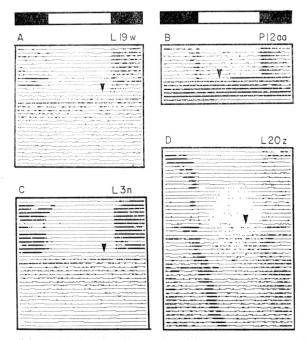

Fig. 10.5. Loss of rhythmicity in *Leucophaea maderae* following the severance of both circum-oesophageal connectives by the head-incision technique (A and C). B — development of extreme hyperactivity in *Periplaneta americana* after cutting both connectives. D—*L. maderae* initially entrained to *LD 12* : 12 then with one circumoesophageal connective cut at time indicated by the arrow; insect was then placed in DD. (From Roberts *et al.*, 1971.)

before the experiment began, and control pairs with the haemocoels unconnected were also included. After being joined in parabiosis the insects were monitored in DD. The results showed that when the lower, headless recipient had its blood system connected to the upper, legless, but rhythmic donor, highly significantly *more* rhythms ensued than when the blood systems were left unconnected. Furthermore, the peaks of activity induced in the recipients corresponded very closely in phase with those of the donor's previous rhythm. At first glance these results might imply a hormonal link between the two insects. It was pointed out, however, that several alternative "explanations" were possible. Between 20 and 40 per cent of the *control* (unconnected) "recipients" showed transferred rhythms, possibly caused by the mechanical struggling of the upper animal.

In addition, the effect of blood transfusion might be either the non-rhythmic restoration of missing cephalic hormones which enable the headless insect to respond to the disturbing stimuli of the type produced *rhythmically* by the upper animal, or a response to blood-borne metabolites or ionic changes. Finally, the results might reflect a response to stress hormones secreted by the upper insect when its central nervous system issues abortive commands; this mechanism, although hormonal, scarcely represents the normal controlling system for the rhythm.

In order to account for the general discrepancies between the work of Harker and later investigators Brady (1967a) turned his attention to the differences between the kind of recorder used. The photocell boxes used by Harker, for example, provided a small static environment whereas the running wheels used by Roberts and later workers provided a moving one which was also "limitless" in the sense that the cockroaches never came to the "end" of it. It is probable, therefore, that the running wheel provides stimulus in the form of "positive feed-back" which reinforces running activity, thereby giving long-term records which do not "fade" so quickly as those from a more static environment. In addition, many operative techniques produce *initial* arrhythmicity or hyperactivity, and long-term records are necessary to determine the real effects of such treatments.

Information on the anatomical location of the driving oscillation in other insects is less complete, but also indicates that the suboesophageal ganglion is not the essential organ. The rhythm of locomotor activity in the grasshopper *Romalea microptera* was unaltered by allatectomy. Removal of the suboesophageal ganglion caused an apparent loss in rhythmicity, but this was not restored by its subsequent reimplantation (Fingerman *et al.*, 1958). Eidmann (1956) found that the activity rhythm of the stick insect *Carausius morosus* could not be changed by removal of the corpora allata, the corpora cardiaca, or the optic lobes, or by implantation of the brain or suboesophageal ganglion. Extirpation of the brain or its protocerebrum, however, abolished the rhythm, as did severance of the circum-oesophageal commissures. However, in view of the strong criticisms of Eidmann's work voiced by Godden (1973), these last results should presumably be accepted with caution.

In the cricket *Teleogryllus commodus*, severance of both optic lobes caused a breakdown in the stridulatory rhythm, suggesting that the clock was located within these structures (Loher, 1972). Section of one optic tract, however, did not affect the rhythm. In this species both the location of the photoreceptors in the compound eyes, and the location of the driving oscillation within the optic lobes, makes it directly comparable to the cockroaches discussed earlier.

B. Clocks Controlling Rhythms of Eclosion

The rhythm of pupal eclosion in *Drosophila* spp. can be initiated and phase-set by light signals at any stage of post-embryonic development (Brett, 1955; Zimmerman and Ives, 1971). Since the compound eyes and ocelli are only fully differentiated in the pharate adult, and the brain is the only organ which is not extensively reorganized during metamorphosis, it would seem, *a priori*, that the photoreceptors for entrainment—and probably also the clock—lie in the brain.

Attempts to locate the photoreceptors in *Drosophila* have been hampered by the insects' small size. Nevertheless, Zimmerman and Ives (1971) have made some progress by painting either the anterior or the posterior ends of the puparium of *D. pseudoobscura* with an opaque black paint. Puparia thus treated were then transferred to DD and exposed to dim monochromatic light pulses (15 minutes at 456 nm and 11.33 to 11.49 log quanta sec^{-1} cm^{-2}) applied 17 hours after the time when dawn would have occurred on day 1, and subsequently examined for the resulting phase-shift ($\Delta\phi$) (Chapter 3). It was found that light signals falling on the anterior half generated as great a $\Delta\phi$ as light falling on the whole organism, whereas light falling on the posterior half alone had no effect. In earlier experiments using cardboard masks Kalmus (1938b) reached a similar conclusion.

Although the compound eyes and ocelli become differentiated in the pharate adult they are not required for photoreception. Pittendrigh (unpublished) and Engelmann and Honegger (1966), for example, showed that an eyeless and ocelliless mutant of *D. melanogaster* was able to entrain normally to a light-cycle. Zimmerman and Ives (1971) have subsequently repeated this observation using dim monochromatic light. These authors have also shown that although the compound eyes of a white-eyed mutant of *D. pseudoobscura* were two log units more sensitive to light than wild-type because of the lack of screening pigments, no such difference was observed with respect to the sensitivity of the circadian oscillator (Chapter 3). These observations constitute strong evidence that the organized photoreceptors are not involved; they also suggest direct (extraoptic) light absorption by the brain.

Direct photoreception by the brain has been demonstrated by Truman and Riddiford (1970) for the silkmoths *Antheraea pernyi* and *Hyalophora cecropia* which, because of their large size, obviate many of the difficulties inherent in *Drosophila*. These authors showed that adults of *H. cecropia* emerge in a well-defined "gate" during the forenoon, whereas adults of *A. pernyi* emerge towards the end of the light-phase (Fig. 3.3). Section of the optic nerves or extirpation of the compound eye anlagen from diapausing pupae was inconsequential, but removal of the brain also removed the "gating" control of eclosion and the moths emerged in a random fashion. Response to photoperiod was fully restored, however, merely by reimplanting the excised brain into the abdomen (Fig. 10.6). The location of the photoreceptors within the brain was demonstrated un-

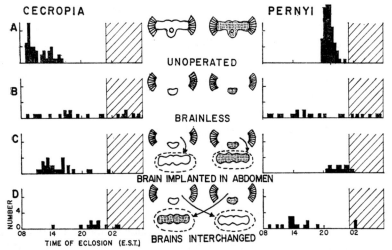

FIG. 10.6. The eclosion of *Hyalophora cecropia* and *Antheraea pernyi* in an *LD 17 : 7* regime showing the effects of brain removal, the transplantation of the brain to the abdomen, and the interchange of brains between the two species. (From Truman, 1971d.)

equivocally by removing the brains from twenty pupae of *H. cecropia* and reimplanting them into either the head (ten insects) or the abdomen (ten insects). These "loose-brain" pupae were then inserted into holes drilled in an opaque board in such a way that the anterior and posterior ends of the pupae received light regimes (*LD 12*: 12) differing only in phase. Figure 10.7 shows that the time of eclosion was determined by the photo-period to which the brain was exposed.

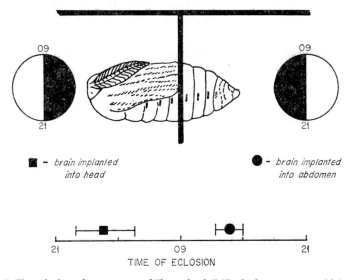

■ – *brain implanted*   ● – *brain implanted*
  into head   *into abdomen*

TIME OF ECLOSION

FIG. 10.7. The eclosion of two groups of "loose-brain" *Hyalophora cecropia* which differed only in the site of brain implantation. The anterior end of each was exposed to light from 09.00 to 21.00; the posterior ends from 21.00 to 09.00. The time of eclosion was determined solely by the photoperiod to which the brain was exposed. The mean and standard deviation of each group is given. (From Truman, 1971d.)

Although the compound eyes are not the photoreceptors for the entrainment of the eclosion rhythm, they are involved in certain *exogenous* effects caused by the lights-on stimulus (Truman, 1971d). Figure 10.8, for example, shows a strong "skew" towards lights-on in *H. cecropia* which was eliminated after transection of the optic nerves or by the transfer of the brain to the abdomen. When the brain was transplanted together with its attached eye discs, however, well-formed (but inverted) compound eyes developed in the pharate adult, and the *immediate* response to light was restored.

Truman and Riddiford (1970) also demonstrated that when the brains of *H. cecropia* and *A. pernyi* pupae were interchanged the timing of eclosion was characteristic for the species of *brain*, but the emergence behaviour was characteristic for the *body* of the recipient. Section of the circum-oesophageal connectives, or extirpation of the sub-oesophageal ganglion, frontal ganglion, or corpora allata-corpora cardiaca complex had no effect. These very beautiful results demonstrated that the clock was also located in the brain and, since "loose-brain" insects functioned in the same way as intact animals, the output of the clock was hormonal. Furthermore, the hormone involved was neither species nor genus-specific.

In a later paper (Truman, 1972b), attempts were made to locate the clock within the brain itself. Surgical bisection of the brain had no effect on gating control. Brains were then excised from *A. pernyi* pupae and the small central wedge containing the median neurosecretory cells separated from the peripheral areas containing the lateral cells.

These pieces were then implanted separately into brainless recipients. In all cases eclosion was random suggesting that neither part of the brain was competent to control the oscillation by itself. Truman pointed out, however, that since each insect emerges only once, one cannot discriminate between the result of implantation of a fragment which lacks a clock and that of one which just lacks a photoreceptor and, thus, is free-running. In both instances a randomized distribution of eclosions would result. The result might also be attributed to the damage caused during the surgical procedures.

It was shown that the hormone responsible for the control of eclosion (the neurotropic eclosion hormone) can be extracted from the brain of *A. pernyi* during the latter two-

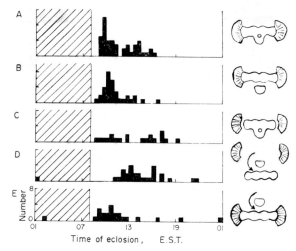

FIG. 10.8. The influence of the eyes in the emergence of *Hyalophora cecropia* in an *LD 17* : 7 regime. A — normal moths; B — circumoesophageal connectives cut; C — optic nerves severed; D — brain transplanted to abdomen; E — brain with attached eye imaginal discs transplanted to abdomen. (From Truman, 1971d.)

thirds of adult development (Truman, 1970). It is secreted by the median neurosecretory cells and passed down their axons to the corpora cardiaca. The release of the hormone into the blood is a gated event dictated by the eclosion clock. Experimental injection of brain homogenates into moths developmentally competent to emerge, initiates a programme of abdominal movements which leads, within 2 hours, to eclosion, escape from the cocoon and the spreading of the wings. The activation of this nervous programme by the hormone has been examined by electrophysiological methods (Truman and Sokolove, 1972). By recording from a nerve cord with severed peripheral nerves it was shown that this pre-eclosion behaviour is pre-patterned in the abdominal ganglia and requires no sensory feedback.

It is interesting that pharate adults of *A. pernyi* exhibit essentially "pupal" behaviour until eclosion, restricting their movements to a simple rotation of the abdomen (Truman, 1971 c, d). This pupal behaviour persists even if the pupal cuticle is removed: "peeled" moths do not, for example, attempt to spread their wings. When the eclosion gate opens, however, the entire eclosion sequence is acted out despite the fact that the insects have neither pupal integument nor cocoon to escape from. If the brain is removed this sequence is destroyed. In some insects, for example, eclosion may occur before the resorption of the moulting fluid is completed and the moths emerge wet; in others the intersegmental abdominal muscles degenerate prematurely so that the moth is trapped within its pupal cuticle. The proper sequence is restored, however, if a brain is implanted into the abdomen.

   Truman (1972a) showed that the larval moults of *A. pernyi* and *Manduca sexta* occur at particular times of the day depending on species, the instar, and the photoperiod. Figure 10.9 shows that the times of the larval ecdyses of *A. pernyi* occur later in each succeeding instar and that the distributions of ecdysis time tend to broaden. The moult from

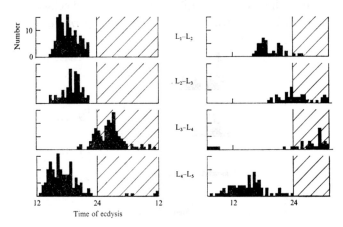

FIG. 10.9. The distribution of the four larval ecdyses of *Antheraea pernyi* reared at 26°C under (left) an *LD 12 : 12* regime, or (right) an *LD 17 : 7* regime. (From Truman, 1972a.)

first to second instar (L₁–L₂) and from L₂–L₃, for example, occurred during the light phase of both *LD 12*: 12 and *LD 17*: 7, but that for L₃–L₄ occurred in the following night, and that for L₄–L₅ was delayed until the light phase of the following day. Series of ligations between the head and thorax at various times during the light cycle demonstrated that the release of brain hormone (PTTH) was a gated event. In *M. sexta*, for example, secretion of PTTH occurred only during a gate in the early to middle portion of the night (Fig. 10.10). The larval ecdyses themselves were *not* gated, however, but occurred a fixed number of hours after the hormone release. This time interval was instar specific and dependent on temperature: a 3° rise (25° to 28°C), for example, caused a 2-hour advance in the L₃–L₄ moult. The proportion of the larval population exploiting a particular gate was also temperature-dependent, as in *D. pseudoobscura* and other gated

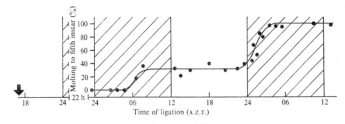

FIG. 10.10. The effects of applying ligatures to fourth instar larvae of *Manduca sexta* at various times after the L₃–L₄ ecdysis. The percentage of the neck-ligated larvae subsequently developing to the fifth instar is plotted against the time of ligation. Each point represents a sample of about fifty ligated larvae. The arrow identifies the median time of the L₃–L₄ ecdysis. (From Truman, 1972a.)

rhythms. As with pupal eclosion, the photoreceptors are probably located within the brain; cautery of the larval ocelli (=stemmata) failed to abolish the insects' ability to "see" the light regime.

C. Photoperiodic Clocks

Available evidence for photoperiodic clocks suggests that the "organized" photoreceptors (compound eyes and ocelli) are not involved in the response to light. Tanaka (1950), for instance, cauterized the lateral ocelli (stemmata) of fourth instar larvae of *Antheraea pernyi*, causing the complete disappearance of these organs in the next instar. Nevertheless, the photoperiodic sensitivity of these "blinded" larvae was unimpaired, short-dayexposure leading to the production of diapausing pupae. De Wilde *et al.* (1959) blinded newly emerged adults of *Leptinotarsa decemlineata* either by cautery or by covering the compound eyes with opaque black paint, and similarly found no interference with the photoperiodic response. Comparable experiments have also been carried out with *Megoura viciae* (Lees, 1964) and the grasshopper *Anacridium aegyptium* (Geldiay, 1971), with similar results. In this respect, therefore, photoperiodic clocks are similar to those controlling eclosion rhythms, but different from those controlling, say, the locomotor activity rhythm in cockroaches.

An alternative approach to the problem of the location of photoreceptors is to illuminate different parts of the body to determine the site of organs which can differentiate between long and short days. This has been done in at least three insects. Working with larvae of the pine lappet moth *Dendrolimus pini*, Geispits (1957) covered either the head or the abdomen with an opaque hood for 12 hours each day and then exposed the larvae to continuous illumination. Those larvae with the abdomen covered showed a long-day response, but those with the head covered behaved as though they were in LD *12*:12 and entered diapause. These experiments, therefore, showed that the photoperiodic receptors were located in the head.

A similar conclusion was reached by Lees (1960, 1964) working with the aphid *Megoura viciae*. Lees maintained parent aphids (virginoparae) at a short daylength (LD *14*: 10) but supplied a 2 hour period of supplementary illumination to different areas of the head and body by means of microilluminators constructed from fine metal capillaries or plastic filaments (light guides) drawn from a viscous solution of polystyrene in benzene. The aphids were attached to the microilluminators for 2 hours each day; the rest of the time they were allowed to feed undisturbed on the host plant. The rationale behind this approach was that if the microilluminator was placed in a position to stimulate the relevant photoreceptor the aphids would react to a long daylength (LD *16*: 8) and produce virginoparous daughters, whereas if the photoreceptors were not illuminated, the aphids would switch to the production of oviparae. The results obtained with capillaries (Fig. 10.11) showed that illumination of the head induced virginoparae, whereas illumination of the thorax or abdomen induced oviparae. This showed that the photoreceptors were in the head. Since the embryos within the abdomen failed to respond, the results also demonstrated that the controlling mechanism was strictly maternal. Smaller bore illuminators revealed that the most sensitive area of the head was in the midline of the dorsum. A smaller proportion of the aphids showed a long-day response when the lateral parts of the head or the compound eyes were illuminated. Beams of light directed at the front of the head, or directly *into* the compound eyes,

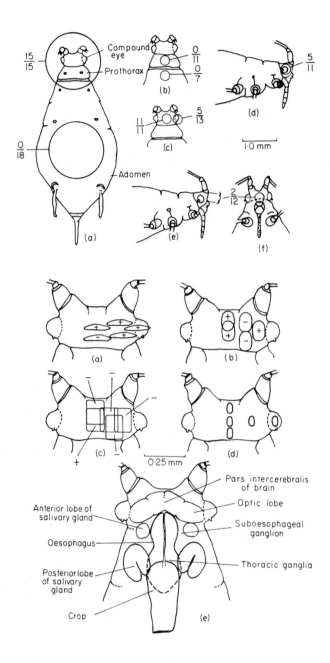

FIG. 10.11. The location of the photoperiodic receptors in *Megoura viciae*. The upper group of figures shows the effect of directing light at the aphids with capillary illuminators. The denominator in each "fraction" indicates the number of aphids tested, the numerator the number responding positively to the supplementary illumination to give a long-day response. The lower group of figures shows the results obtained with light-conducting filaments. Positive and negative responses are indicated by plus and minus signs. (From Lees, 1964.)

were also less effective than a light beam directed vertically on the dorsal midline of the head capsule. Since no dorsal ocelli are present in viviparous aphids and the cuticle in this region is light amber and relatively transparent, Lees concluded that the photoperiodic receptors were probably in the underlying protocerebrum. Furthermore, he suggested that the neurosecretory cells of the pars intercerebralis might be implicated both as receptors and as humoral effectors.

Although illumination of the abdomen of parent aphids had no effect on the determination of progeny-type, such illumination of the grandparents showed that the photoperiodic mechanisms of the (parent) embryos during the last 4 days of prenatal development were fully accessible to light transmitted through the body wall of the grandparent.

Working with *Anacridium aegyptium*, which shows an ovarian diapause of the long-day type, Geldiay (1971) exposed the central part of the head or the abdomen to 6 hours of supplementary illumination after a general body illumination of 9 hours. When the head was thus stimulated the nuclear diameters of the A and B neurosecretory cells, and the oöcytes, were larger than in those insects with the abdomen illuminated. She concluded that long-day illumination activated the neurosecretory cells which have a positive control over oöcyte development.

In the giant silkmoth *Antheraea pernyi*, it has proved possible to demonstrate that the photoperiodic receptors—and perhaps also the clock—are indeed in the brain (Williams, 1963; Williams and Adkisson, 1964). Williams and Adkisson fitted diapausing pupae of *A. pernyi* into holes drilled in an opaque board in such a way that some of them received long-day illumination (*LD 16* : 8) at their head ends and short-day illumination (*LD 8* : 16) at their tails, and others the opposite combination. Out of eighty unchilled pupae maintained in such a "photoperiod gradient" (for 7 weeks), all of those with their head ends exposed to long days developed and emerged as moths, whereas those with their head ends exposed to short days remained firmly in diapause. The photoperiod which the tip of the abdomen "saw" was inconsequential. The fact that the brain was the organ responsible for photoreception was demonstrated in a similar experiment in which brains were removed from twenty-six chilled pupae and then reimplanted under a plastic "window" at the tip of the abdomen. These pupae were then placed in a photoperiodic gradient with a short day (*LD 8* : 16) on one side and continuous darkness (DD) on the other. The results showed that ten (71 per cent) of the fourteen pupae with brainless anteriors exposed to *LD 8* : 16 developed (i.e. a DD response), whereas none of the twelve pupae with "brainy" abdomens exposed to *LD 8* : 16 broke diapause. Although this experiment would have been more compelling had the pupae been placed in a gradient with a strong long daylength on one side and a strong short daylength on the other, it clearly demonstrates that transplantation of the brain to the tip of the abdomen also transplanted sensitivity to photoperiod, and thus identifies the brain as the photoreceptor. In another experiment, twenty-eight brainless pupae failed to develop even after 6 months in a diapause-breaking photoperiod (*LD 17* : 7).

In principle the minimal mechanism for the photoperiodic response must include a photoreceptive pigment, a "clock" to measure daylength and to integrate photoperiodic "information", and an output controlling development or diapause. In silkmoths it is known that the diapause control centres are in the brain, and that the neurosecretory cells in the pars intercerebralis synthesize and secrete brain hormone (PTTH) which, in turn, activates the prothoracic glands to produce ecdysone (Williams, 1952). Since the photoreceptors and the humoral effectors are located in the brain, it follows that the

most likely site for the "clock" is also in the brain. Conclusive evidence for this proposition, however, requires the reciprocal transfer of brains between insects possessing
different clock characteristics (e.g. critical daylengths), and the subsequent demonstration
of these characteristics in the recipient animals.

It is interesting in this context that the richest source of brain hormone in *A. pernyi*
has proved to be the brains of diapausing pupae maintained in short daylength (Williams,
1967). It would thus appear that short daylength inhibits the translocation of the hormone
along the neurosecretory axons, or its release from the corpora cardiaca, rather than its
synthesis within the neurosecretory cells.

By a series of surgical procedures on the brains of diapausing *A. pernyi*, Williams
(1969a) attempted to locate the photoperiodic mechanism within the brain itself. The
strategy of these operations was to do as little a possible to the brain until its ability to
differentiate between a long and a short photoperiod was finally destroyed. Each operation was performed on forty animals, twenty of which were then placed in *LD 17 : 7*
and twenty in *LD 12 : 12*. It was found that cutting the circum-oesophageal connectives,
the tracheal connections to the brain, the nerves to the antennae or eyes, or the nerves
to the corpora cardiaca, had no effect on the insects' ability to discriminate. Neither did
bisection of the brain in the midline, or the dissociation of certain pigmented tips from
the brain lobes. Only when the optic lobes were dissociated from the rest of the brain,
or the entire brain excised and only the cerebral lobes reimplanted, was some damage to
the mechanism evident, although, even then, the insects could still discriminate between
a short day (45 per cent reactivation) and a long day (80 per cent reactivation). A similar
result was obtained when only the dorsal half of the cerebral lobes was returned to the
brainless pupa. Finally, when the whole brain was excised and only the pars intercerebralis (i.e. that part of the brain with the medial but not the lateral neurosecretory cells)
replaced, the pupae then failed to respond to the opposing photoperiods. Williams
pointed out that the proportion of these insects developing (25 per cent in both long and
short daylength) was the same as in the *LL* control. He considered this a "puzzling find-
ing" because one might have anticipated that pupae deprived of their photoperiodic
receptors would have behaved as if in constant dark. No DD control was included,
however, and the proportion of reactivating individuals in DD and *LL* are remarkably
similar in this insect (Williams and Adkisson, 1964). Williams (1969a) concluded that
"the photoperiodic mechanism is located in a tiny mass situated just lateral to the medial
neurosecretory cells. Moreover, this crucial region includes the lateral neurosecretory
cells."

Finally Williams (1969a) attempted to dissociate the hormonal action of the neurosecretory cells from the electrical function of the surrounding neuropile by the use of
puffer-fish Tetrodotoxin which is thought to block the passage of sodium ions through
electrically excitable membranes and therefore to block action potentials. When adminis-
tered to pupae of *A. pernyi* at doses as high as 1 μg/g body weight the insects were paralysed but still underwent normal development to fully formed but flaccid moths which
failed to escape from their pupae. More importantly, diapausing pupae of *A. pernyi* treated
with tetrodotoxin were still able to discriminate between a long and a short photoperiod.
Williams tentatively concluded that nerve impulses conducted in the neuropile played
no part in the photoperiodic mechanism.

Claret (1966 a, b) has also obtained evidence that the photoperiodic receptors lie in
the brain. In the cabbage butterfly *Pieris brassicae*, the sensitive period is apparently

confined to a few days in the third and fourth instars. Claret transferred the brains from larvae just after the fourth moult, and therefore sensitive to photoperiod, into the abdomens of older larvae in which sensitivity was lost. The recipients were then exposed to either short-day (*LD 8 : 16*) or long-day (*LD 16 : 8*) illumination. The results showed that those exposed to short days produced 60 per cent diapause in the pupal instar, whereas those exposed to long days produced only 23 per cent. It was concluded that the brain responded directly to the photoperiodic stimuli. In this connection it is interesting that the head capsule of the larva is entirely black before the third moult. After the third moult, however, the larva develops a translucent yellow triangle on the clypeus which admits light to the brain and sensitivity begins (Claret, 1966b). A similar change in the head capsule was reported for *Bombyx mori* (Bounhiol and Moulinier, 1965) which is sensitive in the early larval instars.

Annotated Summary

1. With a few possible exceptions, the photoreceptors for entrainment of the circadian rhythm of loco-motor or stridulatory activity in cockroaches and crickets are the compound eyes. Occlusion or surgical destruction or isolation of these organs causes the insects to "free-run" in *LL* or *LD* as though they were in DD. The role of the ocelli and/or extraoptic and extracephalic photoreceptors is not satisfactorily established, or is still open to question. For this class of behavioural rhythm there is no sound evidence that the brain itself is involved in photoreception.

2. The clock controlling these locomotory rhythms is in the brain, and in at least two species has been located in the optic lobes. Section of the optic nerves causes cockroaches and crickets to free-run in *LD* suggesting that the photoreceptors (compound eyes) have been isolated from the clock; section of the tissue between the optic lobes and the rest of the protocerebrum causes them to become arrhythmic suggesting that the clock has been isolated from the thoracic locomotory centres.

3. The output from the locomotory clock in cockroaches and crickets appears to be nervous, and cutting the circum-oesophageal commissures (in *Leucophaea maderae*) causes arrhythmia which would not be expected if the output was humoral. Neurosecretory cells in the sub-oesophageal ganglion are not the locus of the driving oscillation.

4. The photoreceptors used for entrainment in the eclosion rhythms of *Drosophila* spp. and in silkmoths are in the brain tissue, and the "organized" photoreceptors are not involved. Transfer of the brain from the head to the abdomen in silkmoth pupae also transfers sensitivity to the environmental light-cycle. Certain exogenous light effects, however, are mediated through the compound eyes of the pharate adult.

5. Removal of the brain from silkmoth pupae did not prevent eclosion but removed the "gated" control. Transfer of the excised brain either into the abdomen of the same individual, or into the abdomen of a decerebrated individual of another species, restored the gated control of eclosion, and in the latter case also determined the *time* of eclosion which was characteristic of the *brain* of the donor species. Since all nerves are cut in these operations the output from the cerebral clock must be humoral rather than nervous.

6. The time of the release of the eclosion hormone in silkmoths is dictated by the circadian clock. Experimental injection of brain homogenates into moths competent to emerge initiates a behavioural sequence, characteristic of the species of *body*, culminating in emergence from the pupal cuticle and cocoon and the spreading of the wings. This sequence is pre-patterned in the CNS.

7. The photoreceptors for "classical" photoperiodism are also in the brain, the eyes and ocelli, where differentiated, not being important. The clock is almost certainly brain-centred, but unequivocal experiments to demonstrate this have yet to be performed. The output of the clock is humoral and involves the decision whether brain hormone (PTTH) is or is not released. Diapausing silkmoth pupae can still be reactivated by long days after injection with tetrodotoxin, although the paralysed moths are unable to emerge.

8. It is likely that the two types of clock (Truman's Type I and Type II) are different in these respects. In Type I, such as eclosion and "developmental" rhythms, the photoreceptors lie in the brain and the output is humoral. Photoperiodism belongs in this category. In Type II clocks, such as those controlling locomotory and other behavioural rhythms, the compound eyes are the principal, if not the only, photoreceptors involved, the brain is insensitive to light, and the clock output is nervous.

APPENDIX A

Glossary and List of Symbols Used to Describe Clock Phenomena

This terminology is based on the *Circadian Vocabulary* of Aschoff, Klotter and Wever (1965), with additions.

Activity time (α). Time in a "sleep–wake" or activity cycle when the animal is active.

Advance phase-shift ($+\Delta\phi$). One or more periods shortened after perturbation by a light or temperature signal causing the overt phase to occur earlier in steady-state than in the control (unperturbed) oscillation.

Aestivation. A summer dormancy (diapause or quiescence). An aestivation diapause is frequently induced by long daylength.

Amplitude. The difference between the maximum (or minimum) value and the mean value in a sinusoidal oscillation. In "population" rhythms it often also refers to the number of individuals emerging or eclosing etc., through a particular gate.

Aschoff's rule ($\tau_{DD} < \tau_{LL}$ nocturnal; $\tau_{DD} > \tau_{LL}$ diurnal). Period of the free-running oscillation (τ) lengthens on transfer from DD to *LL*, or with an increase in light intensity, for dark-active animals, but shortens for light-active animals (Pittendrigh, 1960).

Asymmetrical skeleton photoperiod. A skeleton photoperiod comprising a "main" photoperiod (e.g. 4 to 12 hours) and a short supplementary light pulse which systematically scans the accompanying "night" (Pittendrigh and Minis, 1964).

Bistability phenomenon. Ability of an endogenous oscillation to adopt *either* of two distinct phase relationships to a symmetrical skeleton photoperiod (PP_s), depending on (1) the phase-point in the oscillation which is illuminated first, and (2) the value of the first interval. In *Drosophila pseudoobscura* the bistability phenomenon is observed between PP_s 10.3 and PP_s 13.7 (Pittendrigh, 1966).

Bivoltine. An insect life cycle with two generations per year.

Circadian (rhythm). An endogenous oscillation with a natural period (τ) *close to*, but not necessarily equal to, that of the solar day (24 hours).

Circadian rule. The relationship of the ratio of activity time (α) to rest time (ϱ), and the total amount of activity, to the light intensity. In many vertebrate species α/ϱ and activity increase with light intensity in day-active animals, but decrease in night-active animals (Aschoff, 1965).

Circadian system. The sum total of circadian oscillators (driving oscillators) and driven rhythms in an organism. (These may be independent, loosely coupled or coupled.)

Circadian time (Ct). Time scale (in hours, radians or degrees) covering one full circadian period of an oscillation. The zero point is defined arbitrarily. In *Drosophila pseudoobscura* the point at which the oscillation enters darkness from *LD 12*:12 or from *LL* is considered to be the beginning of the subjective night (Ct 12); the first eclosion peak occurs at Ct 03, 15 hours after the *LL*/DD transition (Pittendrigh, 1966).

Circadian topography. A three-dimensional plot of diapause incidence against the length of the photoperiod and the period of the driving light cycle (T). In species with a circadian photoperiodic clock the topography shows "mountains" of diapause at 24-hour intervals.

Circannual. An endogenous oscillation with a natural period (τ) close to, but not equal to, a year. In the beetle *Anthrenus verbasci* (Blake, 1958) τ is 41 to 44 weeks.

Circasyzygic. An endogenous oscillation with a natural period which approximates to the interval between successive spring or neap tides (14.7 days) (= semilunar rhythm).

Cophase (θ). The interval between the end of the light perturbation and the centroid of the eclosion peaks (Winfree, 1970).

Crepuscular. Twilight active.

Critical daylength (or nightlength). The length of the light (or dark) fraction of the light/dark cycle which separates the "strong" long daylengths from the "strong" short daylengths in the photoperiodic response curve, i.e. a 50 per cent response.

Day-neutral (species, response etc.). Apparently with no reaction to photoperiodic influences.

Delay phase-shift ($-\Delta\phi$). One or more periods lengthened after perturbation by a light or temperature signal causing the overt phase to occur later in steady state than in the control (unperturbed) oscillation.

Diapause. A period of arrest of growth and development which enables the species to overwinter (hibernate) or aestivate, or to synchronize its development cycle to that of the seasons. In most cases diapause involves the cessation of neuro-endocrine activity, and is most frequently induced by photoperiod.

Diurnal. Occurring during the day or light period of the cycle, day-active (cf. *Nocturnal*). An older usage of this term denoting a daily cycle is not used in this book.

Endogenous rhythm (or oscillation). A periodic system which is part of the temporal organization of the organism. It is self-sustaining, i.e. it "free-runs" in the absence of temporal cues such as the daily cycles of light and temperature.

Entrainment. The coupling of a self-sustained oscillation to a *Zeitgeber* (or forcing oscillation) so that both have the same frequency ($\tau = T$)(synchronization) or that the frequencies are integral multiples (frequency demultiplication). Entrainment is possible only within a limited range of frequencies.

Eudiapause. A facultative cessation of development with species-specific sensitive periods and diapausing instars. In favourable conditions development proceeds unchecked; as unfavourable periods approach diapause supervenes. This type of diapause is usually induced by photoperiod and terminated by a period of chilling or by a change in the level of the temperature (Müller, 1970).

Exogenous rhythm. A rhythm of activity which is a direct response to the environmental cycle of light and temperature. In the absence of these variables the rhythm does not persist.

External coincidence. A model for the photoperiodic clock in which light has a dual role: (1) it entrains and hence phase-sets the photoperiodic oscillation, and (2) it controls photoperiodic induction by a temporal coincidence with a photoperiodically-inducible phase (ϕ_i) (Pittendrigh and Minis, 1964).

Fixed point. A point in the circadian oscillation at which a light pulse leaves the oscillation at the *same* circadian time it was when the pulse started, regardless of the duration of the pulse. In *Drosophila pseudoobscura* the theoretical position of the fixed point is at Ct 10.75 (Johnsson and Karlsson, 1972b).

Forcing oscillation (see also *Zeitgeber*). An oscillation or periodic environmental factor capable of synchronizing or entraining another oscillation.

Free-running period (τ). The period of an endogenous oscillator, revealed in the absence of a forcing oscillation or *Zeitgeber* (i.e. in constant temperature and DD, or LL).

Free-running rhythm. A biological rhythm or oscillation in its "free-running" condition (unentrained).

Frequency. The reciprocal of period.

Frequency demultiplication. The entrainment of an endogenous oscillation to a *Zeitgeber* when the frequencies are integral multiples, e.g. the entrainment of a circadian oscillation ($\tau \sim 24$ hours) to environmental light cycles with $T = 4, 6, 8$ or 12 hours, resulting in an entrained steady state with a period of exactly 24 hours.

Gate. The "allowed zone" of the cycle, dictated by the circadian clock, through which flies may emerge, hatch, etc. If a particular insect is not at the "correct" morphogenetic stage to utilize one gate it must wait a full 24 hours for the next (Pittendrigh, 1966).

Hibernation. The state of dormancy (diapause or quiescence) which occurs during the winter months.

Hour-glass. A non-repetitive (i.e. non-oscillatory) timer which is set in motion at, say, dawn or dusk and then runs its allotted time. Such timing devices are to be found in aphid photoperiodism (Lees, 1965) and in certain other clock phenomena.

Intermediate photoperiodic response. The photoperiodic response in some "univoltine" species in which non-diapause development is restricted to a very narrow range of daylengths (18 to 20 hours); with both longer and shorter daylengths diapause occurs (Danilevskii, 1965).

Internal coincidence. A model for the photoperiodic clock in which two or more oscillators are independently phase-set by dawn and dusk, and photoperiodic induction depends on the phase-angle between the two (Pittendrigh, 1972; Tyshchenko, 1966).

Interval timer. (a) A non-repetitive timer or hour-glass (Lees, 1965). (b) A type of oscillatory clock which dictates the times of the day at which a particular event (e.g. eclosion) can occur—in contrast to "pure rhythms" or "continuously consulted" clocks (Pittendrigh, 1958).

Isoinduction surface, see *Circadian topography.*

Long-day (species, response, etc.). The photoperiodic response in which the insects grow and develop during the summer months at long daylength, but enter diapause in the autumn as the days shorten.

Long-day-short-day response. The photoperiodic response which requires short days following long days for its operation.

Multivoltine. An insect life-cycle with many generations per year.

Night-interruption experiment. An experiment in which an insect's photoperiodic response is investigated by asymmetric skeleton photoperiods.

Nocturnal. Occurring during the night or dark period of the cycle, night-active (cf. *Diurnal*).

Oligopause. A facultative arrest of development often with the induction and termination of diapause under photoperiodic control (Müller, 1970).

Oscillator. (A-oscillator). In Pittendrigh's terminology (1960, 1967) the A-oscillator is the self-sustained and light sensitive oscillator whose period is temperature-compensated and which drives the temperature-sensitive B-oscillator (rhythm) which more immediately controls the overt rhythm of activity (e.g. eclosion). The circadian pacemaker.

Parapause. An obligatory diapause observed in univoltine species. Clearly defined inductive periods and diapause supervenes in every generation in a species-specific instar. Onset appears to be independent of the environment (Müller, 1970).

Period. The time after which a definite phase of the oscillation reoccurs. In biological systems it should be stated what overt phase reference point (ϕ_r) has been used to determine the period, e.g. onset of activity, or median of eclosion peak, etc.

Phase. Instantaneous state of an oscillation within a period.

Phase-angle (ψ). Value on the abscissa corresponding to a point of the curve (phase) given either in radians, in degrees, or in other fractions of the whole period. It can be given in time units if the length of the period is stated.

Phase angle difference. Difference between two corresponding phase angles in two coupled oscillators, given either in degrees of angle or in time units.

Phase response curve (PRC). Plot of phase shift $(\Delta\phi)$ (magnitude and sign) caused by a single perturbation at different phases (circadian times) of an oscillator in free-run.

Phase shift $(\Delta\phi)$. A single displacement of an oscillation along the time axis, may involve either an advance $(+\Delta\phi)$ or a delay $(-\Delta\phi)$.

Photonon. The kinetics of a clock after the onset of the light (Truman, 1971b).

Photoperiod. The period of light in the daily cycle (daylength), measured in hours.

Photoperiodic counter. That aspect of the photoperiodic response which consists of a temperature-compensated mechanism which accumulates "information" from successive photoperiodic cycles.

Photoperiodically inducible phase (ϕ_i). A hypothetical phase-point in an oscillator (or perhaps a driven rhythm) which is light-sensitive, and an integral part of the external coincidence model for the photoperiodic clock (Pittendrigh, 1966).

Photoperiodic response curve (PPRC). The response of a population of a particular insect to a range of stationary photoperiods (DD to *LL*) usually including the critical daylength.

Photophase = photoperiod, daylength (Beck, 1968).

Quiescence. A state of dormancy directly imposed by adverse factors in the environment (e.g. cold torpor, dehydration) (cf. *Diapause*).

Range of entrainment. Range of frequencies within which a self-sustained oscillation can be entrained with a *Zeitgeber*. For most organisms the range is from about 18 to 30 hours.

Rest time (ϱ). Time in a "sleep–wake" cycle in which the organism is inactive (Aschoff, 1965).

Rhythm. A periodically reoccurring event. In Pittendrigh's (1967) sense the word is restricted to the driven elements (i.e. the temperature-dependent B-oscillations) directly coupled to the light sensitive driver (A-oscillation).

Required day number (RDN). The temperature-compensated number of inductive photoperiods in the photoperiodic counter, required to raise the incidence of diapause in a particular day's batch to 50 per cent. Equivalent to the "critical day number" of Tyshchenko *et al.* (1972).

Resonance experiment (= *T*-experiment). An experimental design in which the photoperiod is held constant but the period of the driving light cycle (T) varied (e.g. 12 to 72 hours).

Scotonon. The kinetics of a clock after the onset of the dark (Truman, 1971b).

Scotophase. The dark period, or night-time, of the diel-cycle (Beck, 1968).

Semivoltine. An insect life cycle which occupies a 2-year period, i.e. half a generation per year.

Sensitive period (SP). The period of an insect's life-cycle when it is sensitive to photoperiodic control of diapause induction or termination.

Short-day. (Species, response, etc.) The photoperiodic response in which the insects grow and develop at short daylength, but enter diapause (aestivation) during the summer months when days lengthen.

Short-day–long-day response. A photoperiodic response which requires long days following short days for its operation.

Skeleton photoperiod. A light regime using two shorter periods of light to simulate dawn and dusk effects of a longer, complete, photoperiod (PP_c). See *Symmetrical* and *Asymmetrical skeleton*.

Singularity (T^*S^*). A critical annihilating light pulse of a particular duration (S^*) on intensity placed a certain time (T^*) after the *LL*/DD transition (i.e. at a particular circadian time) which puts the clock in a non-oscillatory state (i.e. stops the clock). In *Drosophila pseudoobscura* T^*S^* is a 50-second pulse of dim blue light (10 $\mu W/cm^2$) placed 6.8 hours after the *LL*/DD transition (i.e. at about Ct 18–19) (Winfree, 1970).

Subjective day. The first half of the circadian cycle (Ct 0 to 12) of an oscillation, and that half in which "day" normally occurs.

Subjective night. The second half of the circadian cycle (Ct 12 to 24) of an oscillation, and that half in which "night" normally occurs.

Symmetrical skeleton photoperiod (PP_s). A skeleton photoperiod comprised of two short pulses of equal duration which may simulate a complete photoperiod (PP_c) (Pittendrigh and Minis, 1964).

Synchronization. State in which two or more oscillations have the same frequency due to mutual or unilateral influences. See also *Entrainment*.

T-experiment, see *Resonance experiment*.

Thermoperiod. A daily temperature cycle which may be sinusoidal, "square-wave", etc., and may act as a *Zeitgeber*.

Token stimulus. A seasonal signal which serves to indicate the approach of adverse conditions (e.g. winter) but is itself *not* adverse (e.g. short photoperiod) (Lees, 1955).

Transformation curve. A plot of the circadian time of an oscillation at the end of a light pulse as a function of the circadian time at the beginning of the pulse (Johnsson and Karlsson, 1972b).

Transients. One or more temporary oscillatory states between two steady states caused, for instance, by light or temperature perturbations.

Univoltine. An insect life-cycle with one generation per year.

Voltinism. Referring to the number of generations per year.

Zeitgeber. That forcing oscillation which entrains a biological oscillation, e.g. the environmental cycles of light and temperature.

Zeitgeber time (Zt). Time of environmental cycle measured in hours, usually, in the case of light, after the lights-on signal or "dawn".

Zeitgedächtnis. The "time-memory" of bees.

Zeitsinn. The "time-sense" (= time memory) of bees.

Symbols

L	Light fraction of the cycle (intensity may be specified).
D	Dark fraction of the cycle.
LD	Light/dark cycle. LD 4 : 20 represents four hours of light and twenty hours of darkness in each 24 hour cycle.
LL	Continuous light.
DD	Continuous dark.
τ	Natural period of a biological oscillation as revealed in "free-running" conditions.
T	Period of *Zeitgeber*.
ϕ	Phase point.
ϕ_r	Phase reference point. ϕ_r for environmental light cycle may be beginning of light pulse; ϕ_r for the oscillator in *Drosophila pseudoobscura* is the easily assayed point in the phase response curve where a 360° phase-jump occurs (Ct 18.5); ϕ_r for the rhythm (in *D. pseudoobscura*) is the median of pupal eclosion. In other systems it may be, for instance, the onset of locomotor activity.

ϕ_i The photoperiodically inducible phase.

ψ Phase relation.

$\psi_{R, L}$ Phase angle difference between phase reference point of rhythm and light cycle.

$\psi_{R, O}$ Phase angle difference between phase reference point of rhythm and oscillation.

$\psi_{O, L}$ Phase angle difference between phase reference point of oscillation and light.

$\Delta\phi$ Phase shift.

$+\Delta\phi$ Advance phase shift.

$-\Delta\phi$ Delay phase shift.

α Activity time.

ϱ Rest time.

θ Cophase. The time interval between the end of the light perturbation and the centroid of the eclosion peaks.

Ct   Circadian time.

Zt   *Zeitgeber* time.

APPENDIX B

Lists of Insects Exhibiting Rhythmic Activity or Photoperiodic Control

1. Rhythms of activity exhibited by individual insects
2. Rhythms of activity exhibited by populations of insects
3. Rhythms of a physiological nature
4. Insects species in which photoperiodic control of diapause has been demonstrated or reasonably inferred
5. Insect species in which photoperiodic control of phenomena other than diapause has been demonstrated or proposed

1. Rhythms of Activity Exhibited by Individual Insects

D = diurnal; N = nocturnal; C = crepuscular. Exog. = exogenous

Species	Process controlled	Overt phase	Free-running period (τ)	Reference
EPHEMEROPTERA				
Ecdyonurus torrentis	locomotor activity	D		Harker (1953)
Heptagenia lateralis	locomotor activity	D		Harker (1953)
Baetis rhodani	locomotor activity	D		Harker (1953
ORTHOPTERA				
Schistocerca gregaria	locomotor activity	D	Exog. ?	Odhiambo (1966)
Chorthippus curtipennis	oviposition	D, peak in afternoon		Loher and Chandrashekaran (1970)
Poecilocerus hieroglyphicus	locomotor activity	D, peak in afternoon		Abushama, (1968)
Tettigonia viridissima	singing	N		Neilsen (1938)
Stenopelmatus sp.	locomotor activity	N	<24 in DD	Lutz (1932)
Acheta domestica	locomotor activity	N	>24 in *LL*	Lutz (1932); Nowosielski and Patton (1963)
Acheta domestica	mating			Nowosielski and Patton (1963)
Acheta domestica	spermatophore production		Exog. ?	MacFarlane (1968)
Gryllus campestris	locomotor activity	D		Cloudsley-Thompson (1958)
Ephippiger spp.	stridulation	N-D	23 h 36' in DD	Dumortier (1972)
Teleogryllus conmodus	stridulation	N	25 h 40' in *LL*	Loher (1972)
PHASMIDA				
Carausius morosus	locomotor activity	N	24.3 in DD	Godden (1973)
Carausius morosus	oviposition	N		Eidmann (1956); Godden (1973)

Taxon	Behavior	Type	Period / Remarks	References
DICTYOPTERA				
Leucophaea maderae	locomotor activity	N	23.3–24 in DD; 24–24.75 in *LL*; $Q_{10} = 1.04$	Roberts (1960); Lohmann (1967)
Byrsotria fumigata	locomotor activity	N	23.9–24.4 in DD; 24.4–25.5 in *LL*	Roberts (1960)
Periplaneta americana	locomotor activity	N	23.8 in DD; 24.5 in *LL*	Harker (1956); Roberts (1960)
Blatta orientalis	locomotor activity	N		Gunn (1940)
Blaberus craniifer	locomotor activity	N		Ball (1971, 1972)
Blaberus giganteus	locomotor activity	N		Wobus (1966)
Blaberus discoidalis	locomotor activity	N		Cloudsley-Thompson (1960)
Blatella germanica	locomotor activity	N		Shepherd and Keeley (1972)
Ectobius lapponicus	locomotor activity	N in the field, ♂♂ are active in afternoon	<24 in dim light	Dreisig and Nielsen (1971); Dreisig (1971)
HEMIPTERA				
Velia currens	locomotor activity	N	24.2 in DD	Rensing (1961)
Cimex lectularius	locomotor activity	N	27.5 in 700 lux	Mellanby (1939)
Oncopeltus fasciatus	oviposition	D		Caldwell and Dingle (1967); Rankin *et al.* (1972)
NEUROPTERA				
Myrmeleon obscurus	pit-building activity	N, peak after dusk	24.0–24.25 in DD; 23.75–24.0 in *LL*	Youthed and Moran (1969a)
LEPIDOPTERA				
Anagasta kuhniella	flight	C	Exog. ? free-runs in DD	Edwards (1962)
Crambus teterrellus	oviposition	C, dusk		Crawford (1966)
Crambus topiarus	oviposition	D, peak in late afternoon		Crawford (1967)
Pectinophora gossypiella	flight	N	22.75 in DD	Lappla and Spangler (1971)
Pectinophora gossypiella	oviposition	N		Minis (1965)

TABLE 1 (*cont.*)

Species	Process controlled	Overt phase	Free-running period (τ)	Reference
Halisodota argentata / *Nepytia phantasmaria*	larval activity, flight	N, peak at sunset		Edwards (1964 a, b)
DIPTERA				
Aëdes egypti	flight	D, peak in afternoon	22.5 in DD / 26.0 in *LL* / 23.5 in DD	Taylor and Jones (1969)
Aëdes taeniorhynchus	flight	C, peaks at dusk and dawn		Nayar and Sauerman (1969, 1971)
Anopheles gambiae	flight	C, dusk and dawn	~23.0 in DD	Jones, Ford and Gillett (1966)
Aëdes aegypti	oviposition	D, peak at end of light	~24 in DD	Haddow and Gillett (1957)
Aëdes punctor	flight	C		Taylor (1969)
Anopheles plumbeus	flight	C		Taylor (1969)
Anopheles labranchiae atroparvus	flight	N		Taylor (1969)
Anopheles farauti	flight	N, peaks at dawn, dusk, and 11–13 hours after dusk		Taylor (1969)
Anopheles stephensi	flight	N		Taylor (1969)
Culex pipiens pipiens	flight	N		Taylor (1969)
Dacus tryoni	mating	C, dusk		Tychsen and Fletcher (1971)
Drosophila robusta	flight	D, peak in late afternoon	~28.0 at 0.8 lux/ft² free-runs in *LL*	Roberts (1956)
Drosophila melanogaster	locomotor activity	D		Hardeland and Stange (1971); Konopka and Benzer (1971)
Drosophila melanogaster	courtship	N		Hardeland and Stange (1971)
Drosophila melanogaster	oviposition	N		Rensing and Hardeland (1967); Gruwez *et al.* (1972)
Glossina morsitans	activity	D, bimodal		Brady (1972)
Glossina morsitans	larviposition	D		Phelps and Jackson (1971)
Glossina austeni	larviposition	D		Nash and Trewern (1972)

Species	Activity		Notes	Reference
Phornia regina	locomotor activity	D	24.3 in DD	Green (1964)
HYMENOPTERA				
Camponotus clarithorax	locomotor activity	D		McCluskey (1965)
Apis mellifera	locomotor activity	D	workers 21.8 in DD; drones 23.7 in DD	Spangler (1972, 1973)
COLEOPTERA				
Carabus cancellatus	locomotor activity	N	<24 in DD	Weber (1965, 1966); Lamprecht and Weber (1973)
Various Carabidae	locomotor activity	N		Greenslade (1963)
Boletotherus cornutus	locomotor activity	N	persists in DD	Park and Keller (1932)
Tenebrio molitor	locomotor activity	N	23.3–27.3 at 0.01–100 lux	Lohmann (1964)
Ptinus tectus	locomotor activity	N	persists in LL	Bentley et al. (1941)
Cotinus nitida	locomotor activity			Hintze (1925)
Amphimallon majalis	flight	N		Evans and Gyrisco (1958)
Geotrupes sylvaticus	locomotor activity	C, peaks at dawn and dusk	24.0 in DD	Geisler (1961)
Geotrupes vernalis	locomotor activity	D, bimodal	24.7 in 15 lux becomes unimodal	Warnecke (1966)
Geotrupes stercorarius	locomotor activity	N, bimodal	>24 in DD	Warnecke (1966)
Leptinotarsa decemlineata	feeding	D	<24 in DD	Grison (1943)
Calandra granaria	locomotor activity	N	~24 in DD	Birukow (1964)
Graphognathus leucoloma	feeding	N		Senn and Brady (1973)
Graphognathus peregrinus	feeding	N		Senn and Brady (1973)
Blaps mucronata	locomotor activity	N		Thomas and Finlayson (1970)

2. Rhythms of Activity Exhibited by Populations of Insects

Species	Process controlled	Overt phase	Free-running period (τ)	Reference
ODONATA				
Tetragoneuria cynosura	emergence	D		Lutz (1961)
PHASMIDA				
Carausius morosus	egg hatch	D, peak in morning		Kalmus (1938a)
Carausius morosus	moulting	N		Kalmus (1938a)
HEMIPTERA				
Aphis fabae	final moult	D, peak in morning		Johnson et al. (1957)
Various aphids	moulting and reproduction			Haine (1957)
Cardiaspina densitexta	egg hatch	D, peak in morning	persist in LL	White (1968)
LEPIDOPTERA				
Ascia monuste	pupation	D, peak in afternoon		Nielsen (1961)
Manduca sexta	PTTH secretion	N		Truman (1972a)
Antheraea pernyi	egg hatch	C, peak at dawn	~24 in DD	Riddiford and Johnson (1971)
Antheraea pernyi	pupal eclosion	D, peak in afternoon	~22.0 in DD	Truman and Riddiford (1970)
Hyalophora cecropia	pupal eclosion	D, peak in morning		Truman and Riddiford (1970)
Heliothis zea	pupal eclosion	D, peak in morning		Callahan (1958)
Bombyx mori	egg hatch			Tanaka (1961) Bremer (1926); Scott (1936); Moriarty (1959)
Anagasta kuhniella	pupal eclosion	D, peak in afternoon		

Species	Event	Timing	Period	Reference
Pectinophora gossypiella	egg hatch	C, peak close to dawn	24 in DD	Minis and Pittendrigh (1968)
Pectinophora gossypiella	pupal eclosion	D, peak in morning	22.5 in DD	Pittendrigh and Minis (1971)
Halisodota argentata	pupal eclosion	sunset		Edwards (1964)
Nepytia phantasmaria	egg hatch	D, peak at dawn		Edwards (1964)
DIPTERA				
Aëdes taeniorhynchus	pupation	D, peak in morning	22.5 in DD $Q_{10} \sim 1.0$	Nayer (1967a)
Aëdes vittatus	pupation	N		McClelland and Green (1970)
Mansonia titillans	egg hatch	C, peak at dusk	free-runs in *LL*	Nayer et al. (1973)
Anopheles stephensi	pupal eclosion	N		Coluzzi (1972)
Various Chironomidae	pupal eclosion	D, peak in morning		Palmen (1955)
Pseudosmittia arenaria	pupal eclosion	C, peak close to dawn		Remmert (1955)
Dacus tryoni	pupal eclosion	C, peak in morning		Myers (1952); Bateman (1955)
Ceratitis capitata	pupal eclosion	peak in morning		Myburgh (1963)
Pterandrus rosa	pupal eclosion	D, peak in morning		Myburgh (1963)
Coelopa frigida	puparium formation			Remmert (1961)
Drosophila pseudoobscura	pupal eclosion	C?, peak close to dawn	~ 24 in DD $Q_{10} = 1.02$	Pittendrigh (1954)
Drosophila melanogaster	pupal eclosion	C?, close to dawn		Brett (1955); Clayton and Paietta (1972)
Drosophila victoria	puparium-formation	N	free-runs in *LL*	Rensing and Hardeland (1967)
Scopeuma stercoraria	pupal eclosion	C?, peak close to dawn	persists in DD and *LL*	Lewis and Bletchley (1943)

TABLE 2 *(cont.)*

Species	Process controlled	Overt phase	Free-running period (τ)	Reference
Pegomyia betae	pupal eclosion	D, peak in morning		Dunning (1957)
Glossina morsitans	pupal eclosion	D, peak close to dawn	free-runs in DD	Phelps and Jackson (1971)
Sarcophaga argyrostoma	pupal eclosion			Saunders (unpub).
COLEOPTERA		N		
Amphimallon majalis	pupal eclosion	D, peak in evening	22.5 in DD	Evans and Gyrisco (1958)
Xyleborus ferrugineus	pupal eclosion			Saunders and Knoke (1968)

3. Rhythms of a Physiological Nature

DGL—daily growth layers in the endocuticle; NSC—neurosecretory cells

Species	Process controlled	Overt phase	Free-running period (τ)	Reference
EPHEMEROPTERA				
Aeshna grandis	Daily growth layers in cuticle			Neville (1963c)
Aeshna juncea	DGL			Neville (1967a)
ORTHOPTERA				
Schistocerca gregaria	DGL in endocuticle and in resilin		$\sim$23.0 in DD $Q_{10} = 1.04$	Neville (1963 a, b; 1965)
Schistocerca paranensis	DGL			Neville (1967a)
Locusta migratoria	DGL			Neville (1967a)
Nomadacris septemfasciata	DGL			Neville (1967a)
Cyrtacanthacris tartarica	DGL			Neville (1967a)
Decticus verrucivorus	DGL			Neville (1963c)
Tettigonia viridissima	DGL			Neville (1967a)
Omocestus viridulus	DGL			Neville (1967a)
Gomphocerus maculatus	DGL			Neville (1967a)
Stenobothrus lineatus	DGL			Neville (1967a)
Dolichopoda linderi	DGL			Neville (1965)
Pyrgomorpha sp.	DGL			Neville (1967a)
Chortoicetes terminifera	DGL			Neville (1967a)
Humbe tenuicornis	DGL			Neville (1967a)
Gryllus bimaculatus	DGL			Neville (1967a)
Acheta domestica	DGL			Neville (1967a)
Acheta domestica	trehalose titre	peak 3 hr before dawn		Nowosielski and Patton (1964)
Acheta domestica	sensitivity to insecticides			Nowosielski et al. (1964)
Acheta domestica	RNA synthesis in brain NSC	maximum 30 min after lights-on		Cymborowski and Dutkowski (1969, 1970)

TABLE 3 (cont.)

Species	Process controlled	Overt phase	Free-running period (τ)	Reference
Acheta domestica	impulse frequency in brain			Tyshchenko et al. (1972)
PHASMIDA				
Carausius morosus	DGL			Neville (1967a)
Carausius morosus	colour change		persists in DD	Schliep (1910, 1915); Kalmus (1938a)
Siphyloidea sp.	DGL			Neville (1967a)
DERMAPTERA				
Forficula auricularia	DGL			Neville (1963c)
DICTYOPTERA				
Periplaneta americana	DGL		persist in LL?	Neville (1963c)
Periplaneta americana	haemolymph hydrocarbons			Turner and Acree (1967)
Blaberus discoidalis	DGL	main peak at dusk		Neville (1967a)
Blatella germanica	oxygen consumption sensitivity to insecticides, KCN			Beck (1963, 1964)
Blatella germanica	DGL			Beck (1963)
Sphodromantis tenuidentata	DGL			Neville (1970)
HEMIPTERA				
Oncopeltus fasciatus	DGL		~24 in DD $Q_{10} = 1.0$	Neville (1967a); Dingle et al. (1969)
Hydrocyrius columbiae	DGL			Neville (1965)
Aphrophora alni	DGL			Neville (1963c)

	Process measured	Peak / pattern	DD / LL	Reference
LEPIDOPTERA				
Plusia gamma	pigment movement in eye		persist in DD	Demoll (1911)
Trichoplusia ni	pheromone release and response	maximum at end of night	~24.0 in *LL* (0.3 lux)	Shorey and Gaston (1965); Sower *et al.* (1970)
Heliothis virescens	pheromone response			Shorey and Gaston (1965)
Heliothis zea	sensitivity to insecticides			Bull and Lindquist (1965)
Autographa californica	pheromone response			Shorey and Gaston (1965)
Spodoptera exigua	pheromone response			Shorey and Gaston (1965)
Anagasta kühniella	pigment migration in eye		Exog. ?	Day (1941)
Anagasta kühniella	pheromone release and response	peak at dawn	free-runs in DD	Traynier (1970)
Ostrinia nubilalis	oxygen consumption	peak at dusk	persists in DD	Beck (1963, 1964)
Ostrinia nubilalis	diameters of brain NSC	trimodal rhythm		Beck (1964)
Adoxophyes fasciata	pheromone response	main peak at dawn		Nagata *et al.* (1972)
DIPTERA				
Culex tarsalis	glycogen levels	peak at end of light		Takahashi and Harwood (1964)
Anastrepha ludens	pheromone response	peak at dusk		Flitters (1964)
Drosophila melanogaster	oxygen consumption	bimodal		Rensing *et al.* (1968) Belcher and Brett (1973)
Drosophila melanogaster	nuclear size in c/a and brain NSC	bimodal		Rensing (1964)
Drosophila melanogaster	nuclear sizes in fat body, prothoracic gland	bimodal		Rensing *et al.* (1965)
Drosophila melanogaster	nuclear sizes in salivary gland	bimodal	persists in *LL*	Rensing (1969)
Drosophila melanogaster	level of 5-HT (serotonin)	peaks in adults in middle of night		Fowler *et al.* (1972)

TABLE 3 *(cont.)*

Species	Process controlled	Overt phase	Free-running period (τ)	Reference
Sarcophaga argyrostoma	DGL in thoracic apodemes			Schlein (1972)
HYMENOPTERA *Apis mellifera*	DGL			Menzel *et al.* (1969)
COLEOPTERA *Carabus nemoralis*	nuclear diameters of c/a cells			Klug (1958)
Photinus pyralis	flashing	peak in evening	~24.0	Buck (1937)
Tenebrio molitor	DGL	peak at night	persists, but non-circ.	Caveney (1970); Zelazny (1969)
Tenebrio molitor	oxygen consumption	peak at dusk and dawn		Michal (1931); Campbell (1964) Fondacara and Butz (1970)
Tenebrio molitor	sensitivity to methyl parathion		non-circadian	Zelazny and Neville (1972)
Oryctes rhinoceros	DGL		non-circadian	Zelazny (1969)
Oryctes nasicornis	DGL		non-circadian	Zelazny (1969)
Aegus upoluensis	DGL		non-circadian	Zelazny (1969)
Olethrius insularis	DGL		persists in DD	Jahn and Crescitelli (1940); Jahn and Wulff (1943)
Dytiscus fasciventris	electroretinogram			

4. Insect Species in which Photoperiodic Control of Diapause has been Demonstrated or Reasonably Inferred

This list is based on those of Danilevskii (1965) and Beck (1968), with additions. Type of photoperiodic response: L—long-day response; S—short-day response; I—"intermediate" type of Danilevskii; N—day-neutral; Incr.—increasing daylength; Decr.—decreasing daylength; Oblig.—obligatory; L–S—response to sequence of long followed by short daylengths; S–L—response to sequence of short followed by long daylengths.

Species	Type	Sensitive period	Diapause stage	Latitude	Critical daylength	Reference
ODONATA						
Anax imperator	Decr.		nymph			Corbet (1956)
Lestes sponsa			eggs			Corbet (1956)
Neotetrum pulchellum			nymph			Montgomery and Macklin (1962)
Tetragoneuria cynosura			nymph			Lutz and Jenner (1964)
Aeshna mixta			egg			Schaller (1968)
PLECOPTERA						
Capnia bifrons	Incr.	nymph	nymph			Khoo (1968a)
Diura bicaudata		adult?	egg			Khoo (1968b)
Brachyptera risi			egg			Khoo (1968c)
ORTHOPTERA						
Anacridium aegyptium	Incr.	nymphs and adults	adult			Norris (1965); Geldiay (1971)
Nomadacris septemfasciata	L–S	nymphs	adult			Norris (1959, 1965)
Chorthippus brunneus	Oblig.		egg			Moriarty (1969)
Tetrix arenosa	L		nymph			Sabrosky et al. (1933)
Tetrix undulata						Sicker (1964)
Gryllus campestris	L		nymph			Fuzeau–Braesch (1965)
Nemobius yezoensis			nymph			Masaki and Oyama (1963)
Teleogryllus taiwanemma	L	nymph	nymph			Masaki and Ohmachi (1967)
Chortophaga viridifasciata	L		nymph			Halliburton and Alexander (1964)
Teleogryllus yezoemma	Oblig.		egg			Masaki and Ohmachi (1967)
Teleogryllus emma	Oblig.		egg			Masaki and Ohmachi (1967)

TABLE 4 (cont.)

Species	Type	Sensitive period	Diapause stage	Latitude	Critical daylength	Reference
DICTYOPTERA						
Ectobius lapponicus	?	4th instar nymph				Brown (1973)
HOMOPTERA						
Euscelis plebejus	L		egg			Müller (1957, 1961)
Euscelis lineolatus	L		egg			Müller (1957, 1961)
Nephotettix cincticeps	L		nymph			Kisimoto (1959)
Stenocranus minutus	S		adult			Müller (1957, 1958)
Stenocranus major	L–S		adult			Strübing (1963)
Delphacodes striatella	S		nymph			Kisimoto (1956)
Psylla pyri	L	nymph	adult			Bonnemaison and Missonier (1955); Oldfield (1970)
Psylla peregrina	L		adult			Missonier (1956)
Muellerianella brevipennis	L	adult?	egg			Witsack (1971)
Aleurolobus asari	Oblig.		nymph			Bährmann (1972)
Aleyrodes asari	L	nymph	adult			Bährmann (1972)
Aleurochiton complanatus	L		"pupa"			Müller (1962 a, b)
Aphis forbesi	L		egg			Marcovitch (1923, 1924)
Capitophorus hippophaes	L		egg			Marcovitch (1924)
Dysaphis plantaginea	L		egg			Marcovitch (1923, 1924); Bonnemaison (1965)
Myzus persicae	L		egg			Bonnemaison (1951)
Brevicoryne brassicae	L		egg			Bonnemaison (1951)
Acyrthosiphon pisum	L		egg		$14\frac{1}{2}/24$	Kenten (1955)
Megoura viciae	L		egg			Lees (1959)
Macrosiphum euphorbiae	L		egg			MacGillivray and Anderson (1964)
Periphyllus testudinatus	S		1st instar nymph			Bonnemaison (1956)

Species				Locality	Photoperiod	Reference
Drepanosiphum platanoides	S		adult			Dixon (1963)
HETEROPTERA						
Gerris odontogaster	Decr.	nymph	macropt. adult			Vepsäläinen (1971 a, b, 1974)
Lygus hesperus	L	nymph	adult			Beards and Strong (1966); Leigh (1966)
Exolygus rugulipennis	L	nymph	adult			Boness (1963)
Adelphocoris lineolatus						Ewan (1966)
Ischnodemus sabulalati	S?	adult	egg			Müller (1960)
Pyrrhocoris apterus	L	nymph	nymph			Hodek (1968, 1971)
Aelia acuminata	L		adult	Slovakia	17/24	Hodek (1971)
			adult		15–16/24	
NEUROPTERA						
Chrysopa carnea	L	larva	adult	42°27′N.	13.5–14/24	Tauber and Tauber (1969, 1972b); MacLeod (1967)
	Decr.			33°19′N.	12.5–13/24	
Chrysopa oculata	L	larva	3rd instar in cocoon	Ithaca, N.Y.	12.5–13.5/24	Propp *et al.* (1969)
Chrysopa nigricornis	L	larva	3rd instar larva in cocoon	Ithaca, 42°27′N.	12.5–13.5/24	Tauber and Tauber (1972a)
Chrysopa mohave	L	adult	adult			Tauber and Tauber (1973a)
LEPIDOPTERA						
Papilio podalirius	L		pupa			Wohlfahrt (1957)
Papilio polyxenus	L		larva			Oliver (1969)
Luehdorfia japonica	?		pupa			Hidaka, Ishizoka and Sakagami (1971)
Pieris brassicae	L	larva	pupa	43°N.	~10/24	Way *et al.* (1949); Danilevskii (1965); Claret (1966 a, b)
				50°–60°N.	~15/24	

TABLE 4 (*cont.*)

Species	Type	Sensitive period	Diapause stage	Latitude	Critical daylength	Reference
Pieris rapae	L	larva	pupa	39°N. 35°39′N.	$12\frac{1}{4}/24$ 12h.10m.	Barker *et al.* (1963) Kono (1970)
Pieris napi	L		pupa			Danilevskii (1965)
Danaus plexippus	L		adult			Herman (1973)
Araschnia levana	L		pupa			Danilevskii (1948); Müller (1955)
Vanessa io	L		adult			Danilevskii (1965)
Limenitis archippus	L		larva			Clark and Platt (1969)
Lycaena phlaeas daimio	L		early larva		$13\frac{1}{2}/24$	Sakai and Masaki (1965)
Smerinthus ocellatus	L		pupa			Danilevskii (1965)
Smerinthus populi	L	larva	pupa			Danilevskii (1965)
Manduca sexta	L	larva	pupa			Rabb (1966)
Celerio lineata	L	larva	pupa			Leeley (1970)
Antheraea pernyi	L	larva	pupa		$\sim14/24$	Tanaka (1950a); Williams and Adkisson (1964)
Antheraea polyphemus	L	larva	pupa			Mansingh and Smallman (1967)
Antheraea mylitta	L		pupa			Jolly *et al.* (1971)
Philosamia cynthia	L		pupa			Pammer (1965)
Hyalophora cecropia	L	larva	pupa			Mansingh and Smallman (1966)
Arctia caia	L		larva			Danilevskii (1965)
Arctia aulica	L		larva			Danilevskii (1965)
Spilosoma menthastri	L		pupa			Danilevskii (1965)
Spilosoma lubricipeda	L		larva			Danilevskii (1965)
Parasemia plantaginis	L		larva			Danilevskii (1965)
Lithosia griseola	L		pupa			Danilevskii (1965)
Hylophila prasinana	L		pupa			Danilevskii (1965)
Hyphantria cunea	L	larva	pupa		$14\frac{1}{2}/24$	Jermy and Saringer (1955) Masaki *et al.* (1968)
	S	larva	pupa			Umeya and Masaki (1969)

Species	Type	Sensitive stage	Diapause stage	Latitude	Critical photoperiod	Reference
Acronycta rumicis	L	larva	pupa	43°N. 50°N. 55°N. 60°N.	$14\frac{1}{2}/24$ $16\frac{1}{2}/24$ $18/24$ $19/24$	Danilevskii (1965)
Acronycta megacephala	L	larva	pupa			Danilevskii (1965)
Acronycta leporina	L		pupa			Danilevskii (1965)
Acronycta psi	L		pupa			Danilevskii (1965)
Agrotis C-nigrum	L	larva	larva		$\sim 15/24$	Way and Hopkins (1960)
Diataraxia oleracea	L		pupa			Danilevskii (1965)
Diataraxia dissimilis	L		pupa			Danilevskii (1965)
Diataraxia contigua	L		pupa			Danilevskii (1965)
Trachea atriplicis	L		pupa			Danilevskii (1965)
Demas coryli	L		pupa			Danilevskii (1965)
Chloridia obsoleta	L		pupa			Goryshin (1958)
Chloridia dipsacea	L		pupa			Goryshin (1958)
Melicleptria scutosa	L		pupa			Goryshin (1958)
Athetis ambigua	L		larva			Danilevskii (1965)
Chorizagrotis auxiliaris	S		adult			Jacobson and Blakeley (1959); Jacobson (1960)
Mamestra brassicae	L		pupa	40°35'N. 28°23'N.	$14/24$ $12\frac{1}{2}/24$	Otuka and Santa (1955) Masaki (1968) Masaki and Sakai (1965)
Heliothis zea	S	egg, larva	pupa pupa			Wellso and Adkisson (1966) Adkisson and Roach (1971)
Heliothis virescens	L–S	larva	pupa			Phillips and Newsom (1966) Benschoter (1968)
Lophopteryx camelina	L		pupa			Danilevskii (1965)
Pygaera pigra	L		pupa			Danilevskii (1965)
Dasychira pudibunda	L		pupa			Geispitz and Zarankina (1963)
Leucoma salicis	I		larva			Geispitz (1953)
Euproctis chrysorrhoea	I		larva			Geispitz (1953)
Euproctis similis	I	larva	larva			Geispitz (1953)
Orgyia antiqua	I		egg			Kind (1965)
Dendrolimus pini	L		larva			Geispitz (1953)

INSECT CLOCKS

TABLE 4 (cont.)

Species	Type	Sensitive period	Diapause stage	Latitude	Critical daylength	Reference
Dendrolimus sibiricus	L		larva		~15/24	Danilevskii (1965)
Bombyx mori	S	egg and larva (I)	egg (II)			Kogure (1933)
Abaxas miranda	S		pupa			Masaki (1958)
Harrisina brillians	L		larva?			Smith and Langston (1953)
Loxostege staticalis	L		larva			Danilevskii (1965)
Loxostege verticalis	L		larva			Danilevskii (1965)
Ostrinia nubilalis	L	larva	larva		15/24	Beck and Hanec (1960); Beck (1962a)
Diatraea saccharalis	L	larva	larva		~15/24	Katiyar and Long (1961)
Diatraea grandiosella	L	larva	larva			Chippendale and Reddy (1973)
Plodia interpunctella	L		larva			Tsuji (1963)
Crambus trisectus	L		larva			Kamm (1970)
Crambus leachellus cypridalis	L		larva			Kamm (1970)
Chrysoteuchia topiaria	L		larva			Kamm (1973)
Carpocapsa pomonella	L	larva	larva		13/24 $13\frac{1}{2}$/24	Dickson (1949) Peterson and Hammer (1968)
Grapholitha molesta	L	larva	larva		$13\frac{1}{2}$/24	Dickson (1949)
Grapholitha sinana	L	larva	larva		14–15/24	Saringer and Nagy (1968)
Laspeyresia funebrana	L	larva	larva		14–15/24	Saringer (1970)
Polychrosis botrana	L	embryo and larva	pupa			Komarova (1949)
Pandemis coylana	L	larva	larva			Danilevskii (1965)
Pandemis ribeana	L		larva			Danilevskii (1965)
Capua reticulata	L		larva			Danilevskii (1965)
Cacoecia podana	L		larva			Danilevskii (1965)
Acalla fimbriana	L		adult			Danilevskii (1965)

	Oblig. → facult (L)					
Choristoneura fumiferana			larva			Harvey (1957)
Argyrotaenia velutinana			pupa			Glass (1963)
Adoxophyes reticulana	L	larva	2nd instar larva			Ankersmit (1968)
Carposina niponensis			larva			Toshima *et al.* (1961)
Pectinophora gossypiella	L	larva	larva		13/24	Adkisson *et al.* (1963)
Pectinophora malvella	L	larva	larva			Danilevskii (1965)
Homoeosoma electellum	L		larva		$11\frac{1}{2}$/24	Teetes *et al.* (1969)
TRICHOPTERA						
Linnephilus sp.			adult			Novak and Sehnal (1963, 1965)
DIPTERA						
Anopheles maculipennis	L	larva–adult	adult			Vinogradova (1960)
Anopheles superpictus	L		adult			Vinogradova (1960)
Anopheles hyrcanus	L		adult			Vinogradova (1960)
Anopheles bifurcatus	L		larva			Danilevskii (1965)
Anopheles plumbeus	L		larva			Vinogradova (1962)
Anopheles barberi	L		larva			Baker (1935)
Anopheles freeborni	L	larva–adult	adult	37°N. 47°N.	9–10/24 11–12/24	Depner and Harwood (1966)
Aëdes triseriatus	L		eggs, and/or larva		13/24	Love and Whelchel (1955); Kappus and Venard (1967); Wright and Venard (1971); Clay and Venard (1972)
Aëdes togoi	L	adult	egg			Vinogradova (1965)
Aëdes albopictus			egg			Wang Ren-Lai (1966)
Aëdes atropalpus	L	larva–adult	egg	30°N. 41°N.	13/24 14–15/24	Anderson (1968)

TABLE 4 *(cont.)*

Species	Type	Sensitive period	Diapause stage	Latitude	Critical daylength	Reference
Aëdes dorsalis	L	adult	egg			McHaffey and Harwood (1970)
Culex pipiens	L	pupa–adult	adult			Vinogradova (1960); Sanburg and Larsen (1973); Spielman and Wong (1973)
Culex tarsalis	L		adult			Harwood and Halfhill (1964)
Wyeomyia smithii	L	early larval stages	3rd instar larva		15/24	Evans and Brust (1972); Bradshaw and Lounibos (1972)
Toxorhynchites rutilus	L		adult			McCrary and Jenner (1965)
Chaoborus americanus	L		larva			Bradshaw (1969a, 1972)
Chironomus tentans	L		larva			Engelmann and Shapirio (1965)
Metriocnemus knabi	L		larva			Paris and Jenner (1959)
Culicoides guttipennis	L		larva			Baker (1935)
Rhagoletis cerasi	L		pupa			Legay (1962)
Rhagoletis pomonella	L	larva	pupa			Prokopy (1968)
Drosophila phalerata	L		adult			Geispitz and Simonenko (1970); Tyshchenko *et al.* (1972)
Drosophila deflexa	L?		larva			Basden (1954)
Drosophila nitens	L?		adult			Bazatti-Traverso (1943)
Drosophila robusta	L		larva			Carson and Stalker (1948)
Chymomyza costata	L		larva			Hackman *et al.* (1970)
Musca autumnalis	L	adult	adult			Stoffolano and Matthysse (1967); Valder *et al.* (1969)
Lyperosia irritans	L		pupa			Depner (1962)
Erioischia brassicae	L	adults–eggs and larvae	pupa			Hughes (1960) Read (1969)
Pegomyia hyosciami	L		pupa			Zabirov (1961)

Species					/24	Reference
Pegomyia betae	?		pupa			Missonnier (1963)
Lucilia sericata	L	adult	larva			Cragg and Cole (1952)
Lucilia caesar	L	adult	larva			Ring (1967)
Calliphora vicina	L	adult	larva		15–16/24	Vinogradova and Zinovjeva (1972b)
Phormia terrae novae			?			Danilevskii (1965)
Phormia regina	L	larva and embryo	adult			Fraenkel and Hsiao (1968); Saunders (1971)
Sarcophaga argyrostoma	L		pupa		14/24	
Sarcophaga crassipalpis	L	embryos	pupa			Denlinger (1971)
Sarcophaga bullata	L	embryos	pupa		13½/24	Denlinger (1972)
HYMENOPTERA						
Neodiprion sertifer	S	larva	larva, pre-pupa	45°N.	15–16/24	Sullivan and Wallace (1965, 1967)
Neodiprion rugifrons	L	larva	larva		16/24	King and Benjamin (1965)
Neodiprion swainei	L	larva	larva		14–15/24	Philogene and Benjamin (1971)
Athalia rosae	L	larva	pre-pupa in cocoon			Saringer (1964, 1967); Danilevskii (1965)
Athalia glabricollis	L	larva	pre-pupa in cocoon		14–15/24	Saringer (1966)
Lygaeonematus compressicornis	L		pre-pupa in cocoon			Danilevskii (1965)
Apanteles glomeratus	L		larva	60°N.	16/24	Maslennikova (1958)
Apanteles spurius	L		larva			Geispitz and Kyao (1953)
Apanteles congregatus	L	larva	last instar larva	N. Carolina	12½–13/24	Rabb and Thurston (1969)
Apanteles melanoscelus	L		larva		16/24	Weseloh (1973)
Alysia manducator	L	larva	larva		13–14/24	Vinogradova and Zinovjeva (1972a)
Aphaereta minuta	L	larva	larva			Vinogradova and Zinovjeva (1972a)
Coeloides brunneri	L	adult	larva		15/24	Ryan (1965)
Pimpla instigator	L		larva			Claret (1973)
Caraphractus cinctus	L	larva	larva		14–16/24	Jackson (1961, 1963)

TABLE 4 (cont.)

Species	Type	Sensitive period	Diapause stage	Latitude	Critical daylength	Reference
Trichogramma evanescens	L		larva			Danilevskii (1965)
Pteromalus puparum	L		larva			Maslennikova (1958)
Nasonia vitripennis	L	adult	larva	52°N. 42°N.	$15\frac{1}{4}/24$ $13\frac{1}{2}/24$	Saunders (1965a, 1966a)
Hypopteromalus tabacum	L		larva			McNeil and Rabb (1973)
Catolaccus aeneoviridis	L		pupa			McNeil and Rabb (1973)
COLEOPTERA						
Pterostichus nigrita	S–L		adult			Thiele (1966, 1967, 1971)
Pterostichus augustatus	S–L		adult			Thiele (1967, 1971)
Pterostichus oblongopunctatus	S–L		adult			Thiele (1967)
Pterostichus cupreus	S–L		adult			Krehan (1970)
Pterostichus coerulescens	S–L		adult			Krehan (1970)
Patrobus atrorufus	L–S		hib. as larva, aestiv. as adult			Thiele (1969)
Agonum assimile	L		adult			Thiele (1969)
Nebria brevicollis	L–S		hib. as larva			Thiele (1969)
Tachinus rufipes	S–L		adult			Lipkow (1966)
Tachyporus sp.	S–L		adult			Lipkow (1966)

Species					References
Philonthus fuscipennis	S–L Decr.		adult		Eghtedar (1970)
Anthrenus verbasci			larva		Blake (1958, 1959, 1960)
Coccinella septempunctata	L	adult	adult		Hodek and Cerkasov (1960); Hodek (1962)
Semiadalia undecimnotata	L		adult		Hodek and Cerkasov (1958)
Coccinella novemnotata	L	adult	adult		McMullen (1967)
Coccinella transversoguttata	L	adult	adult		Storch (1973)
Chilochorus bipustulatus			adult	15/24	Tadmor and Applebaum (1971)
Leptinotarsa decemlineata	L	adult	adult		de Wilde (1954); Jermy and Saringer (1955)
Phaedon cochleariae	L		adult		Way et al. (1949)
Chrysomela fastuosa	L		adult		Danilevskii (1965)
Haltica saliceti	L		adult		Danilevskii (1965)
Galeruca tanaceti	S		adult	12–14/24	Siew (1965 a, b, c)
Psylliodes chrysocephala			adult		Ankersmit (1964)
Paropsis atomaria		adult	adult		Carne (1966)
Sitona cylindricollis			adult		Hans (1961)
Anthonomus grandis	L	eggs	adult	12–13/24	Earle and Newsom (1964); Mangum et al. (1968)
Hypera postica	L	larva–adult	adult		Huggans and Blickenstaff (1964); Rosenthal and Koehler (1968); Litsinger and Apple (1973)
Ceutorhynchus pleurostigma	L–S		adult		Ankersmit (1964)
Ceutorhynchus assimilis			adult		Ankersmit (1964)

5. Insect Species in which Photoperiodic Control of Phenomena other than Diapause has been Demonstrated or Proposed

SM — seasonal morphs

Species	Process controlled	Type	Reference
ORTHOPTERA			
Locusta migratoria migratorioides	survival		Albrecht (1965)
Locusta migratoria	egg maturation		Perez et al. (1971)
Locusta migratoria	oviposition		Perez et al. (1971)
Locusta migratoria	male sexual behaviour		Perez et al. (1971)
Nemobius yezoensis	wing length		Masaki and Oyama (1963)
Teleogryllus yezoemma	nymphal development accelerated in short days	S	Masaki and Ohmachi (1967)
Teleogryllus emma	nymphal development accelerated in short days		Masaki and Ohmachi (1967)
HEMIPTERA			
Gerris odontogaster	wing length	L, incr.	Vepsäläinen (1971b)
Euscelis plebejus	SM		Müller (1954, 1957)
Nephotettix cincticeps	SM		Kisimoto (1959)
Nephotettix apicalis	SM		Kisimoto (1959)
Stenocranus minutus	SM		Müller (1954)
Delphacodes striatella	SM		Bonnemaison and Missonnier (1955); Oldfield (1970)
Aleurochiton complanatus	winter and summer "puparia"		Müller (1962 a, b)
Stirellus obtutus	SM		Whitcomb et al. (1972)
Aleyrodes asari	SM		Bährmann (1972)
Aphis forbesi	SM		Marcovitch (1924)
Aphis fabae	SM		Davidson (1929)
Aphis chloris	SM		Wilson (1938)
Capitophorus hippophaes	SM		Marcovitch (1924)

Dysaphis plantaginea	SM	Bonnemaison (1958)
Myzus persicae	SM	Bonnemaison (1951)
Brevicoryne brassicae	SM	Bonnemaison (1951)
Acyrthosiphon pisum	SM	Kenten (1955)
Megoura viciae	SM	Lees (1959)
Macrosiphum euphorbiae	SM	MacGillivray and Anderson (1964)
THYSANOPTERA		
Anaphothrips obscurus	wing length	Köppä (1970)
NEUROPTERA		
Chrysopa carnea	winter coloration	MacLeod (1967); Tauber *et al.* (1970)
LEPIDOPTERA		
Ascia monuste	SM	Pease (1962)
Eurema hecabe mandarina	SM	Aida (1963)
Araschnia levana	SM	Danilevskii (1948); Müller (1955)
Limenitis archippus	larval growth rate	Clark and Platt (1969)
Polygonia C-aureum	SM	Aida and Sakagami (1962)
Kaniska canace no-japonicum	SM	Aida (1963)
Colias eurytheme	SM	Watt (1969); Hoffmann (1973)
Lycaena phlaeas daimio	SM	Sakai and Masaki (1965)
Parasemia plantaginis	larval growth rate	Danilevskii (1965)
Agrotis occulta	larval growth rate	Danilevskii (1965)
Agrotis triangulum	larval growth rate	Danilevskii (1965)
Dasychira pudibunda	larval growth rate and number of instars	Geispitz (1953); Geispitz and Zarankina (1963)
Cosmotriche potatoria	larval growth rate	Danilevskii (1965)
Chilo suppressalis	larval growth rate	Fukaya and Mitsuhashi (1961)
Crambus tutillus	larval growth rate	Kamm (1972)
Peronea fimbriana	SM	Danilevskii (1965)
Pectinophora gossypiella	rate of embryonic development	Minis and Pittendrigh (1968)
Plutella maculipennis	larval growth rate	Atwal (1955)
Plutella maculipennis	fecundity	Atwal (1955); Harcourt and Cass (1966)
Acrolepia assectella	fecundity	Meudec (1966)
Hylophila prasinana	SM	Danilevskii (1965)

TABLE 5 (*cont.*)

Species	Process controlled	Type	Reference
DIPTERA			
Anopheles pulcherrimus	larval growth rate		Vinogradova (1960)
Aëdes triseriatus	larval growth rate		Vinogradova (1967)
Culex tarsalis	deposition of fat		Harwood and Takata (1965)
Culex tarsalis	proportion of		Harwood and Takata (1965)
	unsaturated fatty acids		
	in fat body		
Culex tarsalis	autogeny		Harwood (1966)
Toxorhynchites rutilus	sex ratio		McCrary and Jenner (1965)
	of overwintering stages		Pittendrigh (1961)
Drosophila melanogaster	recovery from heat stress		Lumme *et al.* (1972)
Drosophila littoralis	testis pterin content		Fernandez and Randolph (1966)
Musca domestica	insecticide		
	susceptibility		
Sarcophaga argyrostoma	larval growth rate		Denlinger (1972); Saunders (1972)
HYMENOPTERA			
Diprion similis	abnormal development		Baker and Atwood (1969)
	of legs		
Campoletis perdistinctus	sex ratio		Hoelscher and Vinson (1971)
COLEOPTERA			
Coccinella septempunctata	migration to		Hodek and Cerkasov (1960)
	hibernation site		
Semiadalia undecimnotata	migration to		Hodek and Cerkasov (1960)
	hibernation site		

REFERENCES

ABUSHAMA, F. T. (1968) Rhythmic activity of the grasshopper *Poecilocerus heiroglyphicus* (Acrididae: Pyrogomorphinae). *Entomologia exp. appl.* **11,** 341–347.

ADKISSON, P. L. (1961) Effect of larval diet on the seasonal occurrence of diapause in the pink bollworm. *J. econ. Ent.* **54,** 1107–1112.

ADKISSON, P. L. (1964) Action of the photoperiod in controlling insect diapause. *Am. Nat.* **98,** 357–374.

ADKISSON, P. L. (1966) Internal clocks and insect diapause. *Science, Wash.* **154,** 234–241.

ADKISSON, P. L., BELL, R. A. and WELLSO, S. G. (1963) Environmental factors controlling the induction of diapause in the pink bollworm, *Pectinophora gossypiella* (Saunders). *J. Insect Physiol.* **9,** 299–310.

ADKISSON, P. L. and ROACH, S. H. (1971) A mechanism for seasonal discrimination in the photoperiodic induction of pupal diapause in the bollworm *Heliothis zea* (Boddie). In *Biochronometry* (Ed. MENA-KER, M.), pp. 272–280. National Academy of Sciences, Washington.

AIDA, S. (1963) *Seibutsu Kagaku,* **15,** 163–167. (Quoted from Sakai and Masaki, 1965.) (In Japanese.)

AIDA, S. and SAKAGAMI, Y. (1962) *Kagaku,* **32,** 96–97. (In Japanese.) (Quoted from Sakai and Masaki, 1965.)

ALBRECHT, F. O. (1965) Influence du groupement de l'état hygrométrique et de la photopériode sur la résistance au jeune de *Locusta migratoria migratoriodes* R et F. (Orthoptera acridien). *Bull. biol. Fr. Belg.* **94,** 287–339.

ANDERSON, J. F. (1968) Influence of photoperiod and temperature on the induction of diapause in *Aëdes atropalpus* (Diptera: Culicidae). *Entomologia exp. appl.* **11,** 321–330.

ANKERSMIT, G. W. (1964) Voltinism and its determination in some beetles of cruciferous crops. *Mededelingen Landbouwhogeschool Wageningen,* **64,** 1–60.

ANKERSMIT, G. W. (1968) The photoperiod as a control agent against *Adoxophyes reticulana* (Lepidoptera: Tortricidae). *Entomologia exp. appl.* **11,** 231–240.

ANKERSMIT, G. W. and ADKISSON, P. L. (1967) Photoperiodic responses of certain geographical strains of *Pectinophora gossypiella* (Lepidoptera). *J. Insect Physiol.* **13,** 553–564.

ARTHUR, J. M. and HARVILL, E. K. (1937) Plant growth under continuous illumination from sodium vapor lamps supplemented by mercury arc lamps. *Contrib. Boyce Thompson Inst.* **8,** 433–443.

ASCHOFF, J. (1960) Exogenous and endogenous components in circadian rhythms. *Cold Spring Harb. Symp. Quant. Biol.* **25,** 11–28.

ASCHOFF, J. (1965) Response curves in circadian periodicity. In *Circadian Clocks* (Ed. ASCHOFF, J.), pp. 95–111. North-Holland, Amsterdam.

ASCHOFF, J. (1969) Desynchronization and resynchronization of human circadian rhythms. *Aerosp. Med.* **40** 844–849.

ASCHOFF, J., KLOTTER, K. and WEVER, R. (1965) Circadian vocabulary. In *Circadian Clocks* (Ed. ASCHOFF, J.). North-Holland, Amsterdam.

ASCHOFF, J., SAINT PAUL, U. VON and WEVER, R. (1971) Die Lebensdauer von Fliegen unter dem Einfluss von Zeit-verschiebungen. *Naturwiss.* **58,** 574.

ATWAL, A. S. (1955) Influence of temperature, photoperiod, and food on the speed of development, longevity, fecundity, and other qualities of the diamond-back moth *Plutella maculipennis* (Curtis) (Tineidae, Lepidoptera). *Aust. J. Zool.* **3,** 185–221.

BÄHRMANN, R. (1972) Untersuchungen zur Populationsdynamik von *Aleyrodes asari* Schrank und *Aleurolobus asari* Wünn (Homoptera, Aleyrodina). *Zool. Jb., Syst.* **99,** 82–106.

BAKER, F. C. (1935) The effect of photoperiodism on resting, treehole, mosquito larvae. *Can. Ent.* **67,** 149–153.

BAKER, W. V. and ATWOOD, C. E. (1969) The abnormal development of the meso- and meta-thoracic legs of *Diprion similis* (Hymenoptera: Diprionidae), when reared under certain photoperiods. *Can. Ent.* **101,** 990–994.

BALL, H. J. (1965) Photosensitivity in the terminal abdominal ganglion of *Periplaneta americana* (L.). *J. Insect Physiol.* **11,** 1311–1315.

BALL, H. J. (1971) The receptor site for photic entrainment of a circadian activity rhythm in the cockroach *Periplaneta americana. Ann. ent. Soc. Am.* **64,** 1010–1015.

BALL, H. J. (1972) Photic entrainment of circadian activity rhythms by direct brain illumination in the cockroach *Blaberus craniifer*. *J. Insect Physiol.* **18**, 2449–2455.

BALL, H. J. and CHAUDHURY, M. F. B. (1973) Photic entrainment of circadian rhythms by illumination of implanted brain tissues in the cockroach *Blaberus craniifer*. *J. Insect Physiol.* **19**, 823–830.

BARKER, R. J. (1963) Inhibition of diapause in *Pieris rapae* L. by supplementary photophases. *Experientia* **19**, 185.

BARKER, R. J. and COHEN, C. F. (1965) Light–dark cycles and diapause induction in *Pieris rapae* (L.). *Entomologia exp. appl.* **8**, 27–32.

BARKER, R. J., COHEN, C. F. and MAYER, A. (1964) Photoflashes: a potential new tool for control of insect populations. *Science, Wash.* **145**, 1195–1197.

BARKER, R. J., MAYER, A. and COHEN, C. F. (1963) Photoperiod effects in *Pieris rapae. Ann. ent. Soc. Am.* **56**, 292–294.

BARRY, B. D. and ADKISSON, P. L. (1966) Certain aspects of the genetic factors involved in the control of larval diapause of the pink bollworm. *Ann. ent. Soc. Am.* **59**, 122–125.

BASDEN, E. B. (1954) Diapause in *Drosophila* (Diptera: Drosophilidae). *Proc. R. ent. Soc. Lond.* (A), **29**, 114–118.

BATEMAN, M. A. (1955) The effect of light and temperature on the rhythm of pupal ecdysis in the Queensland fruit-fly, *Dacus* (*Strumeta*) *tryoni* (Frogg.). *Aust. J. Zool.* **3**, 22–33.

BAZATTI-TRAVERSO, A. (1943) Morfologia, citologia e biologia di due nuove specie di *Drosophila* (Diptera Acalyptera). *R.C. ist Lombardo*, **77**, 1–13.

BEARDS, G. W. and STRONG, F. E. (1966) Photoperiod in relation to diapause in *Lygus hesperus* Knight. *Hilgardia*, **37**, 345–362.

BEATTIE, T. M. (1971) Histology, histochemistry, and ultrastructure of neurosecretory cells in the optic lobes of the cockroach, *Periplaneta americana. J. Insect Physiol.* **17**, 1843–1855.

BECK, S. D. (1962a) Photoperiodic induction of diapause in an insect. *Biol. Bull. Mar. Biol. Lab., Woods Hole*, **122**, 1–12.

BECK, S. D. (1962b) Temperature effects on insects: Relation to periodism. *Proc. N. Central Branch, Ent. Soc. Am.* **17**, 18–19.

BECK, S. D. (1963) Physiology and ecology of photoperiodism. *Bull. ent. Soc. Am.* **9**, 8–16.

BECK, S. D. (1964) Time-measurement in insect photoperiodism. *Am. Nat.* **98**, 329–346.

BECK, S. D. (1968) *Insect Photoperiodism*. Academic Press, New York and London.

BECK, S. D. and ALEXANDER, N. (1964) Chemically and photoperiodically induced diapause development in the European corn borer, *Ostrinia nubilalis. Biol. Bull. Mar. Biol. Lab., Woods Hole*, **126**, 175–184.

BECK, S. D. and APPLE, J. W. (1961) Effect of temperature and photoperiod on voltinism of geographical populations of the European corn borer, *Pyrausta nubilalis* (Hbn.). *J. econ. Ent.* **54**, 550–558.

BECK, S. D., CLOUTIER, E. J. and MCLEOD, D. G. R. (1962) Photoperiod and insect development. *Proc. 23rd Biol. Colloq. Oregon State Univ.* 1962, pp. 43–64.

BECK, S. D. and HANEC, W. (1960) Diapause in the European corn borer, *Pyrausta nubilalis* (Hübn.). *J. Insect Physiol.* **4**, 304–318.

BEIER, W. (1968) Beeinflussung der inneren Uhr der Bienen durch Phasenverschiebung des Licht-Dunkel-Zeitgebers. *Z. Bienenforsch.* **9**, 356–378.

BEIER, W. and LINDAUER, M. (1970) Der Sonnenstand als Zeitgeber für die Biene. *Apidologie*, **1**, 5–28.

BELCHER, K. and BRETT, W. J. (1973) Relationship between a metabolic rhythm and emergence rhythm in *Drosophila melanogaster. J. Insect Physiol.* **19**, 277–286.

BELING, I. (1929) Über das Zeitgedächtnis der Bienen. *Z. vergl. Physiol.* **9**, 259–338.

BELL R. A. and ADKISSON, P. L. (1964) Photoperiodic reversibility of diapause induction in an insect. *Science, Wash.* **144**, 1149–1151.

BELOZEROV, V. N. (1964) Larval diapause in the tick *Ixodes ricinus* L. and its relation to external conditions. *Zool. Zh.* **43**, 1626–1637. (In Russian.)

BENNETT, M. F. and RENNER, M. (1963) The collecting performance of honey bees under laboratory conditions. *Biol. Bull. Mar. Biol. Lab., Woods Hole*, **125**, 416–430.

BENSCHOTER, C. A. (1968) Diapause and development of *Heliothis zea* and *H. virescens* in controlled environments. *Ann. ent. Soc. Am.* **61**, 953–956.

BENTLEY, E. W., GUNN, D. L. and EWER, D. W. (1941) The biology and behaviour of *Ptinus tectus* Boie. (Coleoptera, Ptinidae), a pest of stored products. I. The daily rhythm of locomotor activity, especially in relation to light and temperature. *J. exp. Biol.* **18**, 182–195.

BIRUKOW, G. (1953) Menotaxis im polarisierten Licht bei *Geotrupes sylvaticus* Panz. *Naturwissenschaften*, **40**, 611.

BIRUKOW, G. (1956) Lichtkompassorientierung beim Wasserlaufer *Velia currens* F. am Tage und zur Nachtzeit. I. Herbst- und Winterversuche. *Z. Tierpsychol.* **13**, 463–484.

BIRUKOW, G. (1960) Innate types of chronometry in insect orientation. *Cold Spring Harb. Symp. Quant. Biol.* **25**, 403–412.

BIRUKOW, G. (1964) Aktivitäts- und Orientierungsrhythmik beim Kornkäfer (*Calandra granaria* L.). *Z. Tierpsychol.* **21**, 279–301.

BIRUKOW, G. and BUSCH, E. (1957) Lichtkompassorientierung beim Wasserlaufer *Velia currens* F. am Tage und zur Nachtzeit. II. Orientierungs rhythmik in Verschiedenen Lichtbedingungen. *Z. Tierpsychol.* **14**, 184–203.

BLAKE, G. M. (1958) Diapause and the regulation of development in *Anthrenus verbasci* (L.) (Col., Dermestidae). *Bull. Ent. Res.* **49**, 751–775.

BLAKE, G. M. (1959) Control of diapause by an "internal clock" in *Anthrenus verbasci* (L.) (Col., Dermestidae). *Nature, Lond.* **183**, 126–127.

BLAKE, G. M. (1960) Decreasing photoperiod inhibiting metamorphosis in an insect. *Nature, Lond.* **188**. 168–169.

BLAKE, G. M. (1963) Shortening of a diapause-controlled life cycle by means of increasing photoperiod. *Nature, Lond.* **198**, 462–463.

BLANEY, L. T. and HAMNER, K. C. (1957) Inter-relations among the effects of temperature, photoperiod, and dark period on floral initiation of Biloxi soybean. *Bot. Gaz.* **119**, 10–24.

BONESS, M. (1963) Biologisch-ökologische Untersuchungen an *Exolygus* Wagner (Heteroptera, Miridae). *Z. wiss. Zool.* **168**, 374–420.

BONNEMAISON, L. (1951) Contribution à l'étude des facteurs provoquant l'apparition des formes ailées et sexuées chez les Aphidinae. *Annls Epiphyt.* (C) **2**, 1–380.

BONNEMAISON, L. (1956) Determinisme de l'apparition des larves aestivales de *Periphyllus* (Aphidinae). *C. R. Acad. Sci.* **243**, 1166–1168.

BONNEMAISON, L. (1958) Facteurs d'apparition des formes sexupares ou sexuées chez le puceron cendre du pommier (*Sappaphis plantaginea* Pass.). *Annls Epiphyt.* (C) **3**, 331–355.

BONNEMAISON, L. (1965) Action d'une photopériode de duree croissante ou decroissante sur l'apparition des formes sexuées de *Dysaphis plantaginea* Pass. *C. R. Acad. Sci.* **260**, 5138–5140.

BONNEMAISON, L. and MISSONNIER, J. (1955) Influence de photopériodisme sur le déterminisme des formes estivales ou hivernales et de la diapause chez *Psylla pyri* L. (Homopteres). *C. R. Acad. Sci.* **240**, 1277–1279.

BORTHWICK, H. A., HENDRICKS, S. B. and PARKER, M. W. (1952) The reaction controlling floral initiation. *Proc. Nat. Acad. Sci. U.S.A.* **38**, 929–934.

BOUNHIOL, J-J. and MOULINIER, C. (1965) L'opacité cranienne et ses modifications naturelles et experimentelles chez le ver à soie. *C. R. Acad. Sci.* **261**, 2739–2741.

BRADSHAW, W. E. (1969a) Major environmental factors inducing the termination of larval diapause in *Chaoborus americanus* Johannsen (Diptera: Culicidae). *Biol. Bull. Mar. Biol. Lab., Woods Hole*, **136**, 2–8.

BRADSHAW, W. E. (1969b) Dawn and dusk differences in the photoperiodic induction of development in *Chaoborus americanus* (Diptera: Culicidae). *Am. Zool.* **9**, 234.

BRADSHAW, W. E. (1970) Interaction of food and photoperiod in the termination of larval diapause in *Chaoborus americanus* (Diptera: Culicidae). *Biol. Bull. Mar. Biol. Lab., Woods Hole*, **139**, 476–484.

BRADSHAW, W. E. (1972) Photoperiodic control in the initiation of diapause by *Chaoborus americanus* (Diptera: Culicidae). *Ann. Ent. Soc. Am.* **65**, 755–756.

BRADSHAW, W. E. and LOUNIBOS, L. P. (1972) Photoperiodic control of development in the pitcherplant mosquito, *Wyeomyia smithii*. *Can. J. Zool.* **50**, 713–719.

BRADY, J. (1967a) Control of the circadian rhythm of activity in the cockroach. I. The role of the corpora cardiaca, brain and stress. *J. exp. Biol.* **47**, 155–163.

BRADY, J. (1967b) Control of the circadian rhythm of activity in the cockroach. II. The role of the suboesophageal ganglion and ventral nerve cord. *J. exp. Biol.* **47**, 165–178.

BRADY, J. (1967c) Histological observations on circadian changes in the neurosecretory cells of cockroach suboesophageal ganglia. *J. Insect Physiol.* **13**, 201–213.

BRADY, J. (1969) How are insect circadian rhythms controlled? *Nature, Lond.* **223**, 781–784.

BRADY, J. (1970) Characteristics of spontaneous activity in tsetse flies. *Nature, Lond.* **228**, 286–287.

BRADY, J. (1971) The search for an insect clock. In *Biochronometry* (Ed. MENAKER, M.), pp. 517–526. National Academy of Sciences, Washington.

BRADY, J. (1972) Spontaneous, circadian components of tsetse fly activity. *J. Insect Physiol.* **18**, 471–484.

BREMER, H. (1926) Über die tageszeitliche Konstanz im Schlupftermine der Imagines einiger Insekten und ihre experimentelle Beeinflussbarkeit. *Z. wiss. Insektenbiol.* **21**, 209–216.

BRETT, W. J. (1955) Persistent diurnal rhythmicity in *Drosophila* emergence. *Ann. ent. Soc. Am.* **48**, 119–131.

BROWN, F. A. (1960) Response to pervasive geophysical factors and the biological clock problem. *Cold Spring Harb. Symp. Quant. Biol.* **25**, 57–71.

BROWN, F. A. (1965) A unified theory for biological rhythms. In *Circadian Clocks* (Ed. ASCHOFF, J.) pp. 231–261. North-Holland, Amsterdam.

BROWN, F. A., FINGERMAN, M., SANDEEN, M. I. and WEBB, H. M. (1953) Persistent diurnal and tidal rhythms of color change in the fiddler crab, *Uca pugnax*. *J. Exp. Zool.* **123**, 29–60.

BROWN, V. K. (1973) The overwintering stages of *Ectobius lapponicus* (L.) (Dictyoptera: Blattidae). *J. Ent.* (A), **48**, 11–24.

BRUCE, V. G. (1960) Environmental entrainment of circadian rhythms. *Cold Spring Harb. Symp. Quant. Biol.* **25**, 29–48.

BRUCE, V. G. and MINIS, D. H. (1969) Circadian clock action spectrum in a photoperiodic moth. *Science, Wash.* **163**, 583–585.

BRUCE, V. G. and PITTENDRIGH, C. S. (1957) Endogenous rhythms in insects and microorganisms. *Am. Nat.* **91**, 179–195.

BRUN, R. (1914) *Die Raumorientierung der Ameisen und das Orientierungsproblem imm Allgemeinen.* Jena.

BUCK, J. B. (1937) Studies on the firefly. I. The effects of light and other agents on flashing in *Photinus pyralis*, with special reference to periodicity and diurnal rhythm. *Physiol. Zool.* **10**, 45–58.

BULL, D. L. and ADKISSON, P. L. (1960) Certain factors influencing diapause in the pink bollworm, *Pectinophora gossypiella. J. econ. Ent.* **53**, 793–798.

BULL, D. L. and ADKISSON, P. L. (1962) Fat content of the larval diet as a factor influencing diapause and growth rate of the pink bollworm. *Ann. ent. Soc. Am.* **55**, 499–502.

BULL, D. L. and LINDQUIST, D. A. (1965) A comparative study of insecticide metabolism in photoperiod-entrained and unentrained bollworm larvae *Heliothis zea* (Boddie). *Comp. Biochem. Physiol.* **16**, 321–325.

BÜNNING, E. (1935) Zur Kenntniss der endogonen Tagesrhythmik bei Insekten und Pflanzen. *Ber. dt. bot. Ges.* **53**, 594–623.

BÜNNING, E. (1936) Die endogone Tagesrhythmik als Grundlage der Photoperiodischen Reaktion. *Ber. dt. bot. Ges.* **54**, 590–607.

BÜNNING, E. (1959) Zur Analyse des Zeitsinnes bei *Periplaneta americana. Z. Naturf.* **14b**, 1–4.

BÜNNING, E. (1960) Circadian rhythms and time measurement in photoperiodism. *Cold Spring Harb. Symp. Quant. Biol.* **25**, 249–256.

BÜNNING, E. (1964) *The Physiological Clock*, 1st English edition. Springer-Verlag.

BÜNNING, E. (1969) Common features of photoperiodism in plants and animals. *Photochem. Photobiol.* **9**, 219–228.

BÜNNING, E. and JOERRENS, G. (1960) Tagesperiodische antagonistische Schwankungen der Blau-violett und Gelbrot-Empfindlichkeit als Grundlage der photoperiodischen Diapause-Induktion bei *Pieris brassicae. Z. Naturf.* **15**, 205–213.

BÜNSOW, R. C. (1953) Über Tages- und Jahresrhythmische Anderungen der photoperiodischen Lichteropfindlichkeit bei *Kalanchoe blossfeldiana* und ihre Beziehungen zur endogonen Tagesrhythmik. *Z. Bot.* **41**, 257–276.

BUTTEL-REEPEN, H. VON (1900) Sind die Bienen "Reflexmaschinen"? *Biol. Zbl.* **20**, 97.

CALDWELL, R. L. and DINGLE, H. (1967) Regulation of cyclic reproductive and feeding activity in the milkweed bug (*Oncopeltus*) by temperature and photoperiod. *Biol. Bull. Mar. Biol. Lab., Woods Hole*, **133**, 510–525.

CALLAHAN, P. S. (1958) Behavior of the imago of the corn earworm, *Heliothis zea* (Boddie), with special reference to emergence and reproduction. *Ann. ent. Soc. Am.* **51**, 271–283.

CAMPBELL, B. O. (1964) Solar and lunar periodicities in oxygen consumption by the mealworm, *Tenebrio molitor*. Ph.D. thesis, Northwestern Univ. (Quoted from Beck, 1968.)

CARNE, P. B. (1966) Ecological characteristics of the eucalypt-defoliating chrysomelid *Paraopsis atomaria* Ol. *Aust. J. Zool.* **14**, 647–672.

CARSON, H. L. and STALKER, H. D. (1948) Reproductive diapause in *Drosophila robusta. Proc. Nat. Acad. Sci. U.S.A.* **34**, 124–129.

CAVENEY, S. (1970) Juvenile hormone and wound modelling of *Tenebrio* cuticle architecture. *J. Insect Physiol.* **16**, 1087–1107.

CHANDRASHEKARAN, M. K. (1967a) Studies on phase-shifts in endogenous rhythms. I. Effects of light pulses on the eclosion rhythm in *Drosophila pseudoobscura. Z. vergl. Physiol.* **56**, 154–162.

CHANDRASHEKARAN, M. K. (1967b) Studies on phase-shifts in endogenous rhythms. II. The dual effect of light on the entrainment of the eclosion rhythm in *Drosophila pseudoobscura. Z. vergl. Physiol.* **56**, 163–170.

CHANDRASHEKARAN, M. K. and LOHER, W. (1969a) The effect of light intensity on the circadian rhythm of eclosion in *Drosophila pseudoobscura. Z. vergl. Physiol.* **62**, 337–347.

CHANDRASHEKARAN, M. K. and LOHER, W. (1969b) The relationship between the intensity of the light pulses and the extent of phase shifts of the circadian rhythm in the eclosion rate of *Drosophila pseudoobscura. J. exp. Zool.* **172**, 147–152.

CHANDRASHEKARAN, M. K., JOHNSSON, A. and ENGELMANN, W. (1973) Possible "dawn" and "dusk" roles of light pulses shifting the phase of a circadian rhythm. *J. Comp. Physiol.* **82**, 347–356. Berlin–Heidelberg–New York: Springer.

CHIPPENDALE, G. M. and REDDY, A. S. (1973) Temperature and photoperiodic regulation of diapause of the southwestern corn borer, *Diatraea grandiosella. J. Insect Physiol.* **19**, 1397–1408.

CHIPPENDALE, G. M. and YIN, C-M. (1973) Endocrine activity retained in diapause insect larvae. *Nature, Lond.* **246**, 511–513.

CHURCH, N. S. (1955) Hormones and the termination and reinduction of diapause in *Cephus cinctus* Nort. *Can. J. Zool.* **33**, 339–369.

CLAES, H. and LANG, A. (1947) Die Blutenbildung von *Hyoscyamus niger* in 48 stundigen Licht-Dunkel-Zyklen und in Zyklen mit Aufgeteiten Lichtphasen. *Z. Naturf.* **2**, 56–63.

CLARET, J. (1966a) Recherche du centre photorécepteur lors de l'induction de la diapause chez *Pieris brassicae* L. *C. R. Acad. Sci.* **262**, 1464–1465.

CLARET, J. (1966b) Mise en evidence du rôle photorécepteur du cerveau dans l'induction de la diapause chez *Pieris brassicae* (Lepido.). *Annls Endocr.* **27**, 311–320.

CLARET, J. (1972) Sensibilité spectrale des chenilles de *Pieris brassicae* (L.) lors de l'induction photopériodique de la diapause. *C. R. Acad. Sci.* **274**, 1727–1730.

CLARET, J. (1973) Le domaine de photosensibilité du parasite *Pimpla instigator* F. (Hymenoptère, Ichneumonidae) lors de l'entrée et de la levée photopériodiques de la diapause. *C. R. Acad. Sci.* **276**, 3163–3166.

CLARK, S. H. and PLATT, A. P. (1969) Influence of photoperiod on development and larval diapause in the viceroy butterfly *Liminitis archippus*. *J. Insect Physiol.* **15**, 1951–1957.

CLAY, M. E. and VENARD, C. E. (1972) Larval diapause in the mosquito, *Aëdes triseriatus*: effects of diet and temperature on photoperiodic induction. *J. Insect Physiol.* **18**, 1441–1446.

CLAYTON, D. L. and PAIETTA, J. V. (1972) Selection for circadian eclosion time in *Drosophila melanogaster*. *Science, Wash.* **178**, 994–995.

CLOUDSLEY-THOMPSON, J. L. (1953) Studies on diurnal rhythms. III. Photoperiodism in the cockroach *Periplaneta americana* (L.). *Ann. Mag. Nat. Hist.* **6**, 705–712.

CLOUDSLEY-THOMPSON, J. L. (1958) Studies in diurnal rhythms—VIII. The endogenous chronometer i. *Gryllus campestris* L. (Orthoptera: Gryllidae). *J. Insect Physiol.* **2**, 275–280.

CLOUDSLEY-THOMPSON, J. L. (1960) Studies in diurnal rhythms. X. Synchronization of the endogenous chronometer in *Blaberus giganteus* (L.) (Dictyoptera: Blattaria) and in *Gryllus campestris* L. (Orthoptera: Gryllidae). *Entomologist*, **3**, 121–127.

CLOUTIER, E. J., BECK, S. D., MCLEOD, D. G. R. and SILHACEK, D. L. (1962) Neural transplants and insect diapause. *Nature, Lond.* **195**, 1222–1224.

COLUZZI, M. (1972) Inversion polymorphism and adult emergence in *Anopheles stephensi*. *Science, Wash.* **176**, 59–60.

CORBET, P. S. (1955) A critical response to changing length of day in an insect. *Nature, Lond.* **175**, 338–339.

CORBET, P. S. (1956) Environmental factors influencing the induction and termination of diapause in the emperor dragonfly, *Anax imperator* Leach (Odonata: Aeschnidae). *J. exp. Biol.* **33**, 1–14.

CORBET, P. S. (1958) Lunar periodicity of aquatic insects in Lake Victoria. *Nature, Lond.* **182**, 330–331.

CRAGG, J. B. and COLE, P. (1952) Diapause in *Lucilia sericata* (Mg.). *J. exp. Biol.* **29**, 600–604.

CRAWFORD, C. S. (1966) Photoperiod-dependent oviposition rhythm in *Crambus teterrellus* (Lepidoptera: Pyralidae: Crambinae). *Ann. ent. Soc. Am.* **59**, 1285–1288.

CRAWFORD, C. S. (1967) Oviposition rhythm studies in *Crambus topiarius* (Lepidoptera: Pyralidae: Crambinae). *Ann. ent. Soc. Am.* **60**, 1014–1018.

CUMMING, B. G. (1971) The role of circadian rhythmicity in photoperiodic induction in plants. *Proc. Int. Symp. Circadian Rhythmicity* (Wageningen, 1971), pp. 33–85.

CYMBOROWSKI, B. (1973) Control of the circadian rhythm of locomotor activity in the house cricket. *J. Insect Physiol.* **19**, 1423–1440.

CYMBOROWSKI, B. and BRADY, J. (1972) Insect circadian rhythms transmitted by parabiosis—a re-examination. *Nature, New Biology*, **236**, 221–222.

CYMBOROWSKI, B. and DUTKOWSKI, A. (1969) Circadian changes in RNA synthesis in the neurosecretory cells of the brain and sub-oesophageal ganglion of the house cricket. *J. Insect Physiol.* **15**, 1187–1197.

CYMBOROWSKI, B. and DUTKOWSKI, A. (1970) Circadian changes in protein synthesis in the neurosecretory cells of the central nervous system of *Acheta domesticus*. *J. Insect Physiol.* **16**, 341–348.

CYMBOROWSKI, B., SKANGIEL-KRAMSKA, J. and DUTKOWSKI, A. (1970) Circadian changes of acetylcholinesterase activity in the brain of house-crickets (*Acheta domesticus* L.). *Comp. Biochem. Physiol.* **32**, 367–370.

DANILEVSKII, A. S. (1948) The photoperiodic reaction of insects in conditions of artificial light. *Dokl. Akad. Nauk SSSR*, **60**, 481–484. (In Russian.)

DANILEVSKII, A. S. (1965) *Photoperiodism and Seasonal Development of Insects*, 1st English edition. Oliver & Boyd, Edinburgh and London.

DANILEVSKII, A. S. and GLINYANYAYA, E. I. (1949) The effect of the relation between the dark and light periods of the day on insect development. *Dokl. Akad. Nauk SSSR*, **68**, 785–788. (In Russian.)

DANILEVSKII, A. S., GORYSHIN, N. I. and TYSHCHENKO, V. P. (1970) Biological rhythms in terrestrial arthropods. *A. Rev. Ent.* **15**, 201–244.

DAVIDSON, J. (1929) On the occurrence of the parthenogenetic and sexual forms of *Aphis rumicis* L. with special reference to the influence of environmental factors. *Ann. appl. Biol.* **16**, 104–134.

DAY, M. F. (1941) Pigment migration in the eyes of the moth, *Ephestia kuehniella* Zeller. *Biol. Bull. Mar. Biol. Lab., Woods Hole,* **80**, 275–291.

DEHN, M. VON (1967) Über den photoperiodismus heterogoner aphiden. Zur frage der direkten oder indirekten wirkung der tageslange. *J. Insect Physiol.* **13**, 595–612.

DEMOLL, R. (1911) Über die Wanderung des Irispigments im Facettenauge. *Zool. Jb., Physiol.* **30**, 169–180.

DENLINGER, D. L. (1971) Embryonic determination of pupal diapause in the flesh fly *Sarcophaga crassipalpis. J. Insect Physiol.* **17**, 1815–1822.

DENLINGER, D. L. (1972) Induction and termination of pupal diapause in *Sarcophaga* (Diptera: Sarcophagidae). *Biol. Bull. Mar. Biol. Lab., Woods Hole,* **142**, 11–24.

DEPNER, K. R. (1962) Effects of photoperiod and of ultraviolet radiation on the incidence of diapause in the horn fly, *Haematobia irritans* (L.). *Int. J. Bioclimatol. Biometeorol.* **5**, 68–71.

DEPNER, K. R. and HARWOOD, R. F. (1966) Photoperiodic responses of two latitudinally diverse groups of *Anopheles freeborni* (Diptera: Culicidae). *Ann. ent. Soc. Am.* **59**, 7–11.

DICKSON, R. C. (1949) Factors governing the induction of diapause in the oriental fruit moth. *Ann. ent. Soc. Am.* **42**, 511–537.

DINGLE, H., CALDWELL, R. L. and HASKELL, J. B. (1969) Temperature and circadian control of cuticle growth in the bug, *Oncopeltus fasciatus. J. Insect Physiol.* **15**, 373–378.

DIXON, A. F. G. (1963) Reproductive activity of the sycamore aphid, *Drepanosiphum platanoides* (Schr.) (Hemiptera, Aphididae). *J. Anim. Ecol.* **32**, 33–48.

DREISIG, H. (1971) Diurnal activity in the dusky cockroach, *Ectobius lapponicus* L. (Blattodea). *Ent. Scand.* **2**, 132-138.

DREISIG, H. and NIELSEN, E. T. (1971) Circadian rhythm of locomotion and its temperature dependence in *Blatella germanica. J. exp. Biol.* **54**, 187–198.

DUMORTIER, B. (1972) Photoreception in the circadian rhythm of stridulatory activity in *Ephippiger* (Ins., Orthoptera). Likely existence of two photoreceptive systems. *J. comp. Physiol.* **77**, 80–112.

DUNNING, R. A. (1957) A diurnal rhythm in the emergence of *Pegomyia betae* Curtis from the puparium. *Bull. ent. Res.* **47**, 645–653.

DUTKOWSKI, A. B., CYMBOROWSKI, B. and PRZELECKA, A. (1971) Circadian changes in the ultrastructure of the neurosecretory cells of the pars intercerebralis of the house cricket. *J. Insect Physiol.* **17**, 1763–1772.

EARLE, N. W. and NEWSOM, L. D. (1964) Initiation of diapause in the boll weevil. *J. Insect Physiol.* **10**, 131–139.

EDWARDS, D. K. (1962) Laboratory determinations of the daily flight times of separate sexes of some moths in naturally changing light. *Can. J. Zool.* **40**, 511–530.

EDWARDS, D. K. (1964a) Activity rhythms of Lepidopterous defoliators. I. Techniques for recording activity, eclosion, and emergence. *Can. J. Zool.* **42**, 923–937.

EDWARDS, D. K. (1964b) Activity rhythms of Lepidopeterous defoliators. II. *Halisodota argentata* Pack. (Arctiidae), and *Nepytia phantasmaria* Stkr. (Geometridae). *Can. J. Zool.* **42**, 939–958.

EDWARDS, R. L. (1954) The host-finding and oviposition behaviour of *Mormoniella vitripennis* (Walker) (Hym., Pteromalidae), a parasite of Muscoid flies. *Behaviour,* **7**, 88–112.

EGHTEDAR, E. (1970) Zur Biologie und Ökologie der Staphyliniden *Philonthus fuscipennis* Mannh. und *Oxytelus rugosus* Grav. *Pedobiologia,* **10**, 169–179.

EIDMANN, H. (1956) Über rhythmische Erscheinungen bei der Stabheuschrecke *Carausius morosus. Z. vergl. Physiol.* **28**, 370–390.

ENGELMANN, W. and HONEGGER, H. W. (1966) Tagesperiodischer Schlupfrhythmik einer angenlosen *Drosophila melanogaster*-Mutante. *Z. Naturf.* **22B**, 1–2.

ENGELMANN, W. and SHAPPIRIO, D. G. (1965) Photoperiodic control of the maintenance and termination of larval diapause in *Chironomus tentans. Nature, Lond.* **207**, 548–549.

EVANS, K. W. and BRUST, R. A. (1972) Induction and termination of diapause in *Wyeomyia smithii* (Diptera: Culicidae), and larval survival studies at low and subzero temperatures. *Can. Ent.* **104**, 1937–1950.

EVANS, W. G. and GYRISCO, G. G. (1958) The influence of light intensity on the nocturnal emergence of the European chafer. *Ecology,* **39**, 761–763.

EWAN, A. B. (1966) A possible endocrine mechanism for inducing diapause in the eggs of *Adelphocoris lineolatus* (Goeze) (Hemiptera: Miridae). *Experientia,* **22**, 470.

FERNANDEZ, A. T. and RANDOLPH, N. M. (1966) The susceptibility of houseflies reared under various photoperiods to insecticide residues. *J. econ. Ent.* **59**, 37–39.

FINGERMAN, M., LAGO, A. D. and LOWE, M. E. (1958) Rhythm of locomotor activity and oxygen consumption of the grasshopper *Romalea microptera*. *Am. Mid. Nat.* **59**, 58–66.

FLITTERS, N. E. (1964) The effect of photoperiod, light intensity, and temperature on copulation, oviposition, and fertility of the Mexican fruit fly. *J. econ. Ent.* **57**, 811–813.

FONDACARA, J. D. and BUTZ, A. (1970) Circadian rhythm of locomotor activity and susceptibility to methyl parathion of adult *Tenebrio molitor* (Coleoptera: Tenebrionidae). *Ann. ent. Soc. Am.* **63**, 952–955.

FOREL, A. (1910) *Das Sinnesleben der Insekten*. Munich.

FOWLER, D. J. and GOODNIGHT, C. J. (1966) Neurosecretory cells: daily rhythmicity in *Leiobunum longipes*. *Science, Wash.* **152**, 1078–1080.

FOWLER, D. J., GOODNIGHT, C. J. and LABRIE, M. M. (1972) Circadian rhythms of 5-hydroxytryptamine (serotonin) production in larvae, pupae, and adults of *Drosophila melanogaster* (Diptera: Drosophilidae). *Ann. ent. Soc. Am.* **65**, 138–141.

FRAENKEL, G. and HSIAO, C. (1968) Manifestations of a pupal diapause in two species of flies, *Sarcophaga argyrostoma* and *S. bullata*. *J. Insect Physiol.* **14**, 689–705.

FRANK, K. D. and ZIMMERMAN, W. F. (1969) Action spectra for phase shifts of a circadian rhythm in *Drosophila*. *Science, Wash.* **163**, 688–689.

FRASER, A. (1960) Humoral control of metamorphosis and diapause in the larvae of certain Calliphoridae (Diptera: Cyclorrhapha). *Proc. R. Soc. Edin.* B **67**, 127–140.

FRISCH, K. VON (1950) Die Sonne als Kompass im Leben der Bienen. *Experientia*, **6**, 210–221.

FRISCH, K. VON (1967) *The Dance Language and Orientation of Bees*, English Edition. Belknap Press of Harvard University Press. London: Oxford University Press.

FRISCH, K. VON and LINDAUER, M. (1954) Himmel und Erde in Konkurrenz bei der Orientierung der Bienen. *Naturwissenschaften*, **41**, 245–253.

FRYER, G. (1959) Lunar rhythms of emergence, differential behaviour of the sexes, and other phenomena in the African midge, *Chironomus brevibucca* (Kieff.). *Bull. ent. Res.* **50**, 1–8.

FUKAYA, M. and MITSUHASHI, J. (1961) Larval diapause in the rice stem borer with special reference to its hormonal mechanism. *Bull. natn. Inst. Agric. Sci. Japan*, **13**, 1–32.

FUKUDA, S. (1951) Production of the diapause eggs by transplanting the sub-oesophageal ganglion in the silkworm. *Proc. Imp. Acad. Japan*, **27**, 672–677.

FUKUDA, S. (1963) Determinisme hormonale de la diapause chez le ver à soie. *Bull. Soc. zool. Fr.* **88**, 151–179.

FUZEAU-BRAESCH, S. (1966) Étude de la diapause de *Gryllus campestris* (Orthoptera). *J. Insect Physiol.* **12**, 449–455.

GARNER, W. W. and ALLARD, H. A. (1920) Effect of the relative length of the day and night and other factors of the environment on growth and reproduction in plants. *J. Agric. Res.* **18**, 553–606.

GEISLER, M. (1961) Untersuchungen zur Tagesperiodik des Mistkäfers *Geotrupes silvaticus* Panz. *Z. Tierpsychol.* **18**, 389–420.

GEISPITZ, K. F. (1953) The reaction of univoltine Lepidoptera to day-length. *Ent. Obozr.* **33**, 17–31. (In Russian.)

GEISPITZ, K. F. (1957) The mechanism of acceptance of light stimuli in the photoperiodic reaction of Lepidoptera larvae. *Zool. Zh.* **36**, 548–560. (In Russian.)

GEISPITZ, K. F. (1965) *Ent. Obozr.* **44**, 538.

GEISPITZ, K. F. and KYAO, N. N. (1953) The effect of the duration of light on the development of some ichneumonids (Hymenoptera, Braconidae). *Ent. Obozr.* **33**, 32–35. (In Russian.)

GEISPITZ, K. F. and SIMONENKO, N. P. (1970) *Ent. Obozr.* **49**, 83–96. (In Russian.)

GEISPITZ, K. F. and ZARANKINA, A. I. (1963) Some features of the photoperiodic reaction of *Dasychira pudibunda* L. (Lepidoptera, Orgyidae). *Ent. Obozr.* **42**, 29–38. (In Russian.)

GELDIAY, S. (1967) Hormonal control of adult reproductive diapause in the Egyptian grasshopper, *Anacridium aegyptium* L. *J. Endocr.* **37**, 63–71.

GELDIAY, S. (1971) Control of adult reproductive diapause in *Anacridium aegyptium* L. by direct action of photoperiod on the cerebral neurosecretory cells. *Proc. XIII Int. Congr. Ent. Moscow*, 1968, I, 379–380.

GILLETT, J. D. (1962) Contributions to the oviposition cycle by the individual mosquitoes in a population. *J. Insect Physiol.* **8**, 665–681.

GILLETT, J. D. (1972) *The Mosquito. Its Life, Activities, and Impact on Human Affairs*. Doubleday & Co. Inc., Garden City, N.Y.

GILLETT, J. D., CORBET, P. S. and HADDOW, A. J. (1961) Observations on the oviposition-cycle of *Aëdes* (*Stegomyia*) *aegypti* (Linnaeus). VI. *Ann. trop. Med. Parasit.* **55**, 427–431.

GILLETT, J. D., HADDOW, A. J. and CORBET, P. S. (1959) Observations on the oviposition-cycle of *Aëdes* (*Stegomyia*) *aegypti* (Linnaeus) V. *Ann. trop. Med. Parasit.* **53**, 35–41.

GLASS, E. H. (1963) A pre-diapause arrested development period in the red-banded leafroller *Argyrotaenia velutinana*. *J. econ. Ent.* **56**, 634–635.

GODDEN, D. H. (1973) A re-examination of circadian rhythmicity in *Carausius morosus*. *J. Insect Physiol.* **19**, 1377–1386.

GORYSHIN, N. I. (1955) The relation between light and temperature factors in the photoperiodic reaction in insects. *Ent. Obozr.* **34**, 9–14. (In Russian.)

GORYSHIN, N. I. (1958) An ecological analysis of the seasonal cycle of the cotton bollworm (*Chloridea obsoleta* F.) in the northern areas of its range. *Sci. Mem. Lenin. State Univ.* **240**, 3–20.

GORYSHIN, N. I. (1964) The influence of diurnal light and temperature rhythms on diapause in Lepidoptera. *Ent. Obozr.* **43**, 43–46. (In Russian.)

GORYSHIN, N. I. and KOZLOVA, R. N. (1967) Thermoperiodism as a factor in the development of insects. *Zhur. obshch. Biol.* **28**, 278–288. (In Russian.)

GORYSHIN, N. I. and TYSHCHENKO, V. P. (1968) Physiological mechanism of photoperiodic reaction and the problem of endogenous rhythms. In *Photoperiodic Adaptations in Insects* and *Acari* (Ed. DANILEVSKII, A. S.), pp. 192–269. Leningrad University Press. (In Russian.)

GORYSHIN, N. I. and TYSHCHENKO, V. P. (1970) Thermostability of the process of perception of photoperiodic information in the moth *Acronycta rumicis* (Lepidoptera, Noctuidae). *Dokl. Akad. Nauk SSSR*, **193**, 458–461. (In Russian.)

GOSS, R. J. (1969a) Photoperiodic control of antler cycles in deer. I. Phase shift and frequency changes. *J. exp. Zool.* **170**, 311–324.

GOSS, R. J. (1969b) Photoperiodic control of antler cycles in deer. II. Alteration in amplitude. *J. exp. Zool.* **171**, 233–234.

GRABENSBERGER, W. (1934) Experimentelle Untersuchungen über das Zeitgedächtnis von Bienen und Wespen nach Verfutterung von Euchinin und Jodthryeoglobulin. *Z. vergl. Physiol.* **20**, 338–342.

GREEN, G. W. (1964) The control of spontaneous locomotor activity in *Phormia regina* Meigen—I. Locomotor activity patterns of intact flies. *J. Insect Physiol.* **10**, 711–726.

GREENSLADE, P. J. M. (1963) Daily rhythms of locomotor activity in some Carabidae (Coleoptera). *Entomologia exp. appl.* **6**, 171–180.

GRISON, P. (1943) Observations sur le rhythme d'activité nycthéméral chez le Doryphore, *Leptinotarsa decemlineata* Say. *C. R. Acad. Sci.* **217**, 621–622.

GRUWEZ, G., HOSTE, C., LINTS, C. V. and LINTS, F. A. (1972) Oviposition rhythm in *Drosophila melanogaster* and its alteration by a change in the photoperiodicity. *Experientia*, **27**, 1414–1416.

GUNN, D. L. (1940) The daily rhythm of activity of the cockroach, *Blatta orientalis*. *J. exp. Biol.* **17**, 267–277.

GWINNER, E. (1967) Circannuale Periodik der Mauser und der Zugunruhe bei einem Vogel. *Naturwissenschaften*, **54**, 447.

GWINNER, E. (1971) A comparative study of circannual rhythms in warblers. In *Biochronometry* (Ed. MENAKER, M.), pp. 405–427. National Academy of Sciences, Washington.

HACKMAN, W., LAKOVAARA, S., SAURA, A., SORSA, M. and VEPSÄLÄINEN, K. (1970) On the biology and karyology of *Chymomyza costata* Zetterstedt with reference to the taxonomy and distribution of the various species of *Chymomyza* (Dipt., Drosophilidae). *Ann. Ent. Fenn.* **36**, 1–9.

HADDOW, A. J. and GILLETT, J. D. (1957) Observations on the oviposition-cycle of *Aëdes* (*Stegomyia*) *aegypti* (Linnaeus). *Ann. trop. Med. Parasit.* **51**, 159–169.

HADDOW, A. J., GILLETT, J. D. and CORBET, P. S. (1959) Laboratory observations on pupation and emergence in the mosquito *Aëdes* (*Stegomyia*) *aegypti* (Linnaeus). *Ann. trop. Med. Parasit.* **53**, 123–131.

HADDOW, A. J., GILLETT, J. D. and CORBET, P. S. (1961) Observations on the oviposition-cycle of *Aëdes* (*Stegomyia*) *aegypti* (Linnaeus) V. *Ann. trop. Med. Parasit.* **55**, 343–356.

HAINE, E. (1957) Periodicity in aphid moulting and reproduction in constant temperature and light. *Z. angew. Ent.* **40**, 100–124.

HALBERG, F. (1960) Temporal coordination of physiologic function. *Cold Spring Harb. Symp. Quant. Biol.* **25**, 289–310.

HALLIBURTON, W. H. and ALEXANDER, G. (1964) Effect of photoperiod on molting of *Chortophaga viridifasciata* (De Geer) (Orthoptera: Acrididae). *Ent. News*, **75**, 133–137.

HAMNER, K. C. (1960) Photoperiodism and circadian rhythms. *Cold Spring Harb. Symp. Quant. Biol.* **25**, 269–277.

HAMNER, K. C., FLINN, J. C., SIROHI, G. S., HOSHIZAKI, T. and CARPENTER, B. H. (1962) Studies of the biological clock at the south pole. *Nature, Lond.* **195**, 476–480.

HAMNER, W. M. (1963) Diurnal rhythms and photoperiodism in testicular recrudescence of the house finch. *Science, Wash.* **142**, 1294–1295.

HAMNER, W. M. (1964) Circadian control of photoperiodism in the house finch demonstrated by interrupted-night experiments. *Nature, Lond.* **203**, 1400–1401.

HAMNER, W. M. (1969) Hour-glass dusk and rhythmic dawn timers control diapause in the codling moth. *J. Insect Physiol.* **15**, 1499–1504.

HANS, H. (1961) Termination of diapause and continuous laboratory rearing of the sweet clover weevil, *Sitona cylindricollis* Fahr. *Entomologia exp. appl.* **4**, 41–46.

HARCOURT, D. C. and CASS, L. M. (1966) Photoperiodism and fecundity in *Plutella maculipennis* (Curt.). *Nature, Lond.* **210**, 217–218.

HARDELAND, R. and STANGE, G. (1971) Einflüsse von Geschlecht und Alter auf die lokomotorische Aktivität von *Drosophila*. *J. Insect Physiol.* **17**, 427–434.

HARDER, R. and BODE, O. (1943) Über die Wirkung von Zwischenbelichtungen während der Dunkelperiode auf das Blühen, die Verlaubung und die Blattsukkulenz bei der Kurztagpflanze *Kalanchoe blossfeldiana*. *Planta*, **33**, 469–504.

HARKER, J. E. (1953) The diurnal rhythm of mayfly nymphs. *J. exp. Biol.* **30**, 525–533.

HARKER, J. E. (1954) Diurnal rhythm in *Periplaneta americana* L. *Nature, Lond.* **173**, 689–690.

HARKER, J. E. (1955) Control of diurnal rhythms of activity in *Periplaneta americana* L. *Nature, Lond.* **175**, 733.

HARKER, J. E. (1956) Factors controlling the diurnal rhythm of activity in *Periplaneta americana* L. *J. exp. Biol.* **33**, 224–234.

HARKER, J. E. (1958a) Experimental production of midgut tumours in *Periplaneta americana* L. *J. exp. Biol.* **35**, 251–259.

HARKER, J. E. (1958b) Diurnal rhythms in the animal kingdom. *Biol. Rev.* **33**, 1–52.

HARKER, J. E. (1960a) The effect of perturbations in the environmental cycle on the diurnal rhythm of activity of *Periplaneta americana* L. *J. exp. Biol.* **37**, 154–163.

HARKER, J. E. (1960b) Internal factors controlling the suboesophageal ganglion neurosecretory cycle in *Periplaneta americana* L. *J. exp. Biol.* **37**, 164–170.

HARKER, J. E. (1960c) Endocrine and nervous factors in insect circadian rhythms. *Cold. Spring Harb. Symp. Quant. Biol.* **25**, 279–287.

HARKER, J. E. (1964) *The Physiology of Diurnal Rhythms.* Cambridge University Press, London and New York.

HARKER, J. E. (1965a) The effect of a biological clock on the development rate of *Drosophila* pupae. *J. exp. Biol.* **42**, 323–337.

HARKER, J. E. (1965b) The effect of photoperiod on the development rate of *Drosophila* pupae. *J. exp. Biol.* **43**, 411–423.

HARRIS, F. A., LLOYD, E. P., LANE, H. C. and BURT, E. C. (1969) Influence of light on diapause in the boll weevil. II. Dependence of diapause response on various bands of visible radiation and a broad band of infrared radiation used to extend the photoperiod. *J. econ. Ent.* **62**, 854–857.

HARTLAND-ROWE, R. (1955) Lunar rhythm in the emergence of an Ephemeropteran. *Nature, Lond.* **176**, 657.

HARTLAND-ROWE, R. (1958) The biology of a tropical mayfly *Povilla adusta* Navas (Ephemeroptera, Polymitarcidae) with special reference to the lunar rhythm of emergence. *Revue Zool. Bot. afr.* **58**, 185–202.

HARVEY, G. T. (1957) The occurrence and nature of diapause-free development in the spruce budworm, *Choristoneura fumiferana* (Clem.). *Can. J. Zool.* **35**, 549–572.

HARWOOD, R. F. (1966) The relationship between photoperiod and autogeny in *Culex tarsalis* (Diptera: Culicidae). *Entomologia exp. appl.* **9**, 327–331.

HARWOOD, R. F. and HALFHILL, E. (1964) The effect of photoperiod on fat body and ovarian development of *Culex tarsalis*. *Ann. ent. Soc. Am.* **57**, 596–600.

HARWOOD, R. F. and TAKATA, N. (1965) Effect of photoperiod and temperature on fatty acid composition of the mosquito *Culex tarsalis*. *J. Insect Physiol.* **11**, 711–716.

HASEGAWA, K. (1951) Studies in voltinism in the silkworm, *Bombyx mori* L., with special reference to the organs concerning voltinism (a preliminary note). *Proc. Imp. Acad. Japan*, **27**, 667–671.

HASHIMOTO, H. (1966) Discovery of *Clunio takahashii* from Japan. *Jap. J. Zool.* **14**, 13–29.

HAYES, D. K. (1971) Action spectra for breaking diapause and absorption spectra of insect brain tissue. In *Biochronometry* (Ed. MENAKER, M.), pp. 392–402. National Academy of Sciences, Washington.

HELLER, H. C. and POULSON, T. L. (1970) Circannian rhythms—II. Endogenous and exogenous factors controlling reproduction and hibernation in chipmunks (*Eutamias*) and ground squirrels (*Spermophilus*). *Comp. Biochem. Physiol.* **33**, 357–383.

HERAN, H. (1962) Anemotaxis und Fluchtorientierung des Bachlaufers *Velia caprai* Tam. (= *V. currens* F.). *Z. vergl. Physiol.* **46**, 129–149.

HERMAN, W. S. (1973) The endocrine basis of reproductive inactivity in monarch butterflies overwintering in central California. *J. Insect Physiol.* **19**, 1883–1887.

HIDAKA, T., ISHIZAKA, Y. and SAKAGAMI, Y. (1971) Control of pupal diapause and adult differentiation in a univoltine papilionid butterfly, *Luehdorfia japonica*. *J. Insect Physiol.* **17**, 197–203.

HIGHKIN, H. R. and HANSON, L. B. (1954) Possible interaction between light–dark cycles and endogenous daily rhythms on the growth of tomato plants. *Plant Physiol.* **29**, 301–302.

HIGHNAM, K. C. (1958) Activity of the brain/corpora cardiaca system during pupal diapause "break" in *Mimas tiliae* (Lepidoptera). *Q. Jl microsc. Sci.* **99**, 73–88.

HILLMAN, W. S. (1956) Injury of tomato plants by continuous light and unfavorable photoperiodic cycles. *Am. J. Bot.* **43**, 89–96.

HILLMAN, W. S. (1964) Endogenous circadian rhythms and the response of *Lemna perpusilla* to skeleton photoperiods. *Am. Nat.* **98**, 323–328.

HILLMAN, W. S. (1973) Non-circadian photoperiodic timing in the aphid *Megoura. Nature, Lond.* **242**, 128–129.

HINKS, C. F. (1967) Relationship between serotonin and the circadian rhythm in some nocturnal moths. *Nature, Lond.* **214**, 386–387.

HINTON, H. E. (1951) A new chironomid from Africa, the larva of which can be dehydrated without injury. *Proc. zool. Soc. Lond.* **121**, 371–380.

HINTON, H. E. (1960) Cryptobiosis in the larva of *Polypedilum vanderplanki* Hint. (Chironomidae). *J. Insect Physiol.* **5**, 286–300.

HINTZE, A. L. (1925) The behavior of the larvae of *Cotinus nitida* Burmeister (Coleoptera). *Ann. ent. Soc. Am.* **18**, 31–34.

HODEK, I. (1960) Hibernation-bionomics in Coccinellidae. *Acta Soc. ent. Čechoslov.* **57**, 1–20.

HODEK, I. (1962) Experimental influencing of the imaginal diapause in *Coccinella septempunctata* L. (Col., Coccinellidae), 2nd. part. *Acta Soc. ent. Čechoslov.* **59**, 297–313.

HODEK, I. (1967) Bionomics and ecology of predaceous Coccinellidae. *A. Rev. Ent.* **12**, 79–104.

HODEK, I. (1968) Diapause in females of *Pyrrhocoris apterus* L. (Heteroptera). *Acta ent. Bohemoslov.* **65**, 422–435.

HODEK, I. (1971) Sensitivity of larvae to photoperiods controlling the adult diapause of two insects. *J. Insect Physiol.* **17**, 205–216.

HODEK, I. and ČERCASOV, J. (1958) A study of the imaginal diapause of *Semiadalia undecimnotata* Schneid. (Coccinellidae, Col.) in the open. I. *Acta Soc. zool. Bohemoslov.* **22**, 180–192.

HODEK, I. and ČERKASOV, J. (1960) Prevention and artificial induction of the imaginal diapause in *Coccinella 7-punctata* L. *Nature, Lond.* **187**, 345.

HOELSCHER, C. E. and VINSON, S. B. (1971) The sex ratio of a hymenopterous parasitoid, *Campoletis perdistinctus*, as affected by photoperiod, mating, and temperature. *Ann. ent. Soc. Am.* **64**, 1373–1376.

HOFFMANN, K. (1953) Experimentelle Änderung des Richtungsfinden beim Star durch Beeinflussung der "inneren Uhr". *Naturwissenschaften*, **40**, 608–609.

HOFFMANN, K. (1954) Versuche zu der Richtungsfinden der Vögel enthaltenen Zeitschätzung. *Z. Tierpsychol.* **11**, 453–475.

HOFFMANN, K. (1960) Experimental manipulation of the orientational clock in birds. *Cold Spring Harb. Symp. Quant. Biol.* **25**, 379–387.

HOFFMANN, K. (1965) Overt circadian frequencies and circadian rule. In *Circadian Clocks* (Ed. ASCHOFF, J.), pp. 87–94. North-Holland, Amsterdam.

HOFFMANN, K. (1969) Circadiane Periodik bei Tupajas (*Tupaia glis*) in Konstanten Bedingungen. *Zool. Anz. Suppl.* **33**, 171–177.

HOFFMANN, K. (1971) Biological clocks in animal orientation and in other functions. *Proc. Int. Symp. Circadian Rhythmicity* (Wageningen, 1971), pp. 175–205.

HOFFMANN, R. J. (1973) Environmental control of seasonal variation in the butterfly, *Colias eurytheme*. I. Adaptive aspects of a photoperiodic response. *Evolution*, **27**, 387–397.

HOLLINGSWORTH, M. J. (1969) Fluctuating temperatures and the length of life in *Drosophila. Nature, Lond.* **221**, 857–858.

HOUSE, H. L. (1967) The decreasing occurrence of diapause in the fly *Pseudosarcophaga affinis* through laboratory-reared generations. *Can. J. Zool.* **45**, 149–153.

HUGGANS, J. L. and BLICKENSTAFF, C. C. (1964) Effects of photoperiod on sexual development in the alfalfa weevil. *J. econ. Ent.* **57**, 167–168.

HUGHES, R. D. (1960) Induction of diapause in *Erioischia brassicae* Bouché (Dipt., Anthomyidae). *J. exp. Biol.* **37**, 218–223.

ICHIKAWA, M. and NISHIITSUTSUJI-UWO, J. (1955) *Mem. Coll. Sci. Univ. Kyoto* B, **22**, 11–15.

JACKLET, J. W. (1969) Circadian rhythm of optic nerve impulses recorded in darkness from isolated eye of *Aplysia. Science, Wash.* **164**, 562–563.

JACKLET, J. W. (1971) A circadian rhythm in optic nerve impulses from an isolated eye in darkness. In *Biochronometry* (Ed. MENAKER, M.), pp. 351–362. National Academy of Sciences, Washington.

JACKSON, D. L. (1961) Diapause in an aquatic Mymarid. *Nature, Lond.* **192**, 823–824.

JACKSON, D. L. (1963) Diapause in *Caraphractus cinctus* Walker (Hymenoptera: Mymaridae), a parasitoid of the eggs of Dytiscidae (Coleoptera). *Parasitology*, **53**, 225–251.

JACOBSON, L. A. (1960) Influence of photoperiod on oviposition by the army cutworm, *Chorizagrotis auxiliaris* (Lepidoptera: Noctuidae) in an insectary. *Ann. ent. Soc. Am.* **53**, 474–475.

JACOBSON, L. A. and BLAKELY, O. E. (1959) Development and behavior of the army cut-worm in the laboratory. *Ann. ent. Soc. Am.* **52**, 100–105.

JAHN, T. L. and CRESCITELLI, F. (1940) Diurnal changes in the electrical responses of the compound eye. *Biol. Bull. Mar. Biol. Lab., Woods. Hole,* **78**, 42–52.

JAHN, T. L. and WULFF, V. J. (1943) Electrical aspects of a diurnal rhythm in the eye of *Dytiscus fasciventris. Physiol. Zool.* **16**, 101–109.

JANDER, R. (1957). *Z. vergl. Physiol.* **40**, 162–238.

JEGLA, T. C. and POULSON, T. L. (1970) Circannian rhythms—I. Reproduction in the cave crayfish, *Orconectes pellucidus inermis. Comp. Biochem. Physiol.* **33**, 347–355.

JENNER, C. E. and ENGELS, W. L. (1952) The significance of the dark period in the photoperiodic response of male juncos and white-throated sparrows. *Biol. Bull. Mar. Biol. Lab., Woods Hole,* **103**, 345–355.

JERMY, T. and SARINGER, G. (1955) Die rolle der Photoperiode in der Auslösung der Diapause des Kartoffelkäfers (*Leptinotarsa decemlineata* Say) und des Amerikanischen weissen Bärenspinners (*Hyphantria cunea* Drury). *Acta Agron. Acad. Sci. Hung.* **5**, 419–440.

JOHNSON, C. G., HAINE, E., COCKBAIN, A. J. and TAYLOR, L. R. (1957) Moulting rhythm in the alienicolae of *Aphis fabae* Scop. (Hemiptera: Aphididae) in the field. *Ann. appl. Biol.* **45**, 702–708.

JOHNSSON, A. and KARLSSON, H. G. (1972a) A feedback model for biological rhythms. I. Mathematical description and basic properties of the model. *J. theoret. Biol.* **36**, 153–174.

JOHNSSON, A. and KARLSSON, H. G. (1972b) The *Drosophila* eclosion rhythm, the transformation method, and the fixed point theorem. Department of Electrical Measurements, Lund Institute of Technology, Report No. 2/1972, November 15, 1972.

JOLLY, M. S., SINHA, S. S. and RAZDEN, J. L. (1971) Influence of temperature and photoperiod on termination of pupal diapause in the Tasar silkworm, *Antheraea mylitta. J. Insect Physiol.* **17**, 753–760.

JOLY, P. (1945) La fonction ovarienne et son control humoral chez les Dytiscides. *Archs Zool. exp. gen.* **84**, 49–164.

JONES, M. D. R. (1964) The automatic recording of mosquito activity. *J. Insect Physiol.* **10**, 343–351.

JONES, M. D. R., CUBBIN, C. M. and MARSH, D. (1972a) The circadian rhythm of flight activity of the mosquito *Anopheles gambiae:* the light-on response. *J. exp. Biol.* **57**, 337–346.

JONES, M. D. R., CUBBIN, C. M. and MARSH, D. (1972b) Light-on effects and the question of bimodality in the circadian flight activity of the mosquito *Anopheles gambiae. J. exp. Biol.* **57** 347–357.

JONES, M. D. R., FORD, M. G. and GILLETT, J. D. (1966) Light-on and Light-off effects on the circadian flight activity in the mosquito *Anopheles gambiae. Nature, Lond.* **211**, 871–872.

JONES, M. D. R., HILL, M. and HOPE, A. M. (1967) The circadian flight activity of the mosquito *Anopheles gambiae:* phase setting by the light regime. *J. exp. Biol.* **47**, 503–511.

KALMUS, H. (1934) Über die Natur des Zeitgedächtnisses der Bienen. *Z. vergl. Physiol.* **20,** 405–.

KALMUS, H. (1935) Periodizität und Autochronie (Ideochronie) als Zeitregelnde Eigenschaffen der Organismen. *Biologia generalis,* **11**, 93–114.

KALMUS, H. (1938a) Tagesperiodisch verlaufende vorgänge an der Stabhueschrecke (*Dixippus morosus*) und ihre experimentelle beeinflussung. *Z. vergl. Physiol.* **25**, 494–508.

KALMUS, H. (1938b) Die Lage des Aufnahmeorgans für die Schlupfperiodik von *Drosophila. Z. vergl. Physiol.* **26**, 362–365.

KALMUS, H. (1940) Diurnal rhythms in the Axolotl larva and in *Drosophila. Nature, Lond.* **145**, 72–73.

KALMUS, H. (1956) Sun navigation of *Apis mellifica* L. in the southern hemisphere. *J. exp. Biol.* **33**, 554–565.

KAMM, J. A. (1970) Effects of photoperiod and temperature on *Crambus trisectus* and *C. leachellus cypridalis* (Lepidoptera: Crambidae). *Ann. ent. Soc. Am.* **63**, 412–416.

KAMM, J. A. (1972) Photoperiodic regulation of growth in an insect: response to progressive changes in daylength. *J. Insect Physiol.* **18**, 1745–1749.

KAMM, J. A. (1973) Role of environment during diapause on the phenology of *Chrysoteuchia topiaria* (Lepidoptera: Pyralidae). *Entomologia exp. appl.* **16**, 407–413.

KAPPUS, K. D. and VENARD, C. E. (1967) The effects of photoperiod and temperature on the induction of diapause in *Aëdes triseriatus* (Say). *J. Insect Physiol.* **13**, 1007–1019.

KATIYAR, K. P. and LONG. W. H. (1961) Diapause in the sugarcane borer, *Diatraea saccharalis. J. econ. Ent.* **54**, 285–287.

KEELEY, L. L. (1970) Diapause metabolism and rearing methods for the whitelined sphinx, *Celerio lineata* (Lepidoptera, Sphingidae). *Ann. ent. Soc. Am.* **63**, 905–907.

KENTEN, J. (1955) The effect of photoperiod and temperature on reproduction in *Acyrthosiphon pisum* (Harris) and on the forms produced. *Bull. ent. Res.* **46**, 599–624.

KERFOOT, W. B. (1967) The lunar periodicity of *Specodogastra texana,* a nocturnal bee (Hymenoptera, Halictidae). *Anim. Behav.* **15**, 479–486.

KHOO, S. G. (1968a) Experimental studies on diapause in stoneflies. I. Nymphs of *Capnia bifrons* (Newman). *Proc. R. ent. Soc. Lond.* (A), **43**, 40–48.

KHOO, S. G. (1968b) Experimental studies on diapause in stoneflies. II. Eggs of *Diura bicaudata* (L.). *Proc. R. ent. Soc Lond.* (A), **43**, 49–56.

KHOO, S. G. (1968c) Experimental studies on diapause in stoneflies. III. Eggs of *Brachyptera risi* (Morton). *Proc. R. ent. Soc. Lond.* (A), **43**, 141–146.

KIND, T. V. (1965) Neurosecretion and voltinism in *Orgyia antiqua* L. (Lepidoptera, Lymanthriidae). *Ent. Obozr.* **44**, 534–536. (In Russian.)

KING, L. L. and BENJAMIN, D. M. (1965) The effect of photoperiod and temperature on the development of multivoltine populations of *Neodiprion rugifrons* Middleton. *Proc. N. Central Branch, Ent. Soc. Am.* **20**, 129–140.

KING, P. E. (1963) The rate of egg resorption in *Nasonia vitripennis* (Walker) (Hymenoptera: Pteromalidae) deprived of hosts. *Proc. R. ent. Soc. Lond.* (A), **38**, 98–100.

KIRKPATRICK, C. M. and LEOPOLD, A. C. (1952) The role of darkness in sexual activity of the quail. *Science, Wash.* **116**, 280–281.

KISIMOTO, R. (1956) Effect of diapause in the fourth larval instar on the determination of wing form in the adult of the small brown plant hopper, *Delphacodes striatella* Fallen. *Oyo-Kontyu*, **12**, 202–210.

KISIMOTO, R. (1959) Studies on the diapause in the planthoppers and leafoppers. III. Sensitivity of various larval stages to photoperiod and the forms of ensuing adults in the green rice leafhopper, *Nephotettix cincticeps. Japan. J. appl. Ent. Zool.* **3**, 200–207.

KLEBER, E. (1935) Hat das Zeitgedächtnis der Bienen biologische Bedeutung? *Z. vergl. Physiol.* **22**, 221–262.

KLOTTER, K. (1960) Theoretical analysis of some biological models. In *Biological Clocks. Cold Spring Harb. Symp. Quant. Biol.* **25**, 189–196.

KLUG, H. (1958) Histo-physiologische Untersuchungen über die Aktivitätsperiodik bei Carabiden. *Wiss. Z. Humboldt. Univ. Berlin, Math.-Naturw. Reihe*, **8**, 405–434.

KOGURE, M. (1933) The influence of light and temperature on certain characters of the silkworm, *Bombyx mori. J. Dept. Agr. Kyushu Univ.* **4**, 1–93.

KOMAROVA, O. S. (1949) the conditions evoking diapause in the vine leafroller (*Polychrosis botrana* Schiff.). *Dokl. Akad. Nauk SSSR*, **68**, 789–792. (In Russian.)

KONO, Y. (1970) Photoperiodic induction of diapause in *Pieris rapae crucivora* Boisduval (Lepidoptera: Pieridae). *Appl. Ent. Zool.* **5**, 213–224.

KONOPKA, R. and BENZER, S. (1971) Clock mutants of *Drosophila melanogaster. Proc. Natn. Acad. Sci. U.S.A.* **68**, 2112–2116.

KÖPPÄ, P. (1970) Studies on the thrips (Thysanoptera) species most commonly occurring on cereals in Finland. *Ann. Agric. fenn.* **9**, 191–265.

KRAMER, G. (1950) Weitere Analyse der Faktoren, welche die Zugaktivität des gekäfigten Vogels orientieren. *Naturwissenschaften*, **37**, 377–378.

KREHAN, I. (1970) Die Steuerung der Jahresrhythmik und Diapause bei Larval- und Imagoüberwinterern der Gattung *Pterostichus* (Col., Carab.). *Oecologia (Berl.)*, **6**, 58–105.

KRISTENSEN, B. I. (1966) Incorporation of tyrosine into the rubber-like cuticle of locusts studied by autoradiography. *J. Insect Physiol.* **12**, 173–177.

LAMPRECHT, G. and WEBER, F. (1973) Mitnahme, Frequenzdemultiplikation und Maskierung der Laufaktivität von *Carabus*-Arten (Coleoptera) durch Lichtzyklen. *J. Insect Physiol.* **19**, 1579–1590.

LEES, A. D. (1953a) Experimental factors controlling the evocation and termination of diapause in the fruit tree red spider mite *Metatetranychus ulmi* Koch (Acarina: Tetranychidae). *Ann. appl. Biol.* **40**, 449–486.

LEES, A. D. (1953b) The significance of the light and dark phases in the photoperiodic control of diapause in *Metatetranychus ulmi* Koch. *Ann. appl. Biol.* **40**, 487–497.

LEES, A. D. (1955) *The Physiology of Diapause in Arthropods.* Cambridge University Press.

LEES, A. D. (1959) The role of photoperiod and temperature in the determination of parthenogenetic and sexual forms in the aphid *Megoura viciae* Buckton—I. The influence of these factors on apterous virginoparae and their progeny. *J. Insect Physiol.* **3**, 92–117.

LEES, A. D. (1960a) The role of photoperiod and temperature in the determination of parthenogenetic and sexual forms in the aphid *Megoura viciae* Buckton—II. The operation of the "interval timer" in young clones. *J. Insect Physiol.* **4**, 154–175.

LEES, A. D. (1960b) Some aspects of animal photoperiodism. *Cold Spring Harb. Symp. Quant. Biol.* **25**, 261–268.

LEES, A. D. (1961) Clonal polymorphism in aphids. *Symp. R. ent. Soc. Lond.* **1**, 68–79.

LEES, A. D. (1963) The role of photoperiod and temperature in the determination of parthenogenetic and sexual forms in the aphid *Megoura viciae* Buckton—III. Further properties of the maternal switching mechanism in apterous aphids. *J. Insect Physiol.* **9**, 153–164.

LEES, A. D. (1964) The location of the photoperiodic receptors in the aphid *Megoura viciae. J. exp. Biol.* **41**, 119–133.

LEES, A. D. (1965) Is there a circadian component in the *Megoura* photoperiodic clock? In *Circadian Clocks* (Ed. ASCHOFF, J.), pp. 351–356. North-Holland, Amsterdam.

LEES, A. D. (1966a) Photoperiodic timing mechanisms in insects. *Nature, Lond.* **210**, 986–989.

LEES, A. D. (1966b) The control of polymorphism in aphids. *Adv. Insect Physiol.* **3**, 207–277.

LEES, A. D. (1967a) The diversity of biological clocks in aphids. In *Insects and Physiology* (Ed. BEAMENT, J. W. L. and TREHERNE, J. E.), pp. 89–99. Oliver & Boyd, Edinburgh and London.

LEES, A. D. (1967b) Direct and indirect effects of daylength on the aphid *Megoura viciae* Buckton. *J. Insect Physiol.* **13**, 1781–1785.

LEES, A. D. (1968) Photoperiodism in insects. In *Photophysiology*, vol. IV (Ed. GIESE, A. C.), pp. 47–137. Academic Press, New York.

LEES, A. D. (1970) Insect clocks and timers. Inaugural Lecture, Imperial College of Science and Technology, 1st December 1970.

LEES, A. D. (1971a) The relevance of action spectra in the study of insect photoperiodism. In *Biochronometry* (Ed. MENAKER, M.), pp. 372–380. National Academy of Sciences, Washington.

LEES, A. D. (1971b) The role of circadian rhythmicity in photoperiodic induction in animals. *Proc. Int. Symp. Circadian Rhythmicity* (Wageningen, 1971), pp. 87–110.

LEES, A. D. (1973) Photoperiodic time measurement in the aphid *Megoura viciae*. *J. Insect Physiol.* **19**, 2279–2316.

LEGAY, J. M. (1962) Caractère des processus métaboliques au cours de la diapause des insectes. *Annls Nutr., Paris,* **16**, 65–89.

LEIGH, T. F. (1966) A reproductive diapause in *Lygus hesperus* Knight. *J. econ. Ent.* **59**, 1280–1281.

LEPPLA, N. C. and SPANGLER, H. G. (1971) A flight-cage actograph for recording circadian periodicity of pink bollworm moths. *Ann. ent. Soc. Am.* **64**, 1431–1434.

LEUTHOLD, R. (1966) Die Bewegungsaktivität der weiblichen Schabe *Leucophaea maderae* (F.) im Laufe des Fortpflanzungszyklus und ihre experimentelle Beeinflussung. *J. Insect Physiol.* **12**, 1303–1331.

LEWIS, C. B. and BLETCHLEY, J. D. (1943) The emergence rhythm of the dung-fly *Scopeuma* (= *Scatophaga*) *stercoraria* (L.). *J. anim. Ecol.* **12**, 11–18.

L'HELIAS, C. (1962) Corrélations entre les ptérines et le photopériodisme dans la regulation du cycle sexuel chez les pucerons. *Bull. Biol. France et Belge,* **96**, 187–198.

LINDAUER, M. (1957) Sonnenorientierung der Bienen unter der Äquatorsonne und zur Nachzeit. *Naturwissenschaften,* **44**, 1–6.

LINDAUER, M. (1959) Angeborene und erlendte Komponenten in der Sonnenorientierung der Bienen. *Z. vergl. Physiol.* **42**, 43–62.

LINDAUER, M. (1960) Time-compensated sun orientation in bees. *Cold Spring Harb. Symp. Quant. Biol.* **25**, 371–377.

LIPKOW, E. (1966) Biologisch-ökologische Untersuchungen über *Tachyporus*-Arten und *Tachinus rufipes* (Col., Staphyl.). *Pedobiologia,* **6**, 140–177.

LIPTON, G. R. and SUTHERLAND, D. J. (1970) Activity rhythms in the American cockroach, *Periplaneta americana*. *J. Insect Physiol.* **16**, 1555–1566.

LITSINGER, J. A. and APPLE, J. A. (1973) Estival diapause of the alfalfa weevil in Wisconsin. *Ann. ent. Soc. Am.* **66**, 11–20.

LOHER, W. (1972) Circadian control of stridulation in the cricket *Teleogryllus commodus* Walker. *J. comp. Physiol.* **79**, 173–190.

LOHER, W. and CHANDRASHEKARAN, M. K. (1970) Circadian rhythmicity in the oviposition of the grasshopper *Chorthippus curtipennis*. *J. Insect Physiol.* **16**, 1677–1688.

LOHMANN, M. (1964) Der einfluss von Beleuchtungsstärke und Temparatur auf die Tagesperiodische Laufaktivität des Mehlkäfers, *Tenebrio molitor* L. *Z. vergl. Physiol.* **49**, 341–389.

LOHMANN, M. (1967) Ranges of circadian period length. *Experientia,* **23**, 788–790.

LOVE, G. J. and WHELCHEL, J. G. (1955) Photoperiodism and the development of *Aedes triseriatus* (Diptera, Culicidae). *Ecology,* **36**, 340–342.

LUM, P. T. M., NAYAR, J. K. and PROVOST, M. W. (1968) The pupation rhythm in *Aedes taeniorhynchus* III. Factors in developmental synchrony. *Ann. ent. Soc. Am.* **61**, 889–899.

LUMME, J., LAKOVAARA, S. and SAURA, A. (1972) The influence of daylength and temperature on the testis pterin content of *Drosophila littoralis*. *J. Insect Physiol.* **18**, 2043–2053.

LUTZ, F. E. (1932) Experiments with Orthoptera concerning diurnal rhythms. *Am. Mus. Novitates,* **550**, 1–24.

LUTZ, P. E. (1961) Pattern of emergence in the dragonfly *Tetragoneuria cynosura*. *J. Elisha Mitchell Sci. Soc.* **77**, 114–115.

LUTZ, P. E. and JENNER, C. E. (1964) Life-history and photoperiodic response of nymphs of *Tetragoneuria cynosura* (Say). *Biol. Bull. Mar. Biol. Lab., Woods Hole,* **127**, 304–316.

MACFARLANE, J. E. (1968) Diel periodicity in spermatophore formation in the house cricket, *Acheta domesticus* (L.). *Can. J. Zool.* **46**, 695–698.

MACGILLIVRAY, M. E. and ANDERSON, G. B. (1964) The effect of photoperiod and temperature on the production of gamic and agamic forms in *Macrosiphum euphorbiae* (Thomas). *Can. J. Zool.* **42**, 491–510.

MacLeod, E. G. (1967) Experimental induction and elimination of adult diapause and autumnal coloration in *Chrysopa carnea* (Neuroptera). *J. Insect Physiol.* **13**, 1343–1349.

Magnum, C. L., Earle, N. W. and Newsom, L. D. (1968) Photoperiodic induction of diapause in the boll weevil, *Anthonomus grandis*. *Ann. ent. Soc. Am.* **61**, 1125–1128.

Mansingh, A. (1971) Physiological classification of dormancies in insects. *Can. Ent.* **103**, 983–1009.

Mansingh, A. and Smallman, B. N. (1966) Photoperiod control of an "obligatory" pupal diapause. *Can. Ent.* **98**, 613–616.

Mansingh, A. and Smallman, B. N. (1967) Effect of photoperiod on the incidence and physiology of diapause in two Saturniids. *J. Insect Physiol.* **13**, 1147–1162.

Mansingh, A. and Smallman, B. N. (1971) The influence of temperature on the photoperiodic regulation of diapause in Saturniids. *J. Insect Physiol.* **17**, 1735–1739.

Marcovitch, S. (1923) Plant lice and light exposure. *Science, Wash.* **58**, 537–538.

Marcovitch, S. (1924) The migration of the Aphididae and the appearance of sexual forms as affected by the relative length of daily light exposure. *J. Agric. Res.* **27**, 513–522.

Masaki, S. (1956) The local variation in the diapause pattern of the cabbage moth, *Barathra brassicae* Linné, with particular reference to the aestival diapause (Lepidopetera: Noctuidae). *Bull. Fac. Agr. Mie Univ.* **13**, 29–46.

Masaki, S. (1957) Ecological significance of diapause in the seasonal cycle of *Abraxas miranda* Btl. *Bull. Fac. Agr. Mie Univ.* **15**, 15–24.

Masaki, S. (1958) The response of a "short-day" insect to certain external factors: the induction of diapause in *Abraxas miranda* Butl. *Japan. J. appl. Ent. Zool.* **2**, 285–294.

Masaki, S. (1968) Geographic adaptation in the seasonal life cycle of *Mamestra brassicae* (Linné) (Lepidoptera: Noctuidae). *Bull. Fac. Agr. Hirosaki Univ.* **14**, 16–26.

Masaki, S. and Ohmachi, F. (1967) Divergence of photoperiodic response and hybrid development in *Teleogryllus* (Orthoptera: Gryllidae). *Kontyu*, **35**, 83–105.

Masaki, S. and Oyama, N. (1963) Photoperiodic control of growth and wing form in *Nemobius yezoensis* Shiraki (Orthoptera: Gryllidae). *Kontyu*, **31**, 16–26.

Masaki, S. and Sakai, T. (1965) Summer diapause in the seasonal cycle of *Mamestra brassicae* Linné *Japan. J. appl. Ent. Zool.* **9**, 191–205.

Masaki, S., Umeya, K., Sekiguchi, Y. and Kawasaki, R. (1968) Biology of *Hyphantria cunea* Drury (Lepidoptera: Arctiidae) in Japan. III. Photoperiodic induction of diapause in relation to the seasonal life cycle. *Appl. Ent. Zool.* **3**, 55–66.

Masslennikova, V. A. (1958) Conditions determining diapause in the parasitic Hymenoptera *Apanteles glomeratus* L. (Braconidae), and *Pteromalus puparum* (Chalcididae). *Ent. Obozr.* **37**, 538–545. (In Russian.)

Masslennikova, V. A. (1968) The regulation of seasonal development in parasitic insects. In *Photoperiodic Adaptations in Insects and Acari* (Ed. Danilevskii, A. S.), pp. 129–152. Leningrad University Press. (In Russian.)

McClelland, G. A. H. and Green, C. A. (1970) Subtle periodicity of pupation in rapidly developing mosquitoes. *Bull. Wld Hlth Org.* **42**, 951–955.

McCluskey, E. S. (1965) Circadian rhythms in male ants of five diverse species. *Science, Wash.* **150**, 1037–1039.

McCrary, A. B. and Jenner, C. E. (1965) Influence of daylength on sex ratio in the giant mosquito, *Toxorhynchites rutilus*, in nature. *Am. Zool.* **5**, 206.

McHaffey, D. G. and Harwood, R. F. (1970) Photoperiod and temperature influences on diapause in eggs of the floodwater mosquito *Aedes dorsalis* Meigen (Diptera, Culicidae). *J. Med. Ent.* **7**, 631–644.

McLeod, D. G. R. and Beck, S. D. (1963) Photoperiodic termination of diapause in an insect. *Biol. Bull. Mar. Biol. Lab., Woods Hole*, **124**, 84–96.

McMullen, R. D. (1967) The effects of photoperiod, temperature, and food supply on rate of development and diapause in *Coccinella novemnotata*. *Can. Ent.* **99**, 578–586.

McNeil, J. N. and Rabb, R. L. (1973) Physical and physiological factors in diapause initiation of two hyperparasites of the tobacco hornworm, *Manduca sexta*. *J. Insect Physiol.* **19**, 2107–2118.

Meder, E. (1958) Über die Einberechnung der Sonnenwanderung bei der Orientierung der Honigbiene. *Z. vergl. Physiol.* **40**, 610–641.

Medugorac, I. (1967) Orientierung der Bienen in Raum und Zeit nach Dauernarkose. *Z. Bienforsch.* **9**, 105–119.

Medugorac, I. and Lindauer, M. (1967) Das Zeitgedächtnis der Bienen unter dem Einfluss von Narkose und von sozialen Zeitgebern. *Z. vergl. Physiol.* **55**, 450–474. Berlin–Heidelberg–New York: Springer.

Melchers, G. (1956) Die Beteiligung der endonomen Tagesrhythmik am Zustandekommen der photoperiodischen Reaktionen der Kurztagpflanze *Kalanchoë blossfeldiana*. *Z. Naturf.* **11b**, 544–548.

Mellanby, K. (1939) The physiology and activity of the bed bug (*Cimex lectularius* L.) in a natural infestation. *Parasitology*, **31**, 200–211.

MELLANBY, K. (1940) The daily rhythm in the cockroach *Blatta orientalis* II. Observations and experiments on a natural infestation. *J. exp. Biol.* **17**, 278–285.

MENAKER, M. and GROSS, G. (1965) Effects of fluctuating temperature on diapause induction in the pink bollworm. *J. Insect Physiol.* **11**, 911–914.

MENZEL, R., MOCK, K., WLADARZ, G. and LINDAUER, M. (1969) Tagesperiodische ablagerungen in der Endokutikula der Honigbiene. *Biol. Zbl.* **88**, 61–67.

MESSENGER, P. S. (1964) The influence of rhythmically fluctuating temperatures on the development and reproduction of the spotted alfalfa aphid, *Therioaphis maculata*. *J. econ. Ent.* **57**, 71–76.

MEUDEC, M. (1966) Influence de la temperature et de la lumière pendant le développement sur l'état ovarien à l'éclosion chez *Acrolepia assectella* Zeller (Insecte lépidoptère). *C. R. Acad. Sci.* **263**, 554–557.

MICHAL, K. (1931) Oszillation im Sauerstoffverbrauch der Mehlwurmlarven (*Tenebrio molitor*). *Zool. Anz.* **95**, 65–75.

MINIS, D. H. (1965) Parallel peculiarities in the entrainment of a circadian rhythm and photoperiodic induction in the pink bollworm (*Pectinophora gossypiella*). In *Circadian Clocks* (Ed. ASCHOFF, J.), pp. 333–343. North-Holland, Amsterdam.

MINIS, D. H. and PITTENDRIGH, C. S. (1968) Circadian oscillation controlling hatching: its ontogeny during embryogenesis of a moth. *Science, Wash.* **159**, 534–536.

MISSONNIER, J. (1956) Note sur la biologie du psylle de l'aubepine (*Psylla peregrina* Forster). *Annls Epiphyt.* **7**, 253–262.

MISSONNIER, J. (1963) Etude écologique du développement nymphale de deux diptères muscides phytophages: *Pegomyia betae* Curtis et *Chortophila brassicae* Bouché. *Annls Epiphyt.* **14**, 293–310.

MONTGOMERY, B. E. and MACKLIN, J. M. (1962) Rates of development in the later instars of *Neotetrum pulchellum* (Drury). *Proc. N. Central Branch, Ent. Soc. Am.* **17**, 21–23.

MORIARTY, F. (1959) The 24-hr rhythm of emergence of *Ephestia kuhniella* Zell. from the pupa. *J. Insect Physiol.* **3**, 357–366.

MORIARTY, F. (1969) Egg diapause and water absorption in the grasshopper *Chorthippus brunneus*. *J. Insect Physiol.* **15**, 2069–2074.

MÜLLER, H. J. (1954) Der Saisondimorphismus bei Zikaden der Gattung *Euscelis* Brullé. *Beitr. Ent.* **4**, 1–56.

MÜLLER, H. J. (1955) Die Saisonformenbildung von *Araschnia levana*, ein photoperiodisch gesteuerter Diapause-effekt. *Naturwissenschaften*, **42**, 134–135.

MÜLLER, H. J. (1957) Die Wirkung exogener Faktoren auf die Zyklische Formenbildung der Insekten, inbesondere der Gattung *Euscelis* (Hom., Auchenorrhyncha). *Zool. Jb., Syst.* **85**, 317–430.

MÜLLER, H. J. (1958) Über den Einfluss der Photoperiode auf Diapause und Körpergrosse der Delphacide *Stenocranus minutus* Fabr. *Zool. Anz.* **160**, 294–311.

MÜLLER, H. J. (1960) Die Bedeutung der Photoperiode im Lebensablauf der Insekten. *Z. angew. Ent.* **47**, 7–24.

MÜLLER, H. J. (1961) Erster Nachweis Eidiapause bei den Jassiden *Euscelis plebejus* Fall. und *lineolatus* Brullé. *Z. angew. Ent.* **48**, 233–241.

MÜLLER, H. J. (1962a) Über den Saisondimorphen entwicklungszyklus und die Aufhebung der Diapause bei *Aleurochiton complanatus* (Baerensprung) (Homoptera, Aleyrodidae). *Entomologia exp. appl.* **5**, 124–138.

MÜLLER, H. J. (1962b) Über die Induktion der Diapause und der Ausbildung der Saisonformen bei *Aleurochiton complanatus* (Baerensprung). *Z. Morph. Okol. Tiere*, **51**, 575–610.

MÜLLER, H. J. (1964) Über die Wirkung Verschiedener Spektralbereiche bei der photoperiodischen Induktion der Saisonformen von *Euscelis plebejus* Fall. (Homoptera: Jassidae). *Zool. Jb. Abt. Allgem. Zool. Physiol. Tiere*, **70**, 411–426.

MÜLLER, H. J. (1970) Formen der Dormanz bei Insekten. *Nova Acta Leopold*, **35**, 7–27.

MYBURGH, A. C. (1963) Diurnal rhythms in emergence of mature larvae from fruit and eclosion of adult *Pterandrus rosae* (Ksh.) and *Ceratitis capitata* (Wied.). *S. Afr. J. Agric. Sci.* **6**, 41–46.

MYERS, K. (1952) Rhythms in emergence and other aspects of behaviour of the Queensland fruit-fly (*Dacus* (*Strumeta*) *tryoni* Frogg.) and the solanum fruit-fly (*Dacus* (*Strumeta*) *cacuminatus* Hering). *Aust. J. Sci.* **15**, 101–102.

NAGATA, K., TAMAKI, Y., NOGUCHI, H. and YUSHIMA, T. (1972) Changes in sex pheromone activity in adult females of the smaller tea tortrix moth *Adoxophyes fasciata*. *J. Insect Physiol.* **18**, 339–346.

NASH, T. A. M. and TREWERN, M. A. (1972) Hourly distribution of larviposition by *Glossina austeni* Newst. and *G. morsitans morsitans* Westw. (Dipt., Glossinidae). *Bull. ent. Res.* **61**, 693–700.

NAYAR, J. K. (1967a) Endogenous diurnal rhythm of pupation in a mosquito population. *Nature, Lond.* **214**, 828–829.

NAYAR, J. K. (1967b) The pupation rhythm in *Aedes taeniorhynchus* (Diptera: Culicidae). II. Ontogenetic timing, rate of development, and endogenous diurnal rhythm of pupation. *Ann. ent. Soc. Am.* **60**, 946–971.

NAYAR, J. K. (1968) The pupation rhythm in *Aedes taeniorhynchus* IV. Further studies of the endogenous diurnal (circadian) rhythm of pupation. *Ann. ent. Soc. Am.* **61**, 1408–1417.

NAYAR, J. K. (1972) Effects of fluctuating temperatures on life span of *Aedes taeniorhynchus* adults. *J. Insect Physiol.* **18**, 1303–1313.

NAYAR, J. K. and SAUERMAN, D. M. (1969) Flight behaviour and phase polymorphism in the mosquito *Aedes taeniorhynchus*. *Entomologia exp. appl.* **12**, 365–375.

NAYAR, J. K. and SAUERMAN, D. M. (1971) The effect of light regimes on the circadian rhythm of flight activity in the mosquito *Aedes taeniorhynchus*. *J. exp. Biol.* **54**, 745–756.

NAYAR, J. K., SAMARAWICKREMA, W. A. and SAUERMAN, D. M. (1973) Photoperiodic control of egg hatching in the mosquito *Mansonia titillans*. *Ann. ent. Soc. Am.* **66**, 831–835.

NEUMANN, D. (1963) Über die Steuerung der lunaren Schwärmperiodik der Mücke *Clunio marinus*. *Verh. dt. Zool. Ges. Wien* 1962, pp. 275–285.

NEUMANN, D. (1966a) Die intraspezifische Variabilität der lunaren und täglichen Schlüpfzeiten von *Clunio marinus* (Diptera: Chironomidae). *Verh. dt. Zool. Ges. Jena* 1965, pp. 223–233.

NEUMANN, D. (1966b) Die lunare und tägliche Schlüpfperiodik der Mücke *Clunio*. Steuerung und Abstimmung auf die Gezeitenperiodik. *Z. vergl. Physiol.* **53**, 1–61. Berlin–Heidelberg–New York: Springer.

NEUMANN, D. (1967) Genetic adaptation in emergence time of *Clunio* populations to different tidal conditions. *Helgoländer wiss. Meeresunters.* **15**, 163–171.

NEUMANN, D. (1971) Eine nicht-reziproke Kreuzungssterilität zwischen ökologischen Rassen der Mücke *Clunio marinus*. *Oecologia, Berl.* **8**, 1–20.

NEUMANN, D. and HONEGGER, W. (1969) Adaptations of the intertidal midge *Clunio* to arctic conditions. *Oecologia, Berl.* **3**, 1–13.

NEVILLE, A. C. (1963a) Daily growth layers in locust rubber-like cuticle influenced by an external rhythm. *J. Insect Physiol.* **9**, 177–186.

NEVILLE, A. C. (1963b) Growth and deposition of resilin and chitin in locust rubber-like cuticle. *J. Insect Physiol.* **9**, 265–278.

NEVILLE, A. C. (1963c) Daily growth layers for determining the age of grasshopper populations. *Oikos*, **14**, 1–8.

NEVILLE, A. C. (1965) Circadian organization of chitin in some insect skeletons. *Q. Jl microsc. Sci.* **106**, 315–325.

NEVILLE, A. C. (1967a) Daily growth layers in animals and plants. *Biol. Rev.* **43**, 421–441.

NEVILLE, A. C. (1967b) Chitin orientation in cuticle and its control. *Adv. Insect Physiol.* **4**, 213–286.

NEVILLE, A. C. (1970) Cuticle ultrastructure in relation to the whole insect. *Symp. R. ent. Soc. Lond.* **5**, 17–39.

NIELSEN, E. T. (1938) Zur Oekologie der Laubheuschrecken. *Ent. Medd.* **20**, 121–164.

NIELSEN, E. T. (1961) On the habits of the migratory butterfly, *Ascia monuste* L. *Biol. Medd. Kgl. Dansk Viddenskab. Selsk.* **23**, 1–81.

NIELSEN, E. T. and HAEGAR, J. S. (1954) Pupation and emergence in *Aëdes taeniorhynchus* (Wied.). *Bull. ent. Res.* **45**, 757–768.

NISHIITSUTSUJI-UWO, J., PETROPULOS, S. F. and PITTENDRIGH, C. S. (1967) Central nervous system control of circadian rhythmicity in the cockroach. I. Role of the pars intercerebralis. *Biol. Bull. Mar. Biol. Lab., Woods Hole*, **133**, 679–696.

NISHIITSUTSUJI-UWO, J. and PITTENDRIGH, C. S. (1967) The neuroendocrine basis of midgut tumour induction in cockroaches. *J. Insect Physiol.* **13**, 851–859.

NISHIITSUTSUJI-UWO, J. and PITTENDRIGH, C. S. (1968a) Central nervous system control of circadian rhythmicity in the cockroach. II. The pathway of light signals that entrain the rhythm. *Z. vergl. Physiol.* **58**, 1–13. Berlin–Heidelberg–New York: Springer.

NISHIITSUTSUJI-UWO, J. and PITTENDRIGH, C. S. (1968b) Central nervous system control of circadian rhythmicity in the cockroach. III. The optic lobes, locus of the driving oscillation? *Z. vergl. Physiol.* **58**, 14–46. Berlin–Heidelberg–New York: Springer.

NITSCH, I. P. and WENT, F. W. (1959) The induction of flowering in *Xanthium pensylvanicum* under long days. In *Photoperiodism and Related Phenomena in Plants and Animals* (Ed. WITHROW, R. B.), pp. 311–314. Am. Ass. Adv. Sci., Washington.

NORRIS, K. H., HOWELL, F., HAYES, D. K., ADLER, V. E., SULLIVAN, W. N. and SCHECHTER, M. S. (1969) The action spectrum for breaking diapause in the codling moth, *Laspeyresia pomonella* (L.) and the oak silkworm, *Antheraea pernyi* Guer. *Proc. natn. Acad. Sci. U.S.A.* **63**, 1120–1127.

NORRIS, M. J. (1959) The influence of day-length on imaginal diapause in the red locust, *Nomadacris septemfasciata* (Serv.). *Entomologia exp. appl.* **2**, 154–168.

NORRIS, M. J. (1962) Diapause induced by photoperiod in a tropical locust, *Nomadacris septemfasciata* (Serv.). *Ann. appl. Biol.* **50**, 600–603.

NORRIS, M. J. (1965) The influence of constant and changing photoperiods on imaginal diapause in the red locust (*Nomadacris septemfasciata* Serv.). *J. Insect Physiol.* **11**, 1105–1119.

NOVAK, K. and SEHNAL, F. (1963) The development cycle of some species of the genus *Limnephilus* (Trichoptera). *Acta Soc. ent. Cech.* **60**, 67–80.

NOVAK, K. and SEHNAL, F. (1965) Imaginaldiapause bei den in periodischen Gewässern lebenden Trichoptera. *Proc. XIIth Int. Congr. Ent. London*, 1964, p. 434.

NOVAK, V. J. A. (1966) *Insect Hormones*, 3rd edition. Methuen & Co. Ltd., London.

NOWOSIELSKI, J. W. and PATTON, R. L. (1963) Studies on circadian rhythms of the house cricket, *Gryllus domesticus* L. *J. Insect Physiol.* **9**, 401–410.

NOWOSIELSKI, J. W. and PATTON, R. L. (1964) Daily fluctuations in the blood sugar concentration of the house cricket, *Gryllus domesticus* L. *Science, Wash.* **144**, 180–181.

NOWOSIELSKI, J. W., PATTON, R. L. and NAEGELE, J. A. (1964) Daily rhythm of narcotic sensitivity in the house cricket, *Gryllus domesticus* L., and the two-spotted spider mite, *Tetranychus urticae* Koch. *J. cell. comp. Physiol.* **63**, 393–398.

ODHIAMBO, T. R. (1966) The metabolic effects of the corpus allatum hormone in the male desert locust. II. Spontaneous locomotor activity. *J. exp. Biol.* **45**, 51–63.

OHTAKI, T. (1966) On the delayed pupation of the flesh fly, *Sarcophaga peregrina* Robineau-Desvoidy. *Jap. J. Med. Sci. Biol.* **19**, 97–104.

OHTAKI, T., MILKMAN, R. D. and WILLIAMS, C. M. (1968) Dynamics of ecdysone secretion and action in the fleshfly *Sarcophaga peregrina*. *Biol. Bull. Mar. Biol. Lab., Woods Hole* **135**, 322–334.

OLDFIELD, G. N. (1970) Diapause and polymorphism in Californian populations of *Psylla pyri* (Homoptera: Psyllidae). *Ann. ent. Soc. Am.* **63**, 180–184.

OLIVER, C. G. (1969) Experiments on the diapause dynamics of *Papilio polyxenus*. *J. Insect Physiol.* **15**, 1579–1589.

OTUKA, M. and SANTA, H. (1955) Studies on the diapause in the cabbage armyworm *Barathra brassicae* L. III. The effect of the rhythm of light and darkness on the induction of diapause. *Bull. Nat. Inst. Agr. Sci. Japan*, Ser. C, pp. 49–56.

PALMÉN, E. (1955) Diel periodicity of pupal emergence in natural populations of some chironomids (Diptera). *Ann. Zool. Soc. Zool. Bot. Fennicae Vonamo*, **17**, 1–30.

PAMMER, E. (1965) Spinnverhalten und Diapauseinduktion bei *Philosamia cynthia*. *Naturwissenschaften*, **52**, 649.

PARK, O. and KELLER, J. G. (1932) Studies in nocturnal ecology. II. Preliminary analysis of activity in nocturnal forest insects. *Ecology*, **13**, 335–346.

PARKER, M. W., HENDRICKS, S. B., BORTHWICK, H. A., and SCULLY, N. J. (1946) Action spectrum for the photoperiodic control of floral initiation of short-day plants. *Bot. Gaz.* **108**, 1–26.

PARIS, O. H. and JENNER, C. E. (1959) Photoperiodic control of diapause in the pitcher-plant midge, *Metriocnemus knabi*. In *Photoperiodism and Related Phenomena in Plants and Animals* (Ed. WITHROW, R. B.), pp. 601–624. Am. Ass. Adv. Sci., Washington.

PEARL, R. (1928) *The Rate of Living*. University of London Press.

PEASE, R. W. (1962) Factors causing seasonal forms in *Ascia monuste* (Lepidoptera). *Science, Wash.* **137**, 987–988.

PENGELLEY, E. T. and FISHER, K. C. (1963) The effect of temperature and photoperiod on the yearly hibernating behavior of captive golden-mantled ground squirrels (*Citellus lateralis tescorum*). *Can. J. Zool.* **41**, 1103–1120.

PEREZ, Y., VERDIER, M. and PENER, M. P. (1971) The effect of photoperiod on male sexual behaviour in a north adriatic strain of the migratory locust. *Entomologia exp. appl.* **14**, 245–250.

PETERSON, D. M. and HAMNER, W. M. (1968) Photoperiodic control of diapause in the codling moth. *J. Insect Physiol.* **14**, 519–528.

PFLÜGER, W. and NEUMANN, D. (1971) Die Steuerung einer gezeitenparallelen Schlüpfrhythmik nach dem Sanduhr-Prinzip. *Oecologia, Berl.* **7**, 262–266.

PHELPS, R. J. and JACKSON, P. J. (1971) Factors influencing the moment of larviposition and eclosion in *Glossina morsitans orientalis* Vanderplank (Diptera: Muscidae). *J. ent. Soc. Sth Afr.* **34**, 145–157.

PHILLIPS, J. R. and NEWSOM, L. D. (1966) Diapause in *Heliothis zea* and *Heliothis virescens*. *Ann. ent. Soc. Am.* **59**, 154–159.

PHILOGENE, B. J. R. and BENJAMIN, D. M. (1971) Diapause in the Swaine jack-pine sawfly, *Neodiprion swainei*, as influenced by temperature and photoperiod. *J. Insect Physiol.* **17**, 1711–1716.

PITTENDRIGH, C. S. (1954) On temperature independence in the clock system controlling emergence time in *Drosophila*. *Proc. Nat. Acad. Sci. U.S.A.* **40**, 1018–1029.

PITTENDRIGH, C. S. (1958) Perspectives in the study of biological clocks. In *Perspectives in Marine Biology* (Ed. BUZZATI-TRAVERSO, A. A.), pp. 239–268. University of California Press.

PITTENDRIGH, C. S. (1960) Circadian rhythms and the circadian organization of living systems. *Cold Spring Harb. Symp. Quant. Biol.* **25**, 159–184.

PITTENDRIGH, C. S. (1961) On temporal organization in living systems. *Harvey Lectures Ser.* **56**, 93–125.

PITTENDRIGH, C. S. (1965) On the mechanism of entrainment of a circadian rhythm by light cycles. In *Circadian Clocks* (Ed. ASCHOFF, J.), pp. 277–297. North-Holland, Amsterdam.

PITTENDRIGH, C. S. (1966) The circadian oscillation in *Drosophila pseudoobscura* pupae: a model for the photoperiodic clock. *Z. Pflanzenphysiol.* **54**, 275–307.

PITTENDRIGH, C. S. (1967a) Circadian rhythms, space research and manned space flight. In *Life Sciences and Space Research*, pp. 122–134. North–Holland, Amsterdam.

PITTENDRIGH, C. S. (1967b) Circadian systems I. The driving oscillation and its assay in *Drosophila pseudoobscura. Proc. Nat. Acad. Sci. U.S.A.* **58**, 1762–1767.

PITTENDRIGH, C. S. (1972) Circadian surfaces and the diversity of possible roles of circadian organization in photoperiodic induction. *Proc. Nat. Acad. Sci. U.S.A.* **69**, 2734–2737.

PITTENDRIGH, C. S. (1974) Circadian oscillations in cells and the circadian organization of multicellular systems. In *The Neurosciences Third Study Program* (Ed. SCHMITT, F. O. and WORDEN, F. G.), pp. 437-458. M.I.T. Press, Cambridge, Mass.

PITTENDRIGH, C. S. and BRUCE, V. G. (1957) An oscillator model for biological clocks. In *Rhythmic and Synthetic Processes in Growth* (Ed. RUDNICK, D.), pp. 75–109. Princeton.

PITTENDRIGH, C. S. and BRUCE, V. G. (1959) Daily rhythms as coupled oscillator systems and their relation to thermoperiodism and photoperiodism. In *Photoperiodism and Related Phenomena in Plants and Animals* (Ed. WITHROW, R. B.), pp. 475–505. Am. Ass. Adv. Sci., Washington.

PITTENDRIGH, C. S., BRUCE, V. G. and KAUS, P. (1958) On the significance of transients in daily rhythms. *Proc. Nat. Acad. Sci. U.S.A.* **44**, 965–973.

PITTENDRIGH, C. S., EICHHORN, J. H., MINIS, D. H. and BRUCE, V. G. (1970) Circadian systems VI. Photoperiodic time measurement in *Pectinophora gossypiella. Proc. Nat. Acad. Sci. U.S.A.* **66**, 758–764.

PITTENDRIGH, C. S. and MINIS, D. H. (1964) The entrainment of circadian oscillations by light and their role as photoperiodic clocks. *Am. Nat.* **98**, 261–294.

PITTENDRIGH, C. S. and MINIS, D. H. (1971) The photoperiodic time measurement in *Pectinophora gossypiella* and its relation to the circadian system in that species. In *Biochronometry* (Ed. MENAKER, M.), pp. 212–250. National Academy of Sciences, Washington.

PITTENDRIGH, C. S. and MINIS, D. H. (1972) Circadian systems: longevity as a function of circadian resonance in *Drosophila melanogaster. Proc. Nat. Acad. Sci. U.S.A.* **69**, 1537–1539.

PITTENDRIGH, C. S. and SKOPIK, S. D. (1970) Circadian systems, V. The driving oscillation and the temporal sequence of development. *Proc. Nat. Acad. Sci. U.S.A.* **65**, 500–507.

POLCIK, B., NOWOSIELSKI, J. W. and NAEGELE, J. A. (1964) Daily sensitivity rhythm of the two-spotted mite, *Tetranychus urticae*, to DDVP. *Science, Wash.* **145**, 405.

PROKOPY, R. J. (1968) Influence of photoperiod, temperature, and food on initiation of diapause in the apple maggot. *Can. Ent.* **100**, 318–329.

PROPP, G. D., TAUBER, M. J. and TAUBER, C. A. (1969) Diapause in the neuropteran *Chrysopa oculata. J. Insect Physiol.* **15**, 1749–1757.

PROVOST, M. W. and LUM, P. T. M. (1967) The pupation rhythm in *Aëdes taeniorhynchus* (Diptera: Culicidae). I. Introduction. *Ann. Ent. Soc. Am.* **60**, 138–149.

RABB, R. L. (1966) Diapause in *Protoparce sexta* (Lepidoptera: Sphingidae). *Ann. ent. Soc. Am.* **59**, 160–165.

RABB, R. L. and THURSTON, R. (1969) Diapause in *Apanteles congregatus. Ann. ent. Soc. Am.* **62**, 125–128.

RANKIN, M. A., CALDWELL, R. L. and DINGLE, H. (1972) An analysis of a circadian rhythm of oviposition in *Oncopeltus fasciatus. J. exp. Biol.* **56**, 353–359.

READ, D. C. (1969) Rearing the cabbage maggot with and without diapause. *Can. Ent.* **101**, 725–737.

REINHARDT, R. (1969) Über den Einfluss der Temperatur auf den Saisondimorphismus von *Araschnia levana* L. (Lepidopt. Nymphalidae) nach photoperiodischer Diapause-Induktion. *Zool. Jb. Physiol.* **75**, 41–75.

REMMERT, H. (1955) Untersuchungen über das Tageszeitlich gebundene Schlüpfen von *Pseudosmittia arenaria. Z. vergl. Physiol.* **37**, 338–354.

REMMERT, H. (1961) Der Tagesgang im Strandanwurf und seine ökologische Bedeutung. *Verh. dt. Zool. Ges. Saarbrücken*, 1961, p. 438.

REMMERT, H. (1962) *Der Schlupfrhythmus der Insekten*. Franz Steiner Verlag, Wiesbaden.

RENNER, M. (1955) Ein Transozeanversuch zum Zeitsinn der Honigbiene. *Naturwissenschaften*, **42**, 540-541.

RENNER, M. (1957) Neue Versuche über den Zeitsinn der Honigbiene. *Z. vergl. Physiol.* **40**, 85–118.

RENNER, M. (1959) Über ein weiteres Versetzungs-experiment zur Analyse des Zeitsinnes und der Sonnenorientierung der Honigbiene. *Z. vergl. Physiol.* **42**, 449–483.

RENSING, L. (1961) Aktivitätsperiodik des Wasserläufers *Velia currens* F. *Z. vergl. Physiol.* **44**, 292–322.

RENSING, L. (1964) Daily rhythmicity of corpus allatum and neurosecretory cells in *Drosophila melanogaster* (Meig.). *Science, Wash.* **144**, 1586–1587.

RENSING, L. (1969) Die circadiane Rhythmik der Speicheldrüsen von *Drosophila* in vivo, in vitro, und unter dem Einfluss von Ecdyson. *J. Insect Physiol.* **15**, 2285–2303.

RENSING, L., BRUNKEN, W. and HARDELAND, R. (1968) On the genetics of a circadian rhythm in *Drosophila. Experientia*, **24**, 509–510.

RENSING, L. and HARDELAND, R. (1967) Zur Wirkung der circadianen Rhythmik auf die Entwicklung von *Drosophila. J. Insect Physiol.* **13**, 1547–1568.

RENSING, L., THACH, B. T. and BRUCE, V. G. (1965) Daily rhythms in the endocrine glands of *Drosophila* larvae. *Experientia*, **21**, 103–104.

RIDDIFORD, L. M. and JOHNSON, L. K. (1971) Synchronization of hatching of *Antheraea pernyi* eggs. *Proc. XIIIth Int. Congr. Ent., Moscow 1968*, **I**, 431–432.

RING, R. A. (1967) Maternal induction of diapause in the larvae of *Lucilia caesar* L. (Diptera, Calliphoridae). *J. exp. Biol.* **46**, 123–136.

ROBERTS, S. K. DE F. (1956) "Clock" controlled activity rhythm in the fruit fly. *Science, Wash.* **124**, 172.

ROBERTS, S. K. DE F. (1960) Circadian activity in cockroaches. I. The freerunning rhythm in steady-state. *J. cell. Comp. Physiol.* **55**, 99–110.

ROBERTS, S. K. DE F. (1962) Circadian activity in cockroaches. II. Entrainment and phase-shifting. *J. cell. Comp. Physiol.* **59**, 175–186.

ROBERTS, S. K. DE F. (1965a) Photoreception and entrainment of cockroach activity rhythms. *Science, Wash.* **148**, 958–959.

ROBERTS, S. K. DE F. (1965b) Significance of endocrines and central nervous system in circadian rhythms. In *Circadian Clocks* (Ed. ASCHOFF, J.), pp. 198–213. North–Holland, Amsterdam.

ROBERTS, S. K. DE F. (1966) Circadian activity rhythms in cockroaches. III. The role of endocrine and neural factors. *J. Cell. Physiol.* **67**, 473–486.

ROBERTS, S. K. DE F. (1974) Circadian rhythms in cockroaches. Effects of optic lobe lesions. *J. comp. Physiol.* **88**, 21–30.

ROBERTS, S. K. DE F., SKOPIK, S. D. and DRISKILL, R. J. (1971) Circadian rhythms in cockroaches: does brain hormone mediate the locomotor cycle? In *Biochronometry* (Ed. MENAKER, M.), pp. 505–516. National Academy of Sciences, Washington.

ROSENTHAL, S. S. and KOEHLER, C. S. (1968) Photoperiod in relation to diapause in *Hypera postica* from California. *Ann. ent. Soc. Am.* **61**, 531–534.

ROUBAUD, E. (1917) Observations biologiques sur *Nasonia brevicornis* Ashm., Chalcidide parasite des pupes des Muscides. Determinisme physiologique de l'instinct de ponte; adaptation a la lutte contre les glossines. *Bull. scient. Fr. Belg.* **1**, 425–439.

ROWAN, W. (1926) On photoperiodism, reproductive periodicity and the annual migration of birds and certain fishes. *Proc. Boston Soc. nat. Hist.* **38**, 147–189.

RYAN, R. B. (1965) Maternal influence on diapause in a parasitic insect, *Coeloides brunneri* Vier. (Hymenoptera: Braconidae). *J. Insect Physiol.* **11**, 1331–1336.

SABROSKY, C. W., LARSON, I. and NABOURS, R. K. (1933) Experiments with light upon reproduction, growth and diapause in grouse locusts. *Trans. Kansas Acad. Sci.* **36**, 298–300.

SACCA, G. (1964) Comparative bionomics of the genus *Musca. A. Rev. Ent.* **9**, 341–358.

SAKAI, T. and MASAKI, S. (1965) Photoperiod as a factor causing seasonal forms in *Lycaena phlaeas daimio* Seitz. (Lepidoptera: Lycaenidae), *Kontyu*, **33**, 275–283.

SANBURG, L. L. and LARSEN, J. R. (1973) Effect of photoperiod and temperature on ovarian development in *Culex pipiens pipiens. J. Insect Physiol.* **19**, 1173–1190.

SANTSCHI, F. (1911) Observations et remarques critiques sur le mecanisme de l'orientation chez les fourmis. *Revue suisse Zool.* **19**, 303–338.

SANTSCHI, F. (1913) A propos de l'orientation virtuelle chez les fourmis. *Bull. Soc. Hist. nat. Afr. Nord* **4–6**, 231–235.

SÁRINGER, G. (1964) The role of photoperiod in the diapause of the "rape sawfly", *Athalia rosae* L. *Ann. Inst. Prot. Plant Hung., Budapest*, **9**, 107–132.

SÁRINGER, G. (1966) Effect of photoperiod and temperature on the diapause of *Athalia glabricollis* Thomson (Tenthred. Hym.). *Acta Phytopath.* **1**, 139–144.

SÁRINGER, G. (1967) Investigations on the light-sensitive larval instar determining the diapause of *Athalia rosae* L. (= *colibri* Christ, Hym. Tenthred.). *Acta Phytopath.* **2**, 119–126.

SÁRINGER, G. (1970) The diapause of a SW Hungarian plum moth (*Laspeyresia funebrana* Tr.) population. *Acta Phytopath.* **5**, 371–374.

SÁRINGER, G. and NAGY, B. (1968) The effect of photoperiod and temperature on the diapause of the hemp moth (*Grapholitha sinana* Feld.) and its relevance to the integrated control. *Prol. XIIIth Int. Congr. Ent., Moscow 1968*, **I**, 435–436.

SAUNDERS, D. S. (1965a) Larval diapause induced by a maternally-operating photoperiod. *Nature, Lond.* **206**, 739–740.

SAUNDERS, D. S. (1965b) Larval diapause of maternal origin: Induction of diapause in *Nasonia vitripennis* (Walk.) (Hymenoptera: Pteromalidae). *J. exp. Biol.* **42**, 495–508.

SAUNDERS, D. S. (1966a) Larval diapause of maternal origin—II. The effect of photoperiod and temperature on *Nasonia vitripennis. J. Insect. Physiol.* **12**, 569–581.

SAUNDERS, D. S. (1966b) Larval diapause of maternal origin—III. The effect of host shortage on *Nasonia vitripennis. J. Insect Physiol.* **12,** 899–908.

SAUNDERS, D. S. (1967) Time measurement in insect photoperiodism: reversal of a photoperiodic effect by chilling. *Science, Wash.* **156,** 1126–1127.

SAUNDERS, D. S. (1968) Photoperiodism and time measurement in the parasitic wasp, *Nasonia vitripennis. J. Insect Physiol.* **14,** 433–450.

SAUNDERS, D. S. (1969) Diapause and photoperiodism in the parasitic wasp *Nasonia vitripennis,* with special reference to the nature of the photoperiodic clock. *Symp. Soc. exp. Biol.* **23,** 301–329.

SAUNDERS, D. S. (1970) Circadian clock in insect photoperiodism. *Science, Wash.* **168,** 601–603.

SAUNDERS, D. S. (1971) The temperature-compensated photoperiodic clock "programming" development and pupal diapause in the flesh-fly, *Sarcophaga argyrostoma. J. Insect Physiol.* **17,** 801–812.

SAUNDERS, D. S. (1972) Circadian control of larval growth rate in *Sarcophaga argyrostoma. Proc. Nat. Acad. Sci. U.S.A.* **69,** 2738–2740.

SAUNDERS, D. S. (1973a) Thermoperiodic control of diapause in an insect: theory of internal coincidence. *Science, Wash.* **181,** 358–360.

SAUNDERS, D. S. (1973b) The photoperiodic clock in the flesh-fly, *Sarcophaga argyrostoma. J. Insect Physiol.* **19,** 1941–1954.

SAUNDERS, D. S. (1974) Evidence for "dawn" and "dusk" oscillators in the *Nasonia* photoperiodic clock. *J. Insect Physiol.* **20,** 77–88.

SAUNDERS, D. S. (1975a) Spectral sensitivity and intensity thresholds in *Nasonia* photoperiodic clock. *Nature, Lond.* **253,** 732-734.

SAUNDERS, D. S. (1975b) "Skeleton" photoperiods and the control of diapause and development in the flesh-fly, *Sarcophaga argyrostoma. J. comp. Physiol.* **97,** 97-112.

SAUNDERS, D. S. (1975c) Manipulation of the length of the sensitive period, and the induction of pupal diapause in the flesh-fly, *Sarcophaga argyrostoma. J. Ent.* (A), **50,** 107-118.

SAUNDERS, D. S. and SUTTON, D. (1969) Circadian rhythms in the insect photoperiodic clock. *Nature, Lond.* **221,** 559–561.

SAUNDERS, D. S., SUTTON, D. and JARVIS, R. A. (1970) The effect of host species on diapause induction in *Nasonia vitripennis. J. Insect Physiol.* **16,** 405–416.

SAUNDERS, J. L. and KNOKE, J. K. (1968) Circadian emergence rhythm of a tropical scolytid, *Xyleborus ferrugineus. Ann. ent. Soc. Am.* **61,** 587–590.

SCHALLER, F. (1968) Action de la température sur la diapause et le developpement de l'embryon d'*Aeschna mixta* (Odonata). *J. Insect Physiol.* **14,** 1477–1483.

SCHLEIN, Y. (1972) Factors that influence the post-emergence growth in *Sarcophaga falculata. J. Insect Physiol.* **18,** 199–209.

SCHLIEP, W. (1910) Der Farbenwechsel von *Dixippus morosus* (Phasmidae). *Zool. Jb.* (*Physiol.*), **30,** 45–132.

SCHLIEP, W. (1915) Über die Frage nach der Beteiligung des Nervensystems beim Farbenwechsel von *Dixippus. Zool. Jb.* (*Physiol.*) **35,** 225–232.

SCHNEIDERMAN, H. A. and HORWITZ, J. (1958) The induction and termination of facultative diapause in the chalcid wasps *Mormoniella vitripennis* (Walker) and *Tritneptis klugii* (Ratzeburg). *J. exp. Biol.* **35,** 520–551.

SCHWAB, R. G. (1971) Circannian testicular periodicity in the European starling in the absence of photoperiodic change. In *Biochronometry* (Ed. MENAKER, M.), pp. 428–447. National Academy of Sciences. Washington.

SCHWEMMLE, B. (1960) Thermoperiodic effects and circadian rhythms in flowering of plants. *Cold Spring Harb. Symp. Quant. Biol.* **25,** 239–243.

SCOTT, W. N. (1936) An experimental analysis of the factors governing the hour of emergence of adult insects from their pupae. *Trans. R. ent. Soc. Lond.* **85,** 303–329.

SENN, L. H. and BRADY, U. E. (1973) Circadian rhythm of feeding by adult white-fringed beetles, *Graphognathus* spp. (Coleoptera: Curculionidae). *Ann. ent. Soc. Am.* **66,** 719–722.

SHA KHBAZOV, V. G. (1961) The reaction of the length of daylight and the light receptor of the pupa of the Chinese oak silkworm *Antheraea pernyi* G. *Dokl. Akad. Nauk SSSR* **140,** 249–252. (In Russian.)

SHEPARD, M. and KEELEY, L. L. (1972) Circadian rhythmicity and capacity for enforced activity in the cockroach, *Blaberus discoidalis,* after cardiacectomy-allatectomy. *J. Insect Physiol.* **18,** 595–601.

SHOREY, H. H. (1966) The biology of *Trichoplusia ni* (Lepidoptera, Noctuidae), IV. Environmental control of mating. *Ann. ent. Soc. Am.* **59,** 502–506.

SHOREY, H. H. and GASTON, L. K. (1965) Sex pheromones of Noctuid moths. V. Circadian rhythm of pheromone-responsiveness in males of *Autographa californica, Heliothis virescens, Spodoptera exigua,* and *Trichoplusia ni* (Lepidoptera: Noctuidae). *Ann. ent. Soc. Am.* **58,** 597–600.

SICKER, W. (1964) Die Abhängigkeit der Diapause von der Periodizität bei *Tetrix undulata* (Sow.) (Saltatoria, Tetrigidae). *Z. Morph. Okol. Tiere,* **54,** 107–140.

SIEW, Y. C. (1965a) The endocrine control of adult reproductive diapause in the chrysomelid beetle, *Galeruca tanaceti* (L.) I. *J. Insect Physiol.* **11**, 1–10.

SIEW, Y. C. (1965b) The endocrine control of adult reproductive diapause in the chrysomelid beetle, *Galeruca tanaceti* (L.). II. *J. Insect Physiol.* **11**, 463–479.

SIEW, Y. C. (1965c) The endocrine control of adult reproductive diapause in the chrysomelid beetle, *Galeruca tanaceti* (L.). III. *J. Insect Physiol.* **11**, 973–981.

SKOPIK, S. D. and PITTENDRIGH, C. S. (1967) Circadian systems, II. The oscillation in the individual *Drosophila* pupa; its independence of developmental stage. *Proc. Nat. Acad. Sci. U.S.A.* **58**, 1862–1869.

SLAMA, K. (1964) Hormonal control of respiratory metabolism during growth, reproduction and diapause in female adults of *Pyrrhocoris apterus* L. *J. Insect Physiol.* **10**, 283–304.

SMITH, O. J. and LANGSTON, R. L. (1953) Continuous laboratory propagation of western grape leaf skeletonizer and parasites by prevention of diapause. *J. econ. Ent.* **46**, 477–484.

SOMME, L. (1961) On the overwintering of house flies (*Musca domestica* L.) and stable flies (*Stomoxys calcitrans* (L.)) in Norway. *Norsk. Entomol. Tidsskr.* **11**, 191–223.

SOUTHWOOD, T. R. E. (1962) Migration of terrestrial arthropods in relation to habit. *Biol Rev.* **37**, 171–214.

SOWER, L. L., SHOREY, H. H. and GASTON, L. K. (1970) Sex pheromones of noctuid moths, XXI. Light-dark cycle regulation and light inhibition of the sex pheromone release by females of *Trichoplusia ni*. *Ann. ent. Soc. Am.* **63**, 1090–1092.

SPANGLER, H. G. (1972) Daily activity rhythms of individual worker and drone honey bees. *Ann. ent. Soc. Am.* **65**, 1073–1076.

SPANGLER, H. G. (1973) Role of light in altering the circadian oscillations of the honey bee. *Ann. ent. Soc. Am.* **66**, 449–451.

SPIELMAN, A. and WONG, J. (1973) Enviromental control of ovarian diapause in *Culex pipiens*. *Ann. ent. Soc. Am.* **66**, 905–907.

STEBBINS, R. C. (1963) Activity changes in the striped plateau lizard with evidence on influence of the parietal eye. *Copeia*, **1963**, 681–691.

STOFFOLANO, J. G. and MATTHYSSE, J. G. (1967) Influence of photoperiod and temperature on diapause in the face fly, *Musca autumnalis* (Diptera: Muscidae). *Ann. ent. Soc. Am.* **60**, 1242–1246.

STORCH, R. H. (1973) The effect of photoperiod on *Coccinella transversoguttata* (Coleoptera: Coccinellidae). *Entomologia exp. appl.* **16**, 77–82.

STROSS, R. G., and HILL, J. C. (1968) Photoperiod control of winter diapause in the fresh water crustacean, *Daphnia*. *Biol. Bull. Mar. Biol. Lab., Woods Hole*, **134**, 176–198.

STRUBING, H. (1963) Zum diapauseproblem in der Gattung *Stenocranus* (Homoptera, Auchenorrhyncha). *Zool. Beitr.* **9**, 1–119.

STRUMWASSER, F. (1965) The demonstration and manipulation of a circadian rhythm in a single neuron. In *Circadian Clocks* (Ed. ASCHOFF, J.), pp. 442–462. North-Holland, Amsterdam.

SULLIVAN, C. R. and WALLACE, D. R. (1965) Photoperiodism in the development of the European pine sawfly, *Neodiprion sertifer* (Geoff.). *Can. J. Zool.* **43**, 233–245.

SULLIVAN, C. R. and WALLACE, D. R. (1967) Interaction of temperature and photoperiod in the induction of prolonged diapa use in *Neodiprion sertifer*. *Can. Ent.* **99**, 834–850.

SWEENEY, B. M. (1969) *Rhythmic Phenomena in Plants*. Academic Press, New York and London.

TADMOR, U. and APPLEBAUM, S. W. (1971) Adult diapause in the predaceous coccinellid, *Chilochorus bipustulatus*: photoperiodic induction. *J. Insect Physiol.* **17**, 1211–1215.

TAKAHASHI, S. and HARWOOD, R. F. (1964) Glycogen levels of adult *Culex tarsalis* in response to photoperiod. *Ann. ent. Soc. Am.* **57**, 621–623.

TAKIMOTO, A. and HAMNER, K. C. (1964) Effect of temperature and preconditioning on photoperiodic response of *Pharbitis nil*. *Pl. Physiol.* **39**, 1024–1030.

TANAKA, S. J. (1961) *Fac. Text. Sci. Technol. Shinshu Univ. Ser. A. Agric. Seric.* **5**, 69.

TANAKA, Y. (1944). Effect of daylength on hibernation of the Chinese oak silkworm. *Jap. J. Agr. Hort.* **19** (9). (In Japanese.)

TANAKA, Y. (1950a) Studies on hibernation with special reference to photoperiodicity and breeding of the Chinese Tussar silkworm. I. *J. Seric. Sci. Japan.* **19**, 358–371. (In Japanese.)

TANAKA, Y. (1950b) Studies on hibernation with special reference to photoperiodicity and the breeding of the Chinese Tussar silkworm—II. *J. Seric. Sci. Japan*, **19**, 429–446. (In Japanese.)

TANAKA, Y. (1950c) Studies on hibernation with special reference to photoperiodicity and breeding of the Chinese Tussar silkworm—III. *J. Seric. Sci. Japan*, **19**, 580–590. (In Japanese.)

TANAKA, Y. (1951) Studies on hibernation with special reference to photoperiodicity and breeding of the Chinese Tussar silkworm—V. *J. Seric. Sci. Japan*, **20**, 132–138. (In Japanese.)

TAUBER, M. J. and TAUBER, C. A. (1969) Diapause in *Chrysopa carnea* (Neuroptera, Chrysopidae)—I. Effect of photoperiod on reproductively active adults. *Can. Ent.* **101**, 364–370.

TAUBER, M. J. and TAUBER, C. A. (1970) Photoperiodic induction and termination of diapause in an insect: response to changing daylengths. *Science, Wash.* **167**, 170.

TAUBER, M. J. and TAUBER, C. A. (1972a) Larval diapause in *Chrysopa nigricornis:* sensitive stages, critical photoperiod, and termination (Neuroptera: Chrysopidae). *Entomologia exp. appl.* **15**, 105–111.

TAUBER, M. J. and TAUBER, C. A. (1972b) Geographic variation in critical photoperiod and in diapause intensity of *Chrysopa carnea* (Neuroptera). *J. Insect Physiol.* **18**, 25–29.

TAUBER, M. J. and TAUBER, C. A. (1973a) Nutritional and photoperiodic control of the seasonal reproductive cycle in *Chrysopa mohave* (Neuroptera). *J. Insect Physiol.* **19**, 729–736.

TAUBER, M. J. and TAUBER, C. A. (1973b) Quantitative responses to daylength during diapause in insects. *Nature, Lond.* **244**, 296–297.

TAUBER, M. J., TAUBER, C. A. and DENYS, C. J. (1970) Adult diapause in *Chrysopa carnea:* photoperiodic control of duration and colour. *J. Insect Physiol.* **16**, 949–955.

TAYLOR, B. and JONES, M. D. R. (1969) The circadian rhythm of flight activity in the mosquito *Aëdes aegypti* (L.): the phase-setting effects of light-on and light-off. *J. exp. Biol.* **51**, 59–70.

TAYLOR, B. (1969) Geographical range and circadian rhythms. *Nature, Lond.* **222**, 296–297.

TEETES, G. L., ADKISSON, P. L. and RANDOLPH, N. M. (1969) Photoperiod and temperature as factors controlling the diapause of the sunflower moth, *Homoeosoma electellum. J. Insect Physiol.* **15**, 755–761.

THIELE, H-U. (1966) Einflusse der Photoperiode auf die Diapause von carabiden. *Z. angew. Ent.* **58**, 143–149.

THIELE, H-U. (1967) Formen der Diapausesteuerung bei Carabiden. *Verh. dt. Zool. Ges., Heidelberg 1967*, pp. 358–364.

THIELE, H-U. (1969) The control of larval hibernation and adult aestivation in the carabid beetles *Nebria brevicollis* F. and *Patrobus atrorufus. Oecologia, Berl.* **2**, 347–361.

THIELE, H-U. (1971) Die Steuerung der Jahesrhythmik von Carabiden durch exogene und endogene Faktoren. *Zool. Jb. (Syst.),* **98**, 341–371.

THIELE, H-U. (1973) Remarks about Mansingh's and Müller's classifications of dormancies in insects. *Can. Ent.* **105**, 925–928.

THOMAS, R. and FINLAYSON, L. H. (1970) Initiation of circadian rhythms in arrhythmic churchyard beetles (*Blaps mucronata*). *Nature, Lond.* **228**, 577–578.

TOSHIMA, A., HONMA, K. and MASAKI, S. (1961) Factors influencing the seasonal incidence of diapause in *Carposina niponensis* Walshingham. *Japan. J. appl. Ent. Zool.* **5**, 260–269.

TRAYNIER, R. M. M. (1970) Sexual behaviour of the mediterranean flour moth, *Anagasta kuhniella:* some influences of age, photoperiod, and light intensity. *Can. Ent.* **102**, 534–540.

TRUMAN, J. W. (1970) The eclosion hormone: its release by the brain, and its action on the central nervous system of silkmoths. *Am. Zool.* **10**, 511–512.

TRUMAN, J. W. (1971a) Hour-glass behavior of the circadian clock controlling eclosion of the silkmoth *Antheraea pernyi. Proc. Nat. Acad. Sci. U.S.A.* **68**, 595–599.

TRUMAN, J. W. (1971b) The role of the brain in the ecdysis rhythm of silkmoths: comparison with the photoperiodic termination of diapause. In *Biochronometry* (Ed. MENAKER, M.), pp. 483–504. National Academy of Sciences, Washington.

TRUMAN, J. W. (1971c) Physiology of insect ecdysis I. The eclosion behavior of Saturniid moths and its hormonal release. *J. exp. Biol.* **54**, 805–814.

TRUMAN, J. W. (1971d) Circadian rhythms and physiology with special reference to neuroendocrine processes in insects. *Proc. Int. Symp. Circadian Rhythmicity* (Wageningen, 1971), pp. 111–135.

TRUMAN, J. W. (1972a) Physiology of insect rhythms. I. Circadian organization of the endocrine events underlying the moulting cycle of larval tobacco hornworms. *J. exp. Biol.* **57**, 805–820.

TRUMAN, J. W. (1972b) Physiology of insect rhythms. II. The silkworm brain as the location of the biological clock controlling eclosion. *J. comp. Physiol.* **81**, 99–114.

TRUMAN, J. W. and RIDDIFORD, L. M. (1970) Neuroendocrine control of ecdysis in silkmoths. *Science, Wash.* **167**, 1624–1626.

TRUMAN, J. W. and SOKOLOVE, P. G. (1972) Silkmoth eclosion: hormonal triggering of a centrally programmed pattern of behavior. *Science, Wash.* **175**, 1491–1493.

TSUJI, H. (1963) Experimental studies on the larval diapause of the Indian meal moth, *Plodia interpunctella* Hubner (Lepidoptera: Pyralidae). Thesis (Kyushu Univ., Fukuoka), 88 pp.

TURNER, R. B. and ACREE, F. (1967) The effect of photoperiod on the daily fluctuation of haemolymph hydrocarbons in the American cockroach. *J. Insect Physiol.* **13**, 519–522.

TYCHSEN, P. H. and FLETCHER, B. S. (1971) Studies on the rhythm of mating in the Queensland fruit fly, *Dacus tryoni. J. Insect Physiol.* **17**, 2139–2156.

TYSHCHENKO, V. P. (1966) Two-oscillatory model of the physiological mechanism of insect photoperiodic reaction. *Zhur. obshch. Biol.* **27**, 209–222. (In Russian.)

TYSHCHENKO, V. P., GORYSHIN, N. I. and AZARYAN, A. G. (1972) The role of circadian processes in insect photoperiodism. *Zhur. obshch. Biol.* **33**, 21–31. (In Russian.)

UMEYA, K. and MASAKI, S. (1969) Biology of *Hyphantria cunea* Drury (Lepidoptera: Arctiidae) in Japan. VII. Delayed development of summer pupae. *Res. Bull. Pl. Prot. Japan*, **7**, 1–16.

VALDER, S. M., HOPKINS, T. L. and VALDER, S. A. (1969) Diapause induction and changes in lipid composition in diapausing and reproducing face flies, *Musca autumnalis*. *J. Insect Physiol.* **15**, 1199–1214.

VEPSÄLÄINEN, K. (1971a) The roles of photoperiodism and genetic switch in alary polymorphism in *Gerris* (Het., Gerridae) (a preliminary report). *Acta Ent. Fennica*, **28**, 101–102.

VEPSÄLÄINEN, K. (1971b) The role of gradually changing daylength in determination of wing length, alary dimorphism and diapause in a *Gerris odontogaster* (Zett.) population (Gerridae, Heteroptera) in South Finland. *Ann. Acad. Sci. Fenn. A IV Biologica*, **183**, 1–25.

VEPSÄLÄINEN, K. (1974) Lengthening of illumination period is a factor in averting diapause: *Nature, Lond.* **247**, 385–386.

VINOGRADOVA, E. B. (1960) The experimental investigation of ecological factors inducing imaginal diapause in bloodsucking mosquitoes (Diptera, Culidiae). *Ent. Obozr.* **39**, 327–340. (In Russian.)

VINOGRADOVA, E. B. (1962) The role of photoperiodism in the seasonal development of *Anopheles plumbeus* Steph. *Dokl. Akad. Nauk SSSR*, **142**, 481–483.

VINOGRADOVA, E. B. (1965) An experimental study of the factors regulating induction of imaginal diapause in the mosquito *Aëdes togoi* Theob. *Ent. Obozr.* **44**.

VINOGRADOVA, E. B. (1967) The effect of photoperiodism upon the larval development and the appearance of diapausing eggs in *Aëdes triseriatus* Say (Diptera, Culicidae). *Parasitologia*, **1**, 19–26. (In Russian.)

VINOGRADOVA, E. B. and ZINOVJEVA, K. B. (1972a) Experimental investigation of the seasonal aspect of the relationship between blowflies and their parasites. *J. Insect Physiol.* **18**, 1629–1638.

VINOGRADOVA, E. B. and ZINOVJEVA, K. B. (1972b) Maternal induction of larval diapause in the blowfly, *Calliphora vicina*. *J. Insect Physiol.* **18**, 2401–2409.

WAHL, O. (1932) Neue Untersuchungen uber das Zeitgedachtnis der Bienen. *Z. vergl. Physiol.* **16**, 529–.

WANG REN-LAI (1966) Observations on the influence of photoperiod on egg diapause in *Aedes albopictus*. *Acta ent. Sinica*, **15**, 75–77.

WARNECKE, H. (1966) Vergleichende Untersuchungen zur tagesperiodischen Aktivitat von 3 *Geotrupes*-Arten. *Z. Tierpsychol.* **23**, 513–536.

WATT, W. B. (1968) Adaptive significance of pigment polymorphisms in *Colias* butterflies. I. Variation of melanin pigment in relation to thermoregulation. *Evolution*, **22**, 437–458.

WATT, W. B. (1969) Adaptive significance of pigment polymorphisms in *Colias* butterflies. II. Thermoregulation and photoperiodically controlled melanism variation in *Colias eurytheme*. *Proc. Nat. Acad. Sci. U.S.A.* **63**, 767–774.

WAY, M. J. and HOPKINS, B. A. (1950) The influence of photoperiod and temperature on the induction of diapause in *Diataraxia oleracea* L. *J. exp. Biol.* **27**, 365–376.

WAY, M. J., HOPKINS, B. A. and SMITH, P. M. (1949) Photoperiodism and diapause in insects. *Nature, Lond.* **164**, 615.

WEBER, F. (1965) Zur Tagaktivitat von *Carabus*-Arten. *Zool. Anz.* **175**, 354–360.

WEBER, F. (1966) Zur tageszeitlichen Aktivitatsverteilung der *Carabus*-Arten. *Zool. Jb. (Zool. Physiol. Tiere)*, **72**, 136–156.

WELLSO, S. G. and ADKISSON, P. L. (1966) A long-day short-day effect in the photoperiodic control of the pupal diapause of the bollworm, *Heliothis zea* (Boddie) (Lepidoptera: Noctuidae). *J. Insect Physiol.* **12**, 1455–1465.

WENT, F. W. (1959) The periodic aspect of photoperiodism and thermoperiodicity. In *Photoperiodism and Related Phenomena in Plants and Animals* (Ed. WITHROW, R. B.), pp. 551–564. Am. Ass. Adv. Sci., Washington.

WERNER, G. (1954) Tanze und Zeitempfinden der Honigbiene in Abhangigkeit vom Stoffwechsel. *Z. vergl. Physiol.* **36**, 464–.

WESELOH, R. M. (1973) Termination and induction of diapause in the gypsy moth larval parasitoid, *Apanteles melanoscelus*. *J. Insect Physiol.* **19**, 2025–2033.

WEVER, R. (1965) A mathematical model for circadian rhythms. In *Circadian Clocks* (Ed. ASCHOFF, J.), pp. 47–63. North-Holland, Amsterdam.

WHITCOMB, R. F., KRAMER, J. P. and COAN, M. E. (1972) *Stirellus bicolor* and *S. obtutus* (Homoptera: Cicadellidae): winter and summer forms of a single species. *Ann. ent. Soc. Am.* **65**, 797–798.

WHITE, T. C. R. (1968) Hatching of eggs of *Cardiaspina densitexta* (Homoptera, Psyllidae) in relation to light and temperature. *J. Insect Physiol.* **14**, 1847–1859.

WILDE, J. DE (1954) Aspects of diapause in adult insects with special regard to the Colorado beetle, *Leptinotarsa decemlineata*. *Arch. Neerl. Zool.* **10**, 375–385.

WILDE, J. DE (1958) Perception of the photoperiod by the Colorado potato beetle (*Leptinotarsa decemlineata* Say.). *Proc. Xth Int. Congr. Ent. Montreal 1956*, **2**, 213–218.

WILDE, J. DE (1962) Photoperiodism in insects and mites. *A. Rev. Ent.* **7**, 1–26.

WILDE, J. DE and BOER, J. A. DE (1961) Physiology of diapause in the adult Colorado beetle—II. Diapause as a case of pseudoallatectomy. *J. Insect Physiol.* **6**, 152–161.

WILDE, J. DE and BONGA, H. (1958) Observations on threshold intensity and sensitivity of different wave lengths of photoperiodic response in the Colorado beetle (*Leptinotarsa decemlineata* Say). *Entomologia exp. appl.* **1**, 301–307.

WILDE, J. DE, DUINTJER, C. S. and MOOK, L. (1959) Physiology of diapause in the adult Colorado beetle (*Leptinotarsa decemlineata*). I. The photoperiod as a controlling factor. *J. Insect Physiol.* **3**, 75–85.

WILLIAMS, C. M. (1946) Physiology of insect diapause. I. The role of the brain in the production and termination of pupal dormancy in the giant silkworm *Platysamia cecropia*. *Biol. Bull. Mar. Biol. Lab.*, *Woods Hole*, **90**, 234–243.

WILLIAMS, C. M. (1952) Physiology of insect diapause. IV. The brain and prothoracic glands as an endocrine system in the cecropia silkworm. *Biol. Bull. Mar. Biol. Lab.*, *Woods Hole*, **103**, 120–138.

WILLIAMS, C. M. (1963) Control of pupal diapause by the direct action of light on the insect brain. *Science, Wash.* **140**, 386.

WILLIAMS, C. M. (1967) The present status of the brain hormone. In *Insects and Physiology* (Ed. BEAMENT, J. W. L. and TREHERNE, J. E.), pp. 133–139. Oliver & Boyd, Edinburgh and London.

WILLIAMS, C. M. (1969a) Photoperiodism and the endocrine aspects of insect diapause. *Symp. Soc. exp. Biol.* **23**, 285–300.

WILLIAMS, C. M. (1969b) Nervous and hormonal communication in insect development. *Dev. Biol. Suppl.* **3**, 133–150.

WILLIAMS, C. M. and ADKISSON, P. L. (1964) Physiology of insect diapause, XIV. An endocrine mechanism for the photoperiodic control of pupal diapause in the oak silkworm, *Antheraea pernyi*. *Biol. Bull. Mar. Biol. Lab.*, *Woods Hole*, **127**, 511–525.

WILLIAMS, C. M. ADKISSON, P. L. and WALCOTT, C. (1965) Physiology of insect diapause. XV. The transmission of photoperiodic signals to the brain of the oak silkworm, *Antheraea pernyi*. *Biol. Bull. Mar. Biol. Lab.*, *Woods Hole*, **128**, 497–507.

WILSON, F. (1938) Some experiments on the influence of environment upon the forms of *Aphis chloris* Koch. *Trans. R. ent. Soc. Lond.* **87**, 165–180.

WINFREE, A. T. (1970a) The temporal morphology of a biological clock. In *Lectures on Mathematics in the Life Sciences* (Ed. GERSTENHABER, M.), pp. 111–150. American Mathematical Society, Providence, R.I.

WINFREE, A. T. (1970b) Integrated view of resetting a circadian clock. *J. theoret. Biol.* **28**, 327–374.

WITSACK, W. (1971) Experimentelle-ökologische Untersuchungen über Dormanz-formen von Zikaden (Homoptera- Auchenorrhyncha). I—Zur Form und Induktion der Embryonaldormanz von *Muellerianella brevipennis* (Boheman) (Delphacidae). *Zool. Jb. Syst.* **98**, 316–340.

WOBUS, V. (1966) Der Einfluss Lichtintensität auf die Resynchronisation der circadianen Laufaktivität der Schabe *Blaberus craniifer* Burm. (Insecta: Blattida). *Z. vergl. Physiol.* **52**, 276–289.

WOHLFAHRT, T. (1957) Über den Einfluss von Licht, Fütterqualität und Temperatur auf Puppenruhe und Diapause des Mitteleuropäischen Segelfalters *Iphiclides podalirius*. *Tagungsber. Wandervers. dt. Ent.*, Berlin, **2**, 6–14.

WOLF, E. (1927) Über das Heimkehrvermögen der Beinen. *Z. vergl. Physiol.* **6**, 221–254.

WRIGHT, J. E. and VENARD, C. E. (1971) Diapause induction in larvae of *Aëdes triseriatus*. *Ann. ent. Soc. Am.* **64**, 11–14.

YIN, C-M. and CHIPPENDALE, G. M. (1973) Juvenile hormone regulation of the larval diapause of the southwestern corn borer, *Diatraea grandiosella*. *J. Insect. Physiol.* **19**, 2403–2420.

YOUTHED, G. J. and MORAN, V. C. (1969a) The solar-day activity rhythm of Myrmeleontid larvae, *J. Insect Physiol.* **15**, 1103–1116.

YOUTHED, G. J. and MORAN, V. C. (1969b) The lunar-day activity rhythm of Myrmeleontid larvae. *J. Insect Physiol.* **15**, 1259–1271.

ZABIROV, S. M. (1961) Factors governing the seasonal development cycles of the spinach leaf miner (*Pegomyia hyosciami* Panz.) and the cabbage maggot (*Hylemyia brassicae* Bouche) (Diptera, Anthomyidae). *Ent. Obozr.* **40**, 148–151. (In Russian.)

ZEEUW, D. DE (1957) Flowering of *Xanthium* under long-day conditions. *Nature, Lond.* **180**, 558.

ZELAZNY, B. (1969) Doctoral thesis. Quoted from Neville (1970).

ZELAZNY, B. and NEVILLE, A. C. (1972) Endocuticle layer formation controlled by non-circadian clocks in beetles. *J. Insect Physiol.* **18**, 1967–1979.

ZIMMERMAN, W. F. and GOLDSMITH, T. H. (1971) Photosensitivity of the circadian rhythm and of visual receptors in carotenoid depleted *Drosophila*. *Science, Wash.* **171**, 1167–1168.

ZIMMERMAN, W. F. and IVES, D. (1971) Some photophysiological aspects of circadian rhythmicity in *Drosophila*. In *Biochronometry* (Ed. MENAKER, M.), pp. 381–391. National Academy of Sciences, Washington.

ZIMMERMAN, W. F., PITTENDRIGH, C. S. and PAVLIDIS, T. (1968) Temperature compensation of the circadian oscillation in *Drosophila pseudoobscura* and its entrainment by temperature cycles. *J. Insect Physiol.* **14**, 669–684.

INDEX

Only includes those species mentioned in the text; others may be found in Appendix B